APPLIED MATHEMATICS AND OMICS TO ASSESS CROP GENETIC RESOURCES FOR CLIMATE CHANGE ADAPTIVE TRAITS

APPLIED MATHEMATICS AND OMICS TO ASSESS CROP GENETIC RESOURCES FOR CLIMATE CHANGE ADAPTIVE TRAITS

EDITED BY
Abdallah Bari
Ardeshir B. Damania
Michael Mackay
Selvadurai Dayanandan

CRC Press
Taylor & Francis Group
Boca Raton London New York

CRC Press is an imprint of the
Taylor & Francis Group, an **informa** business

CRC Press
Taylor & Francis Group
6000 Broken Sound Parkway NW, Suite 300
Boca Raton, FL 33487-2742

Printed on acid-free paper
Version Date: 20160107

International Standard Book Number-13: 978-1-4987-3013-6 (Hardback)

Visit the Taylor & Francis Web site at
http://www.taylorandfrancis.com

and the CRC Press Web site at
http://www.crcpress.com

Contents

Foreword

This landmark book, *Applied Mathematics and Omics to Assess Crop Genetic Resources for Climate Change Adaptive Traits*, eloquently underscores the importance of global warming and associated climate change that is, and will be, affecting millions of human beings and animals. The challenges we are about to face will have a great impact on agricultural economies worldwide. Areas of inquiry are continually emerging in sciences, and these attract researchers and scientists from a myriad range of disciplines including, as this book so adequately illustrates, from mathematics and omics to conservation and utilization of crop plant genetic resources for food, feed, and fiber.

This book discusses innovative approaches that can be used for mining adaptive traits in genebank collections using mathematics and omics technologies in light of climate-change phenomena. The drylands will be one of the areas that are most affected by any scenarios of climate change and warming trends. These areas represent more than 40% of global land cover and are home to over 2.5 billion of the world's population. The number of livestock that would be affected, with a subsequent impact on the livelihoods of human populations, is almost unfathomable.

Crop cultivars with improved tolerance to heat, drought, and soil salinity and resistance to existing and emerging virulent insect pests and diseases are urgently needed by farmers all over the world, but especially in Asia and Africa, including West Asia and North Africa, to mitigate the effects of climate change and global warming. If we are to increase agricultural productivity and sustainably to meet the future demand for food by an ever-increasing human population, then clearly we need to look at technologies and means beyond the conventional breeding practices and contemporary use of our plant genetic resources, many of which are conserved in over 1700 genebanks around the world.

Identifying stress tolerances and resistant traits from more than seven million accessions currently held in the world's major genebanks is not an easy task. Exploring the holdings of these agricultural genebanks for adaptive traits is more urgent than ever before. Applying novel approaches and innovations to identify for evaluating those genebank accessions most likely to possess the desired traits will improve the effective use of genetic resources in accelerating crop improvement, while at the same time maintaining biodiversity, food security, and livelihoods of agricultural communities existing in drylands and other marginal areas.

This book highlights the outcome of the workshop held from June 24 to 27, 2014, at the Institute of Agronomy and Veterinary Medicine, Hasan II, Rabat, Morocco, where researchers and scientists from five continents met to discuss their own work, along with the problems of farmers, and set forth strategies to mitigate the detrimental effects of climate change using mathematics and omics technologies. The sessions were streamed globally on the Internet, and on the first day, there were over 500 individuals viewing the proceedings. Subsequently, that number has increased to over 7000, showing the great interest that these topics have generated. I am certain that this book will go a long way in introducing novel and innovative practices in identifying the adaptive traits needed to combat climate change and enhance food and nutritional security globally, particularly in developing countries.

Mahmoud Solh
Director General
International Centre for Agricultural Research in the Dry Areas (ICARDA)

Preface

We are at the cusp of global climate change, and the increasing frequency of annual droughts has had significant negative impacts on agricultural production, especially in the drylands. Although seeking higher yield has always been the focus of breeders, there is an urgent need to identify crop cultivars with tolerance to biotic and abiotic stresses for adaptation of crops to the changing climate. Climate change is bound to present new pests and disease challenges for which we urgently need new sources of genetic variation to sustain agriculture.

Crop improvement or modifications to meet the challenges of mitigating the effects of climate change will largely depend on availability of genes for each crop that can address the vagaries of the changing environments. However, finding the required alleles that breeders are seeking is a daunting task, and it has been subject to minimal research mainly due to the requirement of significant manpower and many years of time. We are pressed for time as effects of climate change are already being felt globally from California to Australia. Integrative, multidisciplinary approaches, involving analyses of crop accessions, using emerging mathematical modeling techniques and high throughput omic technologies, including phenomics, genomics, proteomics, and transcriptomics, are urgently needed to develop climate-change resilient crops in a timely fashion. These approaches will provide us with the means to achieve the rapid and cost-efficient identification of genes and genomic regions associated with climate-change adaptive traits and accelerate crop improvement programs to maintain genetic diversity, safeguard food security, and improve livelihoods as a result of global warming and climate change.

To seek answers and to chart the way forward to mitigate the effects of climate change, an international workshop on "Applied Mathematics and Omics Technologies for Discovering Biodiversity and Genetic Resources for Climate Change Mitigation and Adaptation for Sustainable Agriculture in Drylands" was held at Rabat, Morocco, June 24–27, 2014. Twenty-two papers prepared by numerous scientists from five continents (Africa, Asia, Europe, North America, and Australia) were presented. Sponsorship and support were received from 12 international organizations and universities.

This international workshop initiative was an imperative follow-up to other major global initiatives, such as the Mathematics of Planet Earth (MPE2013) initiative launched in 2013, and the Fourteenth Regular Session (April 15–19, 2013) of the UN FAO Commission on Genetic Resources for Food and Agriculture, which emphasized the mobilization of crop genetic resources and the continued exploration of agro-biodiversity for climate change-related traits for adaption to and mitigation of changes in the climate. As highlighted in a recent international conference on "Genetic Resources for Food and Agriculture in a Changing World" held in Lillehammer, Norway, January 27–29, 2014, genetic diversity plays a crucial role for climate-change mitigation and adaptation of agriculture to new climate patterns.

This workshop explored the use of a variety of applied mathematical approaches—including Bayes–Laplace inverse, approximation theorems, and fractal geometry in combination with phenomics and genomics—for rapid and cost-effective identification of crop plants with climate-change adaptive traits. These approaches will help accelerate crop improvement to maintain biodiversity and food security and improve livelihoods in dryland areas under rapidly changing climatic conditions. This book is intended to provide the reader with a broad view of the interplay of biodiversity and answers to some of the latest problems in agriculture. It also highlights some of the ways to provide useful information to those who are in the battle to address the effects of global warming and climate change.

The International Center for Agricultural Research in the Dry Areas (ICARDA), together with its partners, aims to harness the research and capacity-building assets through its new platforms where cutting-edge applied mathematics will be used in combination with different omics technologies. The workshop addressed gender issues by engaging women researchers and addressed the transfer of innovation from and to farming communities. The workshop expanded further on the mathematical conceptual framework of solving practical problems such as drought, heat, diseases, and pest resistance. The chapters of this book reflect, among many other things:

The use of *purposive and knowledge-based* targeted sampling approach
The attempt to apply the *focused identification of germplasm strategy* (FIGS) approach to identify germplasm adaptive to extreme climate-change events, and
The practice of the *double gradient selection technique* (DGST) as a buffer to changing/fluctuating climate conditions.

The participants who took this initiative successfully committed themselves to continue beyond 2014 by conducting a similar workshop every 5 years or so, and expanded the participation to farming communities, so that scientists and farmers are in tune and agreement with the best ways to combat climate change. We believe that sharing this knowledge and experiences of some of the world's well-known scientists and practitioners will provide the readers with a broad appreciation and hopefully a heightened awareness of the stakes involved in the preservation of agricultural systems, vis-à-vis climate change and associated phenomenon.

This book is organized into four sections as follows:

Section I—Climate-Change Implications for Drylands and Farming Communities
Section II—Potential of Using Genetic Resources and Biodiversity to Adapt to and Mitigate Climate Change
Section III—Applied Mathematics (Unlocking the Potential of Mathematical Conceptual Frameworks)
Section IV—Applied Omics Technologies

We thank ICARDA, Concordia University, Institute of Agronomy and Veterinary Medicine Hassan II, National Institute for Agricultural Research (INRA)-Morocco, Crop Biodiversity Research and Resource Center (ARCAD), Consultative Group on International Agricultural Research (CGIAR) Climate Change, Agriculture and Food

Security [CCAFS], International Maize and Wheat Improvement Center (CIMMYT), Grain Research Development Corporation (GRDC), University of Helsinki, and NordGen for their support. Our special thanks to the Communication, Documentation and Information Services Unit (CODIS) of ICARDA. We also thank the staff of ICARDA's Rabat Office who did an outstanding job in logistical support, including transport and financial arrangements. The work of the tireless individuals, who constituted the workshop secretariat and the local organizing committee as well as the international organizing committee that provided back up and conceptual feedback through this endeavor, is also gratefully acknowledged. Finally, we express our gratitude to Patrick McGuire for lending his expertise to format the entire book and bring it to meet the publishing requirements of Taylor & Francis Group, to whom we are thankful for taking up our project without any hesitation.

A. Bari
International Center for Agricultural Research in the Dry Areas, Morocco

A.B. Damania
University of California, Davis

M.C. Mackay
University of Queensland

S. Dayanandan
Concordia University

Editors

Dr. A. Bari is a researcher focusing on applied mathematics in agricultural research. His research involves elaborating on mathematical and theoretical aspects in support for practical solutions, such as the application of fractal geometry to capture complex trait variation in plants. He has held several positions spanning from field surveying to capacity building to assessing genetic resources for useful traits including climate change-related traits. Dr. Bari received his MSc (genes' transfer) from a joint program between the Agronomy and Veterinary Institute Hassan II of Morocco and the University of Minnesota, United States. He received his PhD (on imagining techniques to assess genetic variation for water-use efficiency) from the University of Cordoba, Spain. Dr. Bari has published peer-reviewed articles and chapters and has edited a book on the assessment of genetic resources for water-use efficiency. He works with his colleagues from universities, research institutions, and development organizations to explore these new frontiers of research of applied mathematics in agricultural research to address climate change issues.

Dr. A.B. Damania received his bachelor's degree in botany and zoology and a master's degree in ecology and plant geography from the University of Bombay, India. Later, he was awarded a master's degree and a PhD in crop plant genetic resources from the University of Birmingham, UK.

Dr. Damania's early career was with the Food and Agriculture Organization (FAO) of the United Nations, Rome, Italy, as plant collector for South Asia and the Indian Ocean Islands. Later, he joined the International Center for Agricultural Research in the Dry Areas (ICARDA) as durum germplasm scientist, and subsequently became the cereal curator and was based in Aleppo, Syria.

Since 1994, he has been working at the Genetic Resources Conservation Program (GRCP) at the University of California, Davis. He is currently an associate in the Agricultural Experimental Station (AES) at the University of California, Davis, based in the Department of Plant Sciences. His current research interests include climate change, biodiversity conservation, origins of agriculture, crop domestication, and plants of economic and medicinal values. Dr. Damania has edited a number of books. One of the books he edited, *Biodiversity and Wheat Improvement*, was translated into the Chinese language. He has authored numerous papers in refereed international journals and articles in the popular press on wheat improvement, conservation of genetic resources, and Asian agricultural history.

Dr. M.C. Mackay undertook undergraduate studies in agriculture and genetics at the University of Sydney and Macquarie University in Australia. He received a PhD from the Department of Genetics and Plant Breeding of the Swedish University of Agricultural Sciences. He began his career in plant breeding at the Commonwealth Scientific and Research Organization (CSIRO), initially working with cotton and then with sunflower (1974–1980). He then took the role of senior plant breeder with Cargill Seeds, Australia, for 3 years before moving to Tamworth in New South

Wales, where he headed up the Australian Winter Cereals Collection (AWCC) for some 25 years. It was in this role that Dr. Mackay developed his main research interest in how prebreeders and breeders actually went about identifying and using plant genetic resources. This research eventually led to the development of a different approach known as *focused identification of germplasm strategy* (FIGS) in a collaborative Grains Research and Development Corporation (GRDC) project between ICARDA, the AWCC, and the N. I. Vavilov Research Institute of Plant Industry (VIR) in St. Petersburg. During 2008–2012, he led the collaborative Rome (Italy)-based Global Information on Germplasm (GIG) project for the Global Crop Diversity Trust, the secretariat of the International Treaty on Plant Genetic Resources for Food and Agriculture and Bioversity International, aimed at facilitating access to and use of genetic resources. Dr. Mackay is currently an adjunct associate professor at the Queensland Alliance for Agriculture and Food Innovation (QAAFI) based at the University of Queensland, St. Lucia, Brisbane, Australia. Dr. Mackay has published more than 40 articles, including book chapters, scientific papers, and conference proceedings.

Dr. S. Dayanandan is a professor and graduate program director in the Biology Department, Concordia University, Montréal, Canada. He received his PhD in biology from Boston University, and postdoctoral training in population and conservation genetics from the University of Massachusetts at Boston and the University of Alberta at Edmonton. He was the director of the graduate diploma program in biotechnology and genomics at Concordia University and Izaak W. Killam postdoctoral fellow at the University of Alberta. His current research work is focused on understanding the origin and maintenance of biodiversity in forest and agricultural landscapes using genomics technologies. He has published extensively in the field of biodiversity conservation, ranging from socioeconomic factors and tropical deforestation through population, conservation, and evolutionary genomics of forest trees and crop plants covering numerous countries, including Canada, China, Costa Rica, India, Malaysia, Sri Lanka, and the United States.

Contributors

A.M. Adan
Ministry of Agriculture, Livestock and Fisheries
Djibouti, Republic of Djibouti

S.B. Alaoui
Department of Agronomy and Plant Genetics
Institute of Agronomy and Veterinary Medicine Hassan II
Rabat, Morocco

A. Amri
International Center for Agricultural Research in the Dry Areas
Rabat, Morocco

T.O. Apina
Consulting Group GmbH
Nairobi, Kenya

I. Ayadi
Applied Plant Biotechnology Laboratory
National Agronomic Research Institute of Tunisia
Ariana, Tunisia

L. Bachmann
GFA Consulting Group GmbH
Giessen, Germany

A. Bari
International Center for Agricultural Research in the Dry Areas
Rabat, Morocco

H. Bchini
Applied Plant Biotechnology Laboratory
National Agronomic Research Institute of Tunisia
Ariana, Tunisia

K.Y. Belachew
Department of Agricultural Sciences
University of Helsinki
Helsinki, Finland

S. Ben Abedelaali
Institut Supérieur Agronomique de Chott Mariem
University of Sousse
Sousse, Tunisia

A. Ben Naceur
Applied Plant Biotechnology Laboratory
National Agronomic Research Institute of Tunisia
Ariana, Tunisia

M. Ben Naceur
Applied Plant Biotechnology Laboratory
National Agronomic Research Institute of Tunisia
Ariana, Tunisia

C. Biradar
International Center for Agricultural Research in the Dry Areas
Amman, Jordan

J. Borevitz
Research School of Biology
Australian National University
Canberra, Australian Capital Territory, Australia

R. Chaabane
Applied Plant Biotechnology Laboratory
National Agronomic Research Institute of Tunisia
Ariana, Tunisia

S. Chaabane
Higher Institute of Fine Arts
University Campus Mrezgua
Nabeul, Tunisia

R. Chang
Key Laboratory of Crop Gene Resource and Germplasm Enhancement (Ministry of Agriculture)
Institute of Crop Science
Chinese Academy of Agricultural Science
Beijing, China

Y.P. Chaubey
Department of Mathematics and Statistics
Concordia University
Montréal, Québec, Canada

J. Crossa
International Maize and Wheat Improvement Center
Mexico DF, Mexico

S. Dalmannsdóttir
Norwegian Institute for Agriculture and Environmental Research
Oslo, Norway

A.B. Damania
Department of Plant Sciences
University of California, Davis
Davis, California

S. Dayanandan
Centre for Structural and Functional Genomics
Biology Department
Concordia University
Montréal, Québec, Canada

L. DeHaan
The Land Institute
Salina, Kansas

A. Diederichsen
Saskatoon Research Centre
Saskatoon, Saskatchewan, Canada

R. Djurhuus
The Agriculture Center
Farroe Islands, Denmark

M. El-Bouhssini
International Center for Agricultural Research in the Dry Areas
Rabat, Morocco

I. Elouafi
The International Center for Biosaline Agriculture–Agriculture for Tomorrow
Dubai, United Arab Emirates

A. Frederiksen
Agricultural Consulting Services
House of Agriculture
Qaqortoq, Greenland

D.Z. Habash
Securewheat Research
Securewheat
St. Albans, United Kingdom

Á. Helgadóttir
Faculty of Animal and Land Resources
Agricultural University
Borgarnes, Iceland

L.T. Hickey
Queensland Alliance for Agricultural and Food Innovation
University of Queensland
St. Lucia, Queensland, Australia

B. Humeid
International Center for Agricultural Research in the Dry Areas
Rabat, Morocco

and

International Center for Agricultural Research in the Dry Areas
Terbol, Lebanon

Y. Imani
Department of Agronomy and Plant Genetics
Institute of Agronomy and Veterinary Medicine Hassan II
Rabat, Morocco

M. Inagaki
International Center for Agricultural Research in the Dry Areas
Rabat, Morocco

A. Jilal
National Institute for Agricultural Research
Rabat, Morocco

T. Kantarski
Interdepartmental Genetics Program
Kansas State University
Manhattan, Kansas

M. Karrou
International Center for Agricultural Research in the Dry Areas
Rabat, Morocco

Z. Kehel
International Center for Agricultural Research in the Dry Areas
Rabat, Morocco

H. Khazaei
Department of Plant Sciences/Crop Development Centre
College of Agriculture and Bioresources
University of Saskatchewan
Saskatoon, Saskatchewan, Canada

O. Lahlou
Department of Agronomy and Plant Genetics
Institute of Agronomy and Veterinary Medicine Hassan II
Rabat, Morocco

B. Letty
GFA Consulting Group GmbH
Institute of Natural Resources
Pietermaritzburg, South Africa

Y. Li
The National Key Facility for Crop Gene Resources and Genetic Improvement
Institute of Crop Science
Chinese Academy of Agricultural Science
Beijing, China

M. Maatougui
International Center for Agricultural Research in the Dry Areas
Rabat, Morocco

M.C. Mackay
Queensland Alliance for Agricultural and Food Innovation
University of Queensland
St. Lucia, Queensland, Australia

O.F. Mamluk
Consultant Plant Pathologist
Damascus, Syria

M. Mars
UR Agrobiodiversity
Higher Institute of Agronomy
Institution of Agricultural Research and Higher Education-University of Sousse
Sousse, Tunisia

D.H. Matthews
Department of Geography, Planning and Environment
Concordia University
Montréal, Québec, Canada

S. Mnasri
National Gene Bank
Tunis, Tunisia

A. Morgounov
International Maize and Wheat Improvement Center
Ankara, Turkey

J. Motawaj
International Center for Agricultural Research in the Dry Areas
Rabat, Morocco

S. Moufida
Applied Plant Biotechnology Laboratory
National Agronomic Research Institute of Tunisia
Ariana, Tunisia

M. Nachit
International Center for Agricultural Research in the Dry Areas
Rabat, Morocco

and

University Mohammed VI Polytechnique
Marrakech, Morocco

J.A. Nyemba
GFA Consulting Group GmbH
Douala, Cameroon

H. Ouabbou
International Center for Agricultural Research in the Dry Areas
Rabat, Morocco

M. Pagnotta
Department of Agrobiology and Agrochemistry
University of Tuscia
Viterbo, Italy

J. Poland
Department of Plant Pathology and Agronomy
Kansas State University
Manhattan, Kansas

E. Porceddu
Department of Agrobiology and Agrochemistry
University of Tuscia
Viterbo, Italy

G. Poulsen
Seed Savers of Denmark
Tjele, Denmark

L. Qiu
Key Laboratory of Soybean Biology (Beijing)
Institute of Crop Science
Chinese Academy of Agricultural Science
Beijing, China

C.O. Qualset
Department of Plant Sciences
University of California, Davis
Davis, California

M. Reynolds
International Maize and Wheat Improvement Center
Mexico DF, Mexico

M. Rouissi
Applied Plant Biotechnology Laboratory
National Agronomic Research Institute of Tunisia
Ariana, Tunisia

O. Saddoud Debbabi
National Gene Bank
Tunis, Tunisia

A. Saidi
Applied Plant Biotechnology Laboratory
National Agronomic Research Institute of Tunisia
Ariana, Tunisia

A. Sarker
International Center for Agricultural Research in the Dry Areas
New Delhi, India

S. Sayouri
Applied Plant Biotechnology Laboratory
National Agronomic Research Institute of Tunisia
Ariana, Tunisia

R. Schafleitner
Asian Vegetable Research and Development Center
The World Vegetable Center
Tainan, Taiwan

M.J. Sillanpää
Department of Mathematical Sciences
and
Department of Biology and Biocenter Oulu
University of Oulu
Oulu, Finland

C.T. Simmons
Department of Atmospheric and Oceanic Sciences
McGill University
Montréal, Québec, Canada

M. Singh
International Center for Agricultural Research in the Dry Areas
Amman, Jordan

S.Ø. Solberg
Nordic Genetic Resources Center
Alnarp, Sweden

F.L. Stoddard
Department of Agricultural Sciences
and
Department of Food and Environmental Sciences
University of Helsinki
Helsinki, Finland

K.A. Street
International Center for Agricultural Research in the Dry Areas,
Rabat, Morocco

A.G. Tessema
Institute of Biodiversity
Conservation and Research
Addis Ababa, Ethiopia

G.G. Venderamin
Institute of Genetic Improvment of Forest Plants
National Research Council
Florence, Italy

L. Woltering
GFA Consulting Group GmbH
Hamburg, Germany

F. Yndgaard
Nordic Statistic and Data Consultant
Genarp, Sweden

Section I

Climate-Change Implications for Drylands and Farming Communities

1 Climate Change and Dryland Systems

C.T. Simmons and D.H. Matthews

CONTENTS

GLOBAL TEMPERATURE INCREASE: CAUSES AND FEEDBACKS

Meteorological data and model projections suggest that significant climate changes are ongoing and are likely to continue over the course of the next century, requiring important land management and crop adaptations in order to sustain food supply to growing human populations. Global observations on surface air temperature (Hansen et al. 2010, Morice et al. 2012, and Vose et al. 2012) show that most parts of the Earth's surface have already experienced notable warming over the course of the last century, with the longest observational dataset (HadCRUT4), indicating that global temperatures have risen to 0.72°C–0.85°C in the period from 1850–1900 to 2003–2012. Furthermore, the latest assessment report (AR5) from the Intergovernmental Panel on Climate Change (IPCC) (IPCC 2013) predicts that global surface air temperatures could increase as much as 4.5°C over the course of the present century. This projected rise in temperature is largely due to increasing atmospheric concentrations of greenhouse gases (GHGs), such as carbon dioxide (CO_2), methane (CH_4), and nitrous oxide (N_2O), which absorb energy emitted by the Earth's surface (long-wave radiation) and retain a fraction of this energy in the atmosphere. Faster accumulation (than removal) of these gases in the atmosphere allows progressively more energy to be retained near the Earth's surface, causing an increase in temperature.

The warming initiated by greater concentrations of atmospheric GHGs also influences other components of the climate system, which may lead to an enhancement or a reduction of the original warming. For example, atmospheric water vapor (H_2O), which is itself a potent GHG, may account for up to 50% of the total present-day GHG-related warming (Schmidt et al. 2010). As warmer air is less dense and able to hold significantly more water vapor, increasing surface temperatures may stimulate greater evaporation and more H_2O retention in the atmosphere, thus accentuating the contribution of water vapor to GHG warming. The total contribution of the increase in water vapor to future warming, however, remains uncertain due to the influence of water vapor on cloud cover. Low- and mid-tropospheric clouds, such as stratus, stratocumulus, altostratus, and nimbostratus, reflect and diffuse a large fraction of incoming solar radiation, so an increase in these types of clouds in response to greater water vapor in the atmosphere could lead to less energy absorption at the Earth's surface (a cooling effect). However, high clouds (cirrus types) reflect less solar energy back to space, and a globally significant increase in these types may enhance the Earth's warming (Ruddiman 2008). Currently, this cloud–climate response remains a significant uncertainty in future climate projections (Sherwood et al. 2014).

Another significant contributor to climate change is the reduction of seasonal snow and sea ice at mid and high latitudes with increasing surface temperatures. As snow, glaciers, and sea ice reflect a large fraction of incoming solar radiation back to space, their reduced coverage allows the Earth's surface to absorb more radiation (Crook and Forster 2014). Observational datasets (MLOST and GISS) show that the greatest warming over the past century, in excess of 2.5°C in some regions, has largely occurred over mid- and high-latitude regions (northern Canada, Siberia, and Tibetan Plateau), where the ice and snow cover have a relatively stronger influence on annual temperatures (Hansen et al. 2010 and Vose et al. 2012). Furthermore, the melting of ice in high-latitude regions may amplify the greenhouse effect by leading to faster respiration of organic matter stored in permafrost and peatlands (Koven et al. 2011 and Belshe et al. 2013), thus releasing additional CO_2 to the atmosphere.

Human modification of the land surface for agriculture and pasture also significantly influences global climate by reducing or replacing tree coverage in some regions, leading to a net carbon (CO_2) release to the atmosphere and cooling through an increase in surface reflectivity by supplanting forests with grasses and crop varieties (Pongratz et al. 2010). Furthermore, the replacement of forests with croplands and pasture may reduce local moisture retention for precipitation, which can lead to drying that destabilizes native rainforests in tropical regions (Malhi et al. 2009). Additionally, rice cultivation and bovine pasturage are predominant sources of methane to the atmosphere (Kirschke et al. 2013), whereas fertilization and soil management in agricultural regions lead to the production of N_2O (another important GHG).

In response to total warming from GHGs and associated climate feedbacks, global climate patterns may also change, including the location of storm tracks, the intensity of monsoons, and other features of the general atmospheric circulation that affect precipitation and evaporation. The resulting new range of temperature and precipitation variations for many regions may, in turn, put stress on native ecosystems as well as traditional agricultural methods. In the following sections, we review some of the changes in global patterns of precipitation and general atmospheric circulation that are apparent from the twentieth-century observational data, as well as the projected future evolution from climate model simulations, with a particular emphasis on results for dryland climates. In the second section, we review historical data for the twentieth century and the results from the Coupled Model Intercomparison Project Phase 5 (CMIP5) for the twenty-first century, which is the basis for the IPCC AR5 report released in 2013–2014. In the third section, we discuss some of the uncertainties and limitations of climate models, and where climate models may fall short in simulating dryland climates.

OBSERVATIONS AND FUTURE CLIMATE CHANGE

Observed Changes in Precipitation and Atmospheric Circulation Relevant to Dryland Expansion

While measurements suggest that global temperatures are increasing, observational databases (IPCC 2013) also indicate that there is no discernible global trend in precipitation or changes in precipitation extremes over the course of the twentieth century. The Climate Research Unit (CRN), Global Historical Climatology Network (GHCN), and Global Precipitation Climatology Centre (GPCC) datasets show that there is a drying trend in portions of western Africa—the GHCN gives this trend a strong statistical significance (IPCC 2013). They also agree that the Mediterranean basin has received less precipitation from 1951 to 2010, whereas northern Europe and central North America have become wetter over this time period. Infrequent sampling in many parts of the world, strong local variations in precipitation (especially where convective precipitation is prominent), as well as nonstandardization of the methods used for precipitation measurement, all limit the analysis of regional and global precipitation changes. However, the trend for decreasing precipitation in the Mediterranean region is consistent with the poleward shifts in the Hadley Cell, which causes subtropical semiarid and dryland climates to expand northward (IPCC 2013). In addition to a poleward expansion of the Hadley Cell, Bender et al. (2012)—using cloud data for the period 1983–2008 from the International Satellite Cloud Climatology Project—also documented a significant poleward shift of the mid-latitude storm track over the 25-year period (i.e., away from subtropical and subtropical-adjacent regions). This trend was especially evident over the north Atlantic basin.

Hadley Cell dynamics, which influence the location of the subtropical subsidence, have a significant impact on global precipitation patterns. Trade-wind convergence near the Equator (a region known as the Intertropical Convergence Zone or ITCZ) forces air to rise in the tropics. This rising air is forced to diverge to the north and south of the equator in the upper troposphere, where its upward trajectory is deterred by the warm, stably stratified stratosphere. A mixture of radiative cooling in the upper troposphere and the Coriolis effect causes this poleward-moving air to descend (or subside) around 30° latitude, suppressing precipitation in this region. The descending branch of the Hadley Cell is thus responsible for arid and semiarid climates in the subtropics (Scheff and Frierson 2012). Recent observations suggest that Hadley Cell subsidence is expanding poleward in many regions, by as much as 2°–4.5° latitude since 1979 (Hu and Fu 2007). In the Southern Hemisphere, ozone depletion above Antarctica has led to the most notable poleward shift in the upper-tropospheric jet (because little ozone is present to absorb ultraviolet radiation in the ozone hole, the lack of upper-level stratification from ultraviolet warming leads to a deepening of the polar troposphere), which, in turn, stimulates a poleward expansion of the Hadley Cell (Scheff and Frierson 2012). Polvani et al. (2011), using a climate model, suggested that the influence of stratospheric ozone depletion on the Southern Hemisphere hydrological cycle is robust, with significant drying shifting from subtropics toward mid latitudes (centered around 45°S) and moistening in subpolar regions (60°S). However, expected ozone replenishment in the Southern Hemisphere stratosphere in the twenty-first century may eventually slow the rapid advance of the Hadley Cell poleward (Barnes et al. 2014).

Several climate models also suggest that tropospheric warming (and thickening) due to greater GHG concentrations has contributed to Hadley Cell expansion. In particular, Hu et al. (2013) reported that warming contributed 0.39° (±0.156°) of latitude of the total shift in the Hadley circulation from 1979 to 2005 in a multi-model comparison project. While this represents only a fraction of the observed poleward expansion, the same models suggest that GHGs are expected to have an increasing influence on the Hadley Cell, with the most extreme warming scenario (RCP8.5) for the twenty-first century associated with an average 0.27° (±0.04°)-latitude poleward shift every decade. Over the Northern Hemisphere, poleward shifts from September to May occur in the most of the future warming scenarios (RCP4.5, RCP6.0, and RCP8.5) (Hu et al. 2013).

Projected Changes in Drylands

Twenty Earth-system modeling groups recently employed a suite of climate models in the CMIP5 project to assess potential changes in future climate with increasing GHGs. These models produced four scenarios, called representative concentration pathways (RCP) of radiative forcing, which were created to include the total warming (from GHGs) and cooling effect (from aerosols) of anthropogenic emissions from the pre-Industrial period to the end of the twenty-first century. Each RCP scenario is named for the amount of anthropogenic radiative forcing (also measured in watts/m^2) to be achieved by the year 2100. Up to present, the contribution since the pre-Industrial period has been estimated to be approximately 2.3 watts/m^2 (IPCC 2013). RCP2.6 is the most optimistic scenario, showing a near-term and consistent decrease in fossil fuel emissions, whereas RCP4.5 provides a stabilization scenario, with a gradual increase in the first half of the century, followed by a decrease. Among the less optimistic cases, RCP6.0 denotes a longer increase in emissions followed by a decrease in the last quarter of the twenty-first century, and RCP8.5 represents the *business-as-usual* scenario with continued increasing emissions with rises in global population (the most extreme climate-warming scenario used in the climate models). These four standardized scenarios were applied in the CMIP5 climate models, with the multi-model average of certain parameters (such as precipitation change and temperature change) being used as the basis for technical reports contributing to the most recent IPCC report on climate change (IPCC 2013).

The results from simulating these four scenarios suggest that precipitation changes may become more extreme in many regions of the world as anthropogenic radiative forcing increases. In particular, compared to the 1986–2005 average, decreases in precipitation become more pronounced in subtropical regions (especially southern North America, the Mediterranean basin, and southern Africa) with greater warming (IPCC 2013). In addition, for the most extreme scenarios (RCP6.0 and RCP8.5), there is more statistical significance and model agreement for these trends (IPCC 2013, Feng et al. 2014). At the same time, precipitation was projected to increase in many regions of the world, including central Africa and India, associated with a lengthening monsoon season (Feng et al. 2014), as well as northern Eurasia and central and northern North America.

However, while these regions may receive more precipitation, only tropical monsoon-affected regions are expected to become wetter, with a drier climate expected across much of the globe for the greater warming scenarios (Sherwood and Fu 2014). This is due to the fact that the relatively large temperature increases over land lead to substantial decreases in relative humidity, which increases the potential evapotranspiration (PET, the atmospheric evaporative demand) (Lin et al. 2015). Therefore, while precipitation may increase in some regions, the moisture-holding capacity of the air over most land areas increases even more with higher temperatures, leading to greater evaporation and more land-surface drying (Fu and Feng 2014).

For this reason, when evaluating the projected changes in climate with respect to moisture availability, some recent studies use a ratio of precipitation to PET (a quantity near zero in deserts and one in wet climates), or P/PET. Statistically significant decreases in P/PET occur in the RCP8.5 scenario over much of northern and southern Africa, tropical and subtropical South America, and most of North America and Europe (Sherwood and Fu 2014). Feng and Fu (2013) used a classification of dryland categories using P/PET values (hyperarid: 0–0.05, arid: 0.05–0.2, semiarid: 0.2–0.5, and dry subhumid: 0.5–0.65) from Middleton and Thomas (1997) to track changes among 20 of the CMIP5 models. For the business-as-usual scenario, Feng and Fu (2013) found that drylands increase in global coverage by an average of 3 million km^2 in the RCP4.5 (mitigation) scenario and 7 million km^2 for the RCP8.5 (business-as-usual) case. Their findings also suggest that the greatest expansion of drylands is likely to occur in subtropical regions and poleward of subtropical regions. In particular, hyperarid drylands expand in parts of northern Africa (leeward side of the Atlas Mountains) and Iraq, whereas arid lands expand to the northern coast of Africa and parts of Anatolia, southern Africa, interior and southwestern Australia, and southwestern North America. Furthermore, subhumid climates were projected to be replaced by semiarid conditions across much of the northern Mediterranean basin and Black Sea region, parts of southern Africa, the western Sahel, and subtropical South America (Feng and Fu 2013). The shift to a drier climate regime is most significant (with 80% model agreement) for the Mediterranean and Black Sea regions and southern Africa (Feng and Fu 2013). Feng et al. (2014), classifying global climate shifts according to the Köppen-Trewartha Climate Classification scheme for RCP4.5 and RCP8.5, further suggested that drylands increase 6%–9% by mid-century and 8%–16% by the end of the century (with greater dryland conversion characterizing the RCP8.5 scenario).

Other indices used for determining the dryness of a certain geographic area are also used in global analyses. Because P/PET is a theoretical concept (with maximum potential evaporation only ever taking place in very humid or irrigated conditions), Scheff and Frierson (2015) recently suggested an alternative metric (P/(P+PET)), which provides slightly better inter-model agreement (with less variation based on the averaging period) and clarifies some of the global aridification trends suggested in Feng and Fu (2013). Their study found that drying was magnified in southern Europe, northern Africa, southern Africa, southwestern North America, and the Amazon basin for the RCP8.5 scenario. However, contrary to P/PET estimates, this metric showed a relative humidification of Siberia and north-central Eurasia among many of the models, with several models also demonstrating moistening in parts of east-central Africa and certain regions of mid-latitude and east-coastal South America (Scheff and Frierson 2015). Looking at the evaporation fraction, Cook et al. (2014) confirmed that moisture availability for evaporation in the CMIP5 models occurs both in the subtropics and poleward from the subtropics (central plains of North America and Europe in particular) in the CMIP5 models. Miao et al. (2015) used a correlation between climate (temperature and precipitation) changes and the normalized difference vegetation index (NDVI)—related to vegetation growth, with values of less than 0.1, indicating vegetation-limited desertified regions—to predict future dryland conversion in central Asia and northern China from a CMIP5-weighted model average. Their results suggest that significant drying occurs in most RCP scenarios, with NDVI change becoming increasingly negative across Uzbekistan, Turkmenistan, and the southern half of Kazakhstan in RCP2.6 from 2010 to 2100. These trends were magnified (NDVI between −0.1 and −0.5 in these same regions) for RCP4.5 and RCP8.5 scenarios (Miao et al. 2015, their Figure 6). The total desertified area in their analysis region, however, changes only slightly for RCP2.6, but increases by approximately 2% in RCP4.5 and 5% for the RCP8.5 scenarios, with much of the new desertification according to their classification projected for central and northern central Asia and northwestern China (Miao et al. 2015).

MODEL UNCERTAINTIES AND LIMITATIONS

While state-of-the-art climate models provide a baseline for potential future climate changes and dryland expansion, important caveats to the interpretation of these model experiments merit consideration. The CMIP5 results are characterized by three major forms of uncertainty: internal variability (slightly different initial conditions leading to different results for a single climate model), model uncertainty (different models yielding different results for the same RCP scenario), and scenario uncertainty (characterized by the spread of model solutions created by using different RCP scenarios, i.e., uncertainty in how future anthropogenic factors will evolve). Furthermore, these uncertainties vary for different parameters and in different regions. For example, for projected changes in global temperature, scenario uncertainty is dominant for long timescales (projections beyond 50 years in the future), but model uncertainty is dominant on shorter timescales (Hawkins and Sutton 2011). However, for global annual precipitation, model uncertainty is largest on all timescales. For regional precipitation projections, internal variability can be quite large, explaining a substantial fraction (70% to 30%) of uncertainty in winter precipitation in Europe, and 70% to 10% of summer precipitation in the Sahel, with very little contribution from scenario uncertainty for both regions (Hawkins and Sutton 2011). By contrast, greater scenario uncertainty characterizes summer precipitation in Southeast Asia (Hawkins and Sutton 2011). Such regional patterns in uncertainty for temperature and precipitation represent important limitations to global average statistical analyses. In addition to inherent model uncertainties, many of the models used in CMIP5 share common computer code, and thus the generational link between different models and different versions of the same model can compound uncertainties and skew the multi-model average further from observations (Knutti et al. 2013).

There is also some evidence that uncertainties in multi-model averaging can lead to an underestimation of drying in many regions. In particular, P/PET calculations of observational data from 1955 to 2000 suggest that global drylands have increased in area by 4%, whereas the model average greatly underestimates dryland expansion (~1.3%) over this same period (Feng and Fu 2013). It is thus possible that the projected future dryland expansion in the CMIP5 results may also be relatively modest. Some of this underestimation could be related to the multi-model averaging process itself, which leads to a loss in some of the climate signal obtained in individual model simulations. For example, for the CMIP3 models, Knutti et al. (2010) showed that nearly all models produced a precipitation decrease of more than 15% per 1°C increase in annual grid-cell temperature and more than 20% drying per 1°C during the dry season in certain land regions between 60°N and 60°S, but the multi-model average used alone underestimates these changes (~10% per 1°C for annual precipitation and ~15%–17% per 1°C for the dry season).

CONCLUSIONS

This chapter based on a suite of climate-model simulations from the CMIP5 project shows that global temperature and precipitation are likely to change over the present century with increasing anthropogenic-related warming. From these model results, a general drying trend has been projected over subtropical regions (outside of monsoon-affected areas) and some regions poleward from the subtropics. Dynamic changes to atmospheric circulation contribute to this drying; in particular, a poleward shift of the mid-latitude storm track and a poleward expansion of the subsiding branch of the Hadley Cell lead to less overall precipitation in these regions. These shifts in atmospheric circulation are also accompanied by increases in temperature (largely from GHG forcing), which facilitates dryland expansion by reducing relative humidity, resulting in greater evaporation and land-surface drying. Using a variety of predictive indices (P/PET, evaporative fraction, P/[P+PET], and NDVI), research using the CMIP5 results suggest that the greatest dryland expansion is expected for the Mediterranean and Black Sea basins, southern Africa, and southwestern North America, with some additional desertification possible in parts of central Asia and northwestern China. However, there are significant uncertainties to consider when evaluating multi-model means, and it is possible that present-day and future dryland expansion is being underestimated in the current model projections.

REFERENCES

Barnes, E.A., N.W. Barnes, and L.M. Polvani. 2014. Delayed Southern Hemisphere climate change induced by stratospheric ozone recovery, as projected by the CMIP5 models. *J Climate* 27(2):852–867.

Belshe, E.F., E.A.G. Schuur, and B.M. Bolker. 2013. Tundra ecosystems observed to be CO_2 sources due to differential amplification of the carbon cycle. *Ecol Lett* 16(10):1307–1315.

Bender, F.A., V. Ramanathan, and G. Tselioudis. 2012. Changes in extratropical storm track cloudiness 1983–2008: Observational support for a poleward shift. *Clim Dynam* 38(9–10):2037–2053.

Cook, B.I., J.E. Smerdon, R. Seager, and S. Coats. 2014. Global warming and 21st century drying. *Clim Dynam* 43(9–10):2607–2627.

Crook, J.A. and P.M. Forster. 2014. Comparison of surface albedo feedback in climate models and observations. *Geophys Res Lett* 41(5):1717–1723.

Feng, S. and Q. Fu. 2013. Expansion of global drylands under a warming climate. *Atmos Chem Phys* 13(10):10081–10094.

Feng, S., Q. Hu, W. Huang, C.H. Ho, R. Li, and Z. Tang. 2014. Projected climate regime shift under future global warming from multi-model, multi-scenario CMIP5 simulations. *Global Planet Change* 112:41–52.

Fu, Q. and S. Feng. 2014. Responses of terrestrial aridity to global warming. *J Geophys Res-Atmos* 119(13):7863–7875.

Hansen, J., R. Ruedy, M. Sato, and K. Lo. 2010. Global surface temperature change. *Rev Geophys* 48:RG4004. doi:10.1029/2010RG000345.

Hawkins, E. and R. Sutton. 2011. The potential to narrow uncertainty in projections of regional precipitation change. *Clim Dynam* 37(1–2):407–418.

Hu, Y. and Q. Fu. 2007. Observed poleward expansion of the Hadley circulation since 1979. *Atmos Chem Phys* 7(19):5229–5236.

Hu, Y., L. Tao, and J. Liu. 2013. Poleward expansion of the Hadley circulation in CMIP5 simulations. *Adv Atmos Sci* 30:790–795.

IPCC. 2013. *Climate Change 2013: The Physical Science Basis. Contribution of Working Group I to the Fifth Assessment Report of the Intergovernmental Panel on Climate Change*, eds. T.F. Stocker, D. Qin, G.-K. Plattner, M. Tignor, S.K. Allen, J. Boschung, A. Nauels, Y. Xia, V. Bex, and P.M. Midgley. Cambridge and New York: Cambridge University Press. doi:10.1017/CBO9781107415324.

Kirschke, S., P. Bousquet, P. Ciais et al. 2013. Three decades of global methane sources and sinks. *Nat Geosci* 6(10):813–823. doi:10.1038/NGEO1955.

Knutti, R., R. Furrer, C. Tebaldi, J. Cermak, and G.A. Meehl. 2010. Challenges in combining projections from multiple climate models. *J Climate* 23(10):2739–2758.

Knutti, R., D. Masson, and A. Gettelman. 2013. Climate model genealogy: Generation CMIP5 and how we got there. *Geophys Res Lett* 40(6):1194–1199.

Koven, C.D., B. Ringeval, P. Friedlingstein et al. 2011. Permafrost carbon-climate feedbacks accelerate global warming. *Proc Natl Acad Sci USA* 108(36):14769–14774.

Lin, L., A. Gettelman, S. Feng, and Q. Fu. 2015. Simulated climatology and evolution of aridity in the 21st century. *J Geophys Res-Atmos* 120. doi:10.1002/2014JD022912.

Malhi, Y., L.E. Aragão, D. Galbraith et al. 2009. Exploring the likelihood and mechanism of a climate-change-induced dieback of the Amazon rainforest. *Proc Natl Acad Sci USA* 106(49):20610–20615.

Miao, L., P. Ye, B. He, L. Chen, and X. Cui. 2015. Future climate impact on the desertification in the dry land Asia using AVHRR GIMMS NDVI3g data. *Remote Sens* 7(4):3863–3877.

Middleton, N. and D.S.G. Thomas. 1997. *World Atlas of Desertification*, 2nd ed. London: Hodder Arnold.

Morice, C.P., J.J. Kennedy, N.A. Rayner, and P.D. Jones. 2012. Quantifying uncertainties in global and regional temperature change using an ensemble of observational estimates: The HadCRUT4 data set. *J Geophys Res-Atmos* 117(D8):D08101. doi:10.1029/2011JD017187.

Polvani, L.M., D.W. Waugh, G.J. Correa, and S.W. Son. 2011. Stratospheric ozone depletion: The main driver of twentieth-century atmospheric circulation changes in the Southern Hemisphere. *J Climate* 24(3):795–812.

Pongratz, J., C.H. Reick, T. Raddatz, and M. Claussen. 2010. Biogeophysical versus biogeochemical climate response to historical anthropogenic land cover change. *Geophys Res Lett* 37(8):702. doi:10.1029/2010GL043010.

Ruddiman, W.F. 2008. *Earth's Climate: Past and Future*, 2nd ed. New York: W.H. Freeman and Company.

Scheff, J. and D.M.W. Frierson. 2012. Robust future precipitation declines in CMIP5 largely reflect the poleward expansion of model subtropical dry zones. *Geophys Res Lett* 39(18):704. doi:10.1029/2012GL052910.

Scheff, J. and D.M.W. Frierson. 2015. Terrestrial aridity and its response to greenhouse warming across CMIP5 climate models. *J Climate* 28(14):5583–5600.

Schmidt, G.A., R.A. Ruedy, R.L. Miller, and A.A. Lacis. 2010. Attribution of the present-day total greenhouse effect. *J Geophys Res-Atmos* 115(D20):106. doi:10.1029/2010JD014287.

Sherwood, S. and Q. Fu. 2014. A drier future? *Science* 343:737–739.

Sherwood, S.C., S. Bony, and J.L. Dufresne. 2014. Spread in model climate sensitivity traced to atmospheric convective mixing. *Nature* 505(7481):37–42.

Vose, R.S., D. Arndt, V.F. Banzon et al. 2012. NOAA's merged land-ocean surface temperature analysis. *B Am Meteorol Soc* 93(11) 1677–1685.

2 Plant Genetic Resources and Climate Change

Stakeholder Perspectives from the Nordic and Arctic Regions

S.Ø. Solberg, A. Diederichsen, Á. Helgadóttir, S. Dalmannsdóttir, R. Djurhuus, A. Frederiksen, G. Poulsen, and F. Yndgaard

CONTENTS

With the present increase in the emission of greenhouse gases, the average global temperature is predicted to increase by 4°C by the end of this century (Betts and Hawkins 2014). In the Arctic region, the rise could be higher than the global average and winter temperatures may increase more than summer temperatures (Christensen et al. 2007). Precipitation is predicted to rise by 10%–20% in the region (Tebaldi et al. 2011), in particular because of greater rainfall in autumn and winter. The changes will lead to a declining snow cover, which is expected to decrease by 20% by the end of this century, with storms likely to be more frequent, resulting in serious river flows and flooding (Hassol 2004). The expected changes will have a serious impact on food production globally and regionally. For the Arctic region, there are opportunities with regard to growing new crops or cultivars with a higher yield potential because of a longer growing season. However, challenging these opportunities are the negative effects of extreme weather with rapid fluctuations in temperature and reduced snow cover, increased winter damage, and risk of new pests and diseases as likely consequences. Arctic agriculture needs to be prepared for the climate of the future and plant breeding should be a component of such preparations. Omics technologies add a powerful tool to the analysis of genetic diversity and the identification of genetic resources that may come to play a role in adapting crops for climate change. The question is whether sufficient infrastructure and expertise are available locally to bring such knowledge into practical application. The structure in plant breeding enterprises has changed considerably in recent years, both in the Nordic region and globally. If the consolidation and centralization process continues, important crops can be left without breeding programs. Especially in the periphery (like our region), there is a risk that there will be no remaining recipients in a position to transform advanced knowledge into tangible benefits. We propose that climate change requires new ways of interaction, and the inclusion of lay expertise and farmer communities are especially relevant in this context. First, this contribution provides an overview of today's picture of agriculture and plant breeding in the Arctic region. Second, it presents results from a survey that focused on plant breeding and farmers' access to plant cultivars. Finally, it presents a discussion on how we can move forward by better understanding the importance of adaptation of crops to future climates, and how our acquired knowledge can be linked to local initiatives and communities. This is relevant for crop adaptation not only for the Arctic and Nordic regions but also for other marginal regions of the world.

TODAY'S AGRICULTURAL SECTOR AND PLANT BREEDING

Agricultural Production

Agricultural activities vary among the five Nordic countries. In the Arctic region, forage crops are of primary importance, but berries, vegetables, and potatoes also play a role. Greenland covers 2,166,086 km^2 but only 410,449 km^2 of the country is ice free, of which a mere 2420 km^2 is arable agricultural land (Statistics Greenland 2013). Sheep farming is by far the most important agricultural activity, but there is also some dairy production, and there are small areas with potato and vegetable production. A total of 1071 ha are used for the production of winter fodder in the form of hay (Rambøll Management Consulting 2014). In the Faroe Islands, rough grazing areas cover 1181 km^2 or 85% of the total land surface (Fosaa et al. 2006). The sheep graze the outfields the entire year; they keep to the mountainous areas during the summer and during the winter, they graze in the lowlands. The Faroe Islands are self-sufficient with dairy products where the production is based on 1080 ha of cultivated land and supplemented feeding with imported concentrates. The country's potato production covers about 20% of the Faroe Islands' requirement. Vegetables and berries are grown partly for the local market but mostly for personal consumption.

Large permanent natural grazing areas also form part of the production system in Iceland, where a population of 320,000 has a total land area of 103,000 km^2 and the average farm size is 600 ha. This is six times the size of an average farm in the United Kingdom. Iceland is self-sufficient as regards dairy and meat products, and local agriculture provides 75% of the country's demand for potatoes, 38% of the demand for vegetables, 55% of barley requirements, but only 0.1% of the wheat needs. Around two-thirds of all agricultural land is on drained bog land, most of which was turned into agricultural production between 1945 and 1980. Timothy grass (*Phleum pratense*) is the most important sown forage crop, but indigenous species, such as meadow grass (*Poa pratensis*), red fescue (*Festuca rubra*), bent grass (*Agrostis capillaris*), tufted hair grass (*Deschampsia cespitosa*), and white clover (*Trifolium repens*), prevail in the permanent grasslands. Winter survival is a key issue in the production of forages.

In northern Scandinavia, the emphasis is on meat and dairy production from sheep, cattle, and goats, but production of fresh vegetables, potatoes, and berries is important for the local market. Despite the shortness of the season in the north, 24 h daylight makes production possible; however, the cool climate and the short growing season restrict the yield potential. For the Sami people, reindeer herding is especially important, and they use the natural grazing and mountainous areas in both summer and winter. In Norway, the national policy of promoting local food production is motivated by the desire to enhance self-sufficiency. In addition, local agriculture also maintains the cultural landscape and remote settlements. The farmers are paid subsidies from national authorities, and the average farm size is small compared to the farm size in northern Sweden. In northern Scandinavia, timothy and meadow grass are the most important forage species for short-term leys, but other grasses and forage legumes are also used. Winter survival is a key issue for production. In southern Scandinavia, cereals (barley, oats, rye, and wheat) are of greater importance, but short-term leys, oil crops, potato, vegetables, and fruits and berries are also produced. In southern Scandinavia, the agricultural conditions are more similar to those in northern parts of central Europe, and cultivars from Germany and the Netherlands can be used.

In Canada, only 7.3% of the total land mass is used for agricultural production (Statistics Canada 2012). While most of the country is of arctic and subarctic climate, most agriculture production is done in the semiarid climate of western Canada and in the humid continental climate of eastern Canada. Farm numbers have dropped, and now there are about 230,000 farms in Canada. Farm size has been increasing since the middle of the twentieth century, with the average farm size now being about 300 ha. Western Canadian farms are much larger than those in eastern Canada. Large proportions of cereals, pulses, oilseeds, and meat are exported. Vegetable production has decreased in total volume, but greenhouse production in the proximity of major urban centers has expanded. Fruit production is on the rise, with blueberries accounting for nearly 50% of fruit production. The country's agricultural sector is shaped by market forces. High standards in wheat grain quality were formerly supported by national legislation; the emphasis is now more on yield. Centralized governmental structures for cereal marketing created to provide remote and small farmers with market access similar to that of large producers have been removed. All this suggests that the socioeconomic parameters for agriculture in Canada are changing at an accelerating rate.

Plant Breeding Structures: Review

Over the past few decades, the organizational and economical structures in plant breeding have changed from being part of national efforts to increase agricultural production to becoming a function of developments in global enterprises and private markets (Helgadóttir 2014). In Sweden, there are currently three enterprises engaged in plant breeding for food and agriculture, with a further two in Finland, four in Denmark, and one each in Norway and Iceland (Table 2.1). During the past few decades, plant-breeding businesses have undergone significant structural change. In some of the countries, some governmental ownership or backing for plant breeding continues, but most of the breeding activities have now been privatized. The situation in Sweden can serve as an example. In the beginning of the twentieth century, there were 8–10 different breeding companies in Sweden alone: the Swedish Seed Association, W. Weibull AB, Sockerbolaget Hilleshög AB, Carl Engström Algot Holmberg & Söner AB, Otto J. Olsson & Sons, J.E. Ohlsens Enke Malmö, L. Daehnfeldts Fröhandel Helsinborg, and Statens Trädgårdsforsök Alnarp. For decades, these businesses carried out plant breeding for almost all important agricultural and horticultural crops in the country. The emphasis was on the local or national market, and the need to provide cultivars adapted to the country's various geographic and climatic conditions. In the 1950s, the Swedish Seed Association had no less than 23 research stations or substations in Sweden, from Svalöv in the south to Hjälmarsborg and Luleå in the north (Fröier 1951). In the 1960s and 1970s, the structures changed rapidly. Swedish plant breeding expanded to other countries in Europe, with W. Weibull AB establishing

TABLE 2.1
Overview of Plant Breeding for Food and Agriculture in Sweden, Denmark, Finland, Norway, and Iceland in 2013

Country	Company/Institute[a]	Crops
Sweden	Lantmännen	Barley, oat, winter wheat, spring wheat, spring oilseed rape, forage grasses, red clover, white clover, and alfalfa
	Syngenta Seeds	Sugar beet
	Swedish University of Agricultural Sciences	Potato, fruit, and berries
Denmark	Sejet Plant Breeding	Barley (both spring and winter types) and winter wheat
	Nordic Seed	Barley and winter wheat
	DLF Trifolium	Forage grasses, red clover, white clover, and fodder beet
	LKF Vandel	Potato
Finland	Boreal	Barley, oat, rye, spring wheat, winter wheat, spring turnip rape, spring oilseed rape, pea, fava bean, potato, forage grasses, and red clover
Norway	Graminor	Barley, oat, spring wheat, potato, forage grasses, red clover, white clover, alfalfa, fruit (apple, plum), and berries (strawberry, raspberry)
Iceland	Agricultural University of Iceland	Barley, timothy, and white clover

[a] Only breeding programs carried out in the Nordic area are included in the table. A few of the companies are international with programs in other countries as well.

a breeding station in Germany and representatives in Finland, Denmark, Norway, France, Austria, Switzerland, Spain, Poland, and the USSR (Farjesson 1997). The competition in the European seed market became extremely strong. In 1971, the Swedish Seed Association and Otto J. Olsson & Sons (Hammenhög) merged into Svalöf AB. In addition, the Swedish subsidiary of Ohlsens Enke was acquired in this merger, and a few years later also the Danish Ohlsens Enke (Olsson 1997). One consequence of this consolidation process is that the number of Danish and Swedish enterprises with their own vegetable cultivars fell from 12 in the 1950s to a mere 2 in 2013, with only Findus still actively breeding vegetables.

The official variety lists from the National Plant Variety Board in Sweden (NPVB 1952, 1960, 1970, 1980, 1990, 2000, 2007, and 2013) document that since the 1960s, plant breeding in the following crops has ceased: white mustard (*Sinapis alba*), flax (*Linum usitatissimum*), hemp (*Cannabis sativa*), false flax (*Camelina sativa*), poppy (*Papaver* spp.), swede (*Brassica napus* subsp. *rapifera*), turnips (*Br. rapa* ssp. *rapa*), fodder beet (*Beta vulgaris* ssp. *vulgaris*), fava bean (*Vicia faba* var. *equina*), soybean (*Glycine max*), common vetch (*V. sativa*), blue lupine (*Lupinus angustifolius*), common bean (*Phaseolus vulgaris*), gray pea (*Pisum sativum* var. *arvense*), soup pea (*P. sativum* ssp. *sativum*), smooth brome (*Bromus inermis*), field brome (*B. arvensis*), swamp meadow-grass (*Poa palustris*), late meadow grass (*Poa* spp.), hybrid ryegrass (*Lolium* × *boucheanum*), and redtop (*Ag. gigantea*). Changes are even more dramatic for vegetable crops. Since the 1960s, breeding of the following vegetable crops has been discontinued in Sweden: vegetable types of common bean (*Ph. vulgaris*), broad bean (*V. faba* var. *faba*), cabbage (*Br. oleracea*), onion (*Allium cepa*), leek (*Al. ampeloprasum*), carrot (*Daucus carota*), parsnip (*Pastinaca sativa*), parsley (*Petroselinum crispum*), radish (*Raphanus sativus*), garden beet (*Be. vulgaris* var. *conditiva*), celery (*Apium graveolens*), dill (*Anethum graveolens*), lettuce (*Lactuca sativa*), spinach (*Spinacia oleracea*), tomato (*Solanum lycopersicum*), cucumber (*Cucumis sativus*), melon (*C. melo*), and sweet pepper (*Capsicum annuum*). In 2013, a few Swedish-bred cultivars continue to be listed, but wrinkle pea (*P. sativum* var. *sativum*) is the only vegetable species still being bred.

The situation as described for Sweden is not unique. In other Nordic countries, breeding programs for minor crops have ceased and research stations have been centralized or closed. The same trend is evident in Canada, Russia, and the entire circumpolar area, albeit with a few exceptions, such as the breeding program for barley in Iceland, where cultivars such as *Skegla*, *IsKría*, *IsLomur*, and *IsSkumur* have been released. Key breeding objectives for barley breeding in Iceland were early maturity, wind resistance, and maturity at low temperature. In Sweden, Lantmännen has breeding activities for forages at Lännäs in the northern part of the country, and the Norwegian company Graminor has some activities in Tromsø in northern Norway. Apart from these exceptions, breeding and research for new cultivars for the Arctic are limited. A few ongoing or recently finished research projects have concentrated on prebreeding objectives. A Norwegian national research project VARCLIM (Understanding the genetic and physiological basis for adaption of Norwegian perennial forage crops to future climates) examined the basis for adaptation to future climates by investigating important traits in timothy (*Phleum pratense*), perennial ryegrass (*Lolium perenne*), festulolium (*Festuca* × *Lolium*), red clover

(*T. pratense*), and alfalfa (*Medicago sativa*) in existing breeding populations (Østrem et al. 2014). In 2003, the Norwegian Genetic Resource Center initiated a project concerning on-farm methods for conservation and further genetic adaptation of forage crops (Asdal 2008). The Nordforsk's Nordic Forage Crops Genetic Resource Adaptation Network (NOFOCGRAN) has formed a platform for researchers and students to develop knowledge, methods, and germplasm as the basis for future development of cultivars of perennial forage grasses and legumes in the Nordic region (Barua et al. 2014).

In Canada, plant breeding was established in the first instance by the national government and later by provincial governments and universities (Slinkard and Knott 1995). While some agricultural production of maize, garden beans, squash, sunflowers, and tobacco was historically carried out by Iroquoian people in eastern Canada, European settlement has shaped modern Canadian agriculture. Landraces of the above-mentioned crops may continue to exist as relicts, but this has not been explored systematically (Gros-Louis and Gariépy 2013). Moreover, landraces brought by the early peasant immigrants from different regions of Europe have mostly been lost and were often poorly adapted to the new environment. As no structure existed to meet the very urgent demand for such adapted crops, Canada recognized that it was a governmental task to establish plant introduction, evaluation, and breeding activities to ensure the sustainability of the country's agriculture. The public plant-breeding efforts assumed that seeds of the released cultivars would be produced on farms and that the primary client was the farmer. Resistance to seed-borne diseases has therefore always remained one of the breeding objectives. Until the late twentieth century, Canadian government was engaged in plant-breeding efforts in most horticultural and agricultural crops grown in the country, and newly founded universities also took part in this public sector effort. Governmental entities even engaged in ornamental plant breeding, as it was recognized that beauty was a necessity for emotional survival under the harsh conditions faced in remote areas. Today, Canada is part of the globalized market economy and the nation's agricultural settlement has been completed. The crops now being given breeding attention are those that are highly profitable. The governmental sector is retracting from producing and releasing cultivars to the market but continues to act in germplasm development and maintains the national genebank: Plant Gene Resources of Canada. In western Canada, plant breeding is shifting to the private sector, and the focus is on main crops such as barley, oat, wheat, flax, Brassicaceae oilseeds, and mustards. The number of plant breeders has not fallen, but the number of crops that are worked on has decreased (Clarke 2011). Breeding of forage crops and vegetables has almost completely ceased in Canada.

Stakeholder Survey in the Arctic

To attain a better understanding of how key actors in the Arctic food system view the situation, an online survey was conducted in 2013. In the survey, we presented a number of different claims related to plant breeding and farmers' access to varieties in the region (Table 2.2). Key institutions covering plant-breeding enterprises, research institutes, farmers' organizations, authorities, NGOs, and others were invited to participate. The survey was sent via an email link to 95 key persons of these institutions

TABLE 2.2
Overview of the Various Claims from the Survey and Their Abbreviations

Description	Abbreviation
Lack of good cultivars constrains Arctic food production	LACK
Cultivars for the Arctic are better now than 20 years ago	BT20
There is a need for new cultivars in the Arctic	NEWN
There is a proper testing of cultivars in the Arctic	PROT
There is official will/support for variety testing	OWIL
Breeding enterprises will/support to variety testing	WTSU
Nordic cooperation on variety testing is important	NCOI
Information for Arctic cultivars is appropriate	IAPP
Old varieties might be of interest for the Arctic	OLDI
The Nordic Gene Bank (NordGen) is important for the Arctic	NGBI
There is proper plant breeding for the Arctic	PROB
A Nordic cooperation on plant breeding for the Arctic is important	NPBI
A small market limits the breeding of new cultivars	MLIM

using a Google Form. A total of 42 respondents completed the survey. Despite the low number of respondents, the survey was regarded useful because of the key roles of those who did respond. The response to the various claims was given on a scale from 1 = strongly agree to 5 = strongly disagree.

The survey results are shown in Figure 2.1. Around 90% of the responses strongly agreed or agreed with the claim that there is a need for new cultivars for the Arctic (NEWN, median score 1). The majority disagreed with the claim that there is proper plant breeding in the Arctic (PROB, median score 4). Furthermore, the majority strongly agreed with the claims that a Nordic cooperation on plant breeding is important for the Arctic and that the Nordic Gene Bank/NordGen is important for the Arctic (NPBI and NGBI, both with median score 1). Many agreed that reintroducing old cultivars (which are no longer on the variety lists) might be of interest (OLDI, median score 2).

Regarding variety testing, around 65% of the responses strongly disagreed or disagreed with the claim that there is proper variety testing in the region, that there is an official will/support for variety testing in the region, and that there is a will/support from breeding enterprises/institutions for variety testing (all with median score 4). Farmers and farmers' organizations were more positive than others to the claim that there is a proper variety testing for the Arctic and that there is a will/support for variety testing. The results show that there is great potential for cooperation across national borders, and that the NordGen is considered important for the region.

A principal component analysis was carried out for assessing whether there were patterns in the responses based on region or profession, but no clear pattern was detected. Three clusters of respondents could be identified, but these did not relate to country/region or other identifiers such as age, sex, education, occupation, and crop expertise (results not shown).

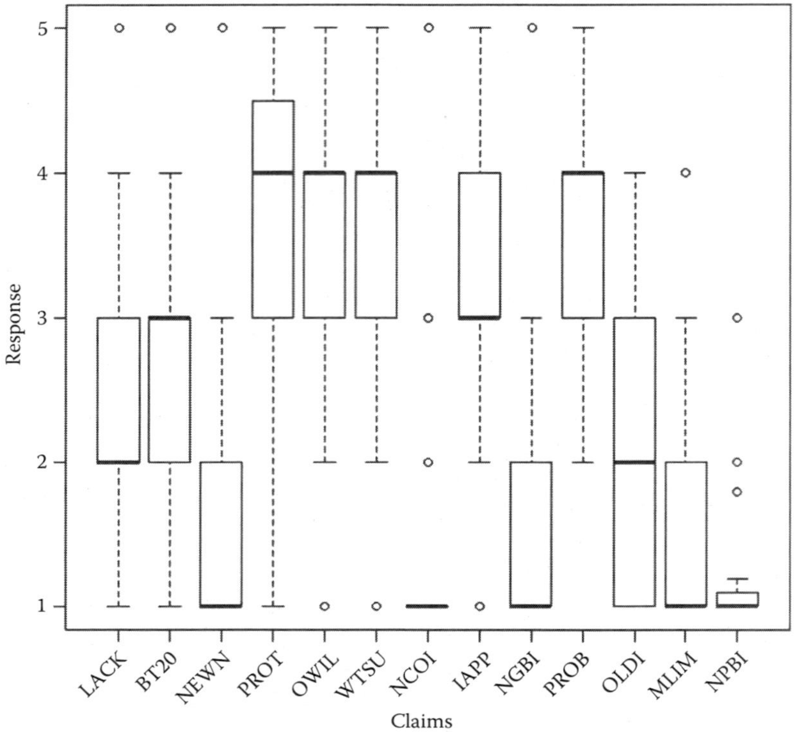

FIGURE 2.1 Box plots of the results on a 1–5 scale, where 1 = strongly agree and 5 = strongly disagree with the various claims (for abbreviations, see Table 2.2).

KEY ISSUES AND FUTURE PERSPECTIVES

The review given above and the stakeholder survey show that there are only a few active plant-breeding programs in the Arctic at present and that access to adapted cultivars is a concern. The number of research stations in the region is also limited. Facing the challenges of climate change and given existing structures, there is a need for new ideas and new types of interaction. Subbarao et al. (2005) brought up the need for decentralized breeding for climate-change adaptation, when discussing breeding for drought tolerance in semiarid parts of the world. The same approach may be valid for other places. The agro-climatic conditions and production systems are very diverse in the Nordic and Arctic regions. Genetic resources are central in crop improvement, but simply finding a magical phenotype in these regions that can solve all the problems is not at all realistic. Genebanks are one of the major access points for genetic resources for breeders and researchers. Omics technologies may play an important role in discovery of genetic resources for climate-change adaptation, but if desirable attributes are found, these must then be transferred into high-yielding locally adapted cultivars. This is not a straightforward process, especially if we relate this to the breeding structures described above. There is a need for an

Arctic bridge—a network across countries and institutions to support collaboration in research and breeding for new cultivars and implementing them into farmers' fields. New structures need to be developed, allowing for participation and recognition of different forms of expertise (Waterton and Wynne 1996 and Turnhout et al. 2012). Moreover, it would be desirable to include lay expertise and farmer communities in the process of obtaining plants adapted to particular environments. This is especially relevant for marginal regions in the world from which both public research and private breeding enterprises have been withdrawing for many years. From our perspective, the chain *genebank–research/breeding–seed/companies–farmers* need to be supplemented by new networks and linkages. Technologies should be discussed and implemented in a transdisciplinary fashion (Wickson and Wynne 2012), producing robust knowledge and long-term social sustainability. Some recent examples on how genebanks can interact to identify valuable genetic resources are given below. These are only small-scale activities in an area filled with numerous opportunities.

Interactions with Civil Society and Local Authorities

The genebank at NordGen is a Nordic cooperation involving five countries. It maintains more than 30,000 seed samples and distributes around 6000 accessions annually. Around 8% of these are delivered for breeding, 36% for research, and close to 50% for private use (farmers and home gardeners). The termination of vegetable breeding in the Nordic region may explain for farmers' and home growers' high level of interest, but such requests do not necessarily lead to any crop improvement or new developments. However, the inclusion of farmers and gardeners in this process can bring genetic resources and evaluation projects into remote regions. One recent project is a cooperation of NordGen with the Danish Seed Savers (SSD), an NGO with 450 active members. The aim of the project is to reintroduce local varieties by cultivating and evaluating 26 Danish pea and bean varieties. Workshops were organized for the stakeholders and the most suitable types were multiplied, marketed, and thus distributed to the general public. Videos were made on how to describe and maintain unique varieties. Another project focuses on asparagus, with *Danish Giant*, a Danish asparagus landrace, being reintroduced. The landrace has a good reputation, but finding it outside the genebank was difficult.

In Sweden and Norway, NordGen has been active in promoting conservation of varieties. These are landraces and old cultivars adapted to local conditions. They are no longer part of agricultural communities and can only be found in genebanks. NordGen has therefore been listing such varieties and taken on the role as the official maintainer of these old varieties and lists them in its inventory of accessions. Farmers can use and sell seeds from these varieties according to national seed legislation. Farmers have joined with micro-companies in order to develop local varieties and on-farm breeding. All such genebank interaction with civil society and local businesses may become part of an innovative approach and an integrative solution, both for evaluation of genebank accessions and for adaptation to climate change. Omics technology may play a role in such interactions. There is not one single solution to the challenges of the future, and different actors and stakeholders must be included in addressing these issues. Genebanks can play an important role in this context.

Genebanks, other than NordGen, with materials highly relevant for the Arctic region are the national genebanks in Canada and Russia. The Canadian genebank preserves and provides access to plant genetic resources of more than 100,000 accessions and is located at three different sites in Canada. Cooperation with newly emerging not-for-profit organizations to share know-how in diversity conservation and utilization is on the rise in Canadian genebank activity. The genebank works with groups such as Seeds of Diversity and the Bauta Family Initiative on Canadian Seed Security. Increasing attention is being given in Canada to Aboriginal agriculture, increased self-sustainability in food production in northern areas, and the concept of relating socially marginalized people in remote areas to food production.

Public–Private Partnerships

A public–private partnership for prebreeding has been established in the Nordic region. The purpose of the partnership is to support the development of Nordic plant breeding to meet the long-term needs of the agricultural and horticultural industries, especially regarding adaptation to climate change. Projects involving prebreeding of ryegrass (*L. perenne*), barley (*Hordeum vulgare*), and apple (*Malus domestica*) have already been started. There is a need to move forward in discovering genetic resources for climate-change adaptation and to include these in breeding cooperation across disciplines, borders, and enterprises. One of the key challenges in this context is the building of trust; although such partnerships require a long-term commitment, funding is often only secured for one or a few years.

CONCLUDING REMARK

Climate change will have an enormous impact on the food system. There is a contradiction between the need for new cultivars identified in our study and the closing down of, or reduction in, breeding efforts that has occurred over recent decades. To meet future challenges, new initiatives must be established that accomplish and sustain access to and use of plant genetic resources. Active, efficient, and comprehensive collaborative networks of institutions must be maintained across borders to facilitate joint solutions in the production of cereals, potatoes, vegetables, fodder crops, berries, and other crops in the Arctic and Nordic regions as well as in other marginal regions of the world.

REFERENCES

Asdal, Å. 2008. *State of plant genetic resources for food and agriculture in Norway*. Second Norwegian National Report on conservation and sustainable utilisation of Plant Genetic Resources for Food and Agriculture. Commissioned report from Skog oglandskap 19/2008, Ås, Norway.

Barua, S.K., P. Berg, A. Bruvoll et al. 2014. *Climate change and primary industries: Impacts, adaptation and mitigation in the Nordic countries*. TemaNord 2014:552. Copenhagen, Denmark: Nordic Council of Ministers.

Betts, R.A. and E. Hawkins. 2014. Climate projections. In *Plant genetic resources and climate change*, eds. M. Jackson, B. Ford-Lloyd, and M. Parry, 38–60. Wallingford, CT: CABI.

Christensen, J.H., B. Hewitson, A. Busuioc et al. 2007. Regional climate projections. In *Climate change 2007: The physical science basis. Contribution of Working Group I to the Fourth Assessment Report of the Intergovernmental Panel on Climate Change*, eds. S. Solomon, D. Qin, M. Manning et al., 849–940. Cambridge: Cambridge University Press.

Clarke, J.M. 2011. *Trends in crop breeding research capacity in western Canada*. Saskatoon, Canada: Ag-West Bio, Inc.

Farjesson, F. 1997. Weibullsholm. Familieföretakets utveckling 1870–1980. In *Den svenska växtförädlingens historia*, ed. G. Olsson, 35–50. Stockholm, Sweden: Kungl. Skogs- och Lantbruksakademien.

Fosaa, A.M., M. Gaard, E. Gaard, and J. Dalsgarð. 2006. *Føroya náttúra: lívfrøðiligt margfeldi*. Torshavn, Faroe Islands: Nám.

Fröier, K. 1951. Institutioner och privatföretag, som bedriva växtförädling i Sverige. In *Svensk växtförädling*, eds. Å. Åkerman, F. Nilsson, N. Sylvèn, and K. Fröier, 699–729. Stockholm, Sweden: Natur och Kultur.

Gros-Louis, M. and S.G. Gariépy. 2013. *Aboriginal agriculture and agri-food in Quebec, status report and considerations for developing a knowledge and technology transfer strategy*. Ottawa, Canada: Ministry of Agriculture and Agri-Food.

Hassol, S.J. 2004. *Impacts of a warming Arctic. Arctic climate impact assessment*. Cambridge: Cambridge University Press.

Helgadóttir, Á. 2014. Why is public plant breeding important in the Nordic region? *J Swedish Seed Assoc* 123:23–28.

NPVB. 1952. *Förteckning over sorter och stammar, som Orginalutsädesnämden som äro berättigad till statsplombering 1952–1953 (Official Swedish list of cultivars)*. Stockholm, Sweden: Emil Kihlströms Tryckeri AB.

NPVB. 1960. *Rikssortlista—omfattande sorter och stammar, som Orginalutsädesnämden förklarats berättigade till statsplombering sesongen 1960–1961 (Official Swedish list of cultivars)*. Helsingborg, Sweden: Schmidts Boktryckeri AB.

NPVB. 1970. *Official Swedish list of cultivars 1970–1971*. Helsingborg, Sweden: Schmidts Boktryckeri AB.

NPVB. 1980. *Rikssortlista 1980–1981*. Meddelande från Statens Växtsotnämnd (National Plant Variety Board, Sweden) 1980(2).

NPVB. 1990. *Sortlista 1990–1991 sorter godkända för statsplombering, Official Swedish list of cultivars*. Meddelande från Statens Växtsotnämnd (National Plant Variety Board, Sweden) 1990(2).

NPVB. 2000. *Sortlista 2000, National list of plant varieties*. Meddelande från Statens Växtsotnämnd (National Plant Variety Board Gazette, Sweden) 2000(2).

NPVB. 2007. *Sortlista 2007, National list of plant varieties*. Meddelande från Statens Växtsotnämnd (National Plant Variety Board Gazette, Sweden) 2007(2).

NPVB. 2013. *Sortlista 2013, National list of plant varieties*. Plant Variety Gazette from the Swedish Board of Agriculture 2013(2).

Olsson, G. 1997. Otto J Olsson & Son i Hammenhög. In *Den svenska växtförädlingens historia*, ed. G. Olsson, 77–80. Stockholm, Sweden: Kungl. Skogs- och Lantbruksakademien.

Østrem, L., P. Marum, and A. Larsen. 2014. Framtidige plantegen—kor finst dei? *Bioforsk FOKUS* 9(2):66.

Rambøll Management Consulting. 2014. *Where can development come from? Potentials and pitfalls in Greenland's economic sectors towards 2025*. Copenhagen, Denmark: Rambøll.

Slinkard, A.E. and D.R. Knott. 1995. *Harvest of gold, the history of plant breeding in Canada*. Saskatoon, Canada: University Press.

Statistics Canada. 2012. *Snapshot of Canadian agriculture*. http://www.statcan.gc.ca/pub/95-640-x/2012002/00-eng.htm.

Statistics Greenland. 2013. Greenland in figures 2013. In *Greenland: Greenland statistics*, ed. D. Michelsen, 21. Nuuk, Greenland: Statistics Greenland.

Subbarao, G.V., O. Ito, R. Serraj et al. 2005. Physiological perspectives on improving adaptation to drought—Justification for a systemic component-based approach. In *Handbook of photosynthesis*, ed. M. Pessaraki, 2nd edn, 577–594. Boca Raton, FL: CRC Press/Taylor & Francis Group.

Tebaldi, C., J.M. Arblaster, and R. Knutti. 2011. Mapping model agreement on future climate projections. *Geophys Res Lett* 38:L23701.

Turnhout, E., B. Bloomfield, M. Hulme et al. 2012. Conservation policy: Listen to the voices of experience. *Nature* 488:454–455.

Waterton, C. and B. Wynne. 1996. Building the European Union: Science and the cultural dimensions of environmental policy. *J Eur Public Policy* 3:421–440.

Wickson, F. and B. Wynne. 2012. The anglerfish deception. *EMBO Reports* 16:100–105.

3 Adaptation of Farmers to Climate Change

A Systems Approach to Cereal Production in Benslimane Region, Morocco

S.B. Alaoui, A.M. Adan, Y. Imani, and O. Lahlou

CONTENTS

Recent model results suggest that not only is precipitation in North Africa likely to decrease between 10% and 20% but also temperatures are likely to rise between 2°C and 3°C by 2050 (Schilling et al. 2012). This is detrimental to the agricultural sector in Morocco, which accounts for 15% of GDP and 40% of employment, and which is particularly volatile and dependent on weather conditions. The dependency on agriculture and on water in particular is making Morocco one of the most sensitive and vulnerable countries to climate change (Schilling et al. 2012).

A recent study by Rochdane et al. (2014) combining climate satellite data with socioeconomic data has shown that the already high vulnerability of Morocco for food and wood production is likely to be exacerbated with climate change and population increase. The study used a vulnerability index (VI) to assess whether the population needs (demand) in terms of food and wood products, including fiber, are produced by the landscape (supply):

$$VI(D, S) = \frac{D(Pt, Pd, Te, Af)}{S(T, P)}$$

where:

D represents demand
S is supply
Pt is the population total
Pd is the population distribution (urban vs. rural)
Te is a technological development indicator representing the efficiency with which the landscape supply is produced and includes harvest and processing efficiencies as well as transport and crop-residue losses (Rochdane et al. 2014)

The vulnerability of regions to climate change is also expressed by the relative reduction in the availability of land for agriculture in 2050 compared to what is available in the current period.

The overall population is vulnerable and agriculture is more susceptible to changing conditions in Morocco; for example, the drought of 1994–1995 reduced the agricultural GDP by 45% and the national GDP by 8%. Agricultural production is mainly rainfed, accounting for 85% of useful agricultural area (UAA) versus 15% for irrigated land. Almost three quarters of arable land are used to grow cereals (Achy 2012a,b). Agricultural production, therefore, goes through significant annual fluctuations primarily related to weather conditions and erratic rainfall. In addition, the UAA is dominated by cereals (57% of the UAA [MAPM 2012]), whose performances are highly correlated to the amount and seasonal distribution of rainfall.

Wheat consumption per capita in Morocco, estimated at 258 kg annually, is among the highest in the world. In the past decade, consumption has been driven by rising population coupled with diversification of bread products, especially in major cities where the higher income population tends to be concentrated (Ahmed and El Honsali 2012). However, Moroccan wheat production does not fully cover the national needs, and massive importation is common. Indeed, wheat alone accounted for up to 38% of total agricultural imports in 2002 (HCP 2003).

GENERAL CHARACTERISTICS OF THE MOROCCAN CLIMATE

Geographically, Morocco lies between two climate zones: the anticyclone of the Azores (to the west) and the Saharan depression (to the southeast). Due to its geographical location in an arid to semi-arid region of the globe, Morocco for millennia has been strongly linked to regional climate variability and change. This climate is characterized by an inter-annual variability of rainfall, with lower rainfall in the southern part, a number of days of very limited rainfall (less than 50 days over a large part of the country), and episodes of periodic and frequent droughts (Yessef and Saltani 2009).

Since 1973, which coincided with the great drought of the Sahel, the Maghreb in general, and Morocco in particular, experienced an abrupt climate change (Tabet-Aoul 2008), with greater occurrence of drought: one year out of three. This drought has been especially felt in the 1980s (1981, 1983, and 1987), it persisted in the 1990s (1992, 1993, 1995, and 1999), and in the first decade of the third millennium (2000, 2002, 2005, and 2007). The rainfall and its intra- and inter-annual distribution are random (Barakat and Handoufe 1998, Balaghi 2000). Intense floods occurred during 1995–2000–2002–2006–2008. Temperature increase has ranged between 1°C and 1.4°C, with an accentuation of heat waves in all seasons at the expense of cold waves.

Climate Projections for Morocco: Increased Risk of Drought

Although drought is not a recent problem, its frequency has been increasing with recent climate change. Indeed, according to the latest Intergovernmental Panel on Climate Change (IPCC), the leading international body for the assessment of climate change, the climate is predicted to become even hotter and drier in most of the Middle East and North Africa (MENA) region. Higher temperatures and reduced precipitation will increase the frequency and duration of droughts, an effect that is already materializing in the Maghreb.

Indeed, according to Morocco's National Communication to the United Nations Framework Convention on Climate Change (UNFCCC) (MATUHE 2001), climate projections anticipate for 2020 the continuation and acceleration of these trends: (1) increase in annual average temperature by 0.6°C (2015), from 2.1°C to 2.9°C (2045), and 3.2°C to 4.1°C (2075); (2) decrease in rainfall by 6% (2015), 13% (2045), and 19% (2075); (3) disruption of seasonal rainfall (winter rainfall concentrated in a short period) with reduced duration of snow cover and withdrawal of the snowpack (migration altitude of 0°C isotherm and accelerated snowmelt); (4) increase in frequency and intensity of heat waves (*chergui*, etc.); and (5) increase in frequency and intensity of droughts in the south and east of the Atlas mountains with the possibility of increased salinization of groundwater and soil. The level of oceans will rise by 0.6–1 m by 2100, with a rise in temperature and increased salinity of seawater. All these threats combined already place Morocco among the countries most threatened by climate change (Yessef and Saltani 2009).

Since 1980, Morocco has experienced a 25% decrease in average rainfall, accompanied by an increase in the frequency and severity of droughts (Barakat and Handoufe 1998, Skees 2001, Azzam and Sekkat 2005). The expected decline of the agricultural sector's contribution to growth in Morocco has numerous repercussions on economic and social development. Thus, drought directly affects the level of grain production, which is the essential ingredient in the diet of Moroccan families, particularly middle- to low-income families (Achy 2012a,b).

The vast majority of farmers who are affected by changing climate conditions have a small area of land. In fact, 7 out of 10 farmers have no more than 2.1 ha of land and suffer recurring droughts in the absence of protection mechanisms that meet their needs and suit their means (Achy 2012a,b). A decrease in cereal yields by 50% in dry years (yielding only 14 million quintals) and 10% in normal years (yielding only 51 million quintals) is threatening the 60-million-quintals level of the security food program set by the Moroccan Department of Agriculture. The need for cereals in 2020 would be in the order of 130 million quintals, 85 million of which is for human consumption (MATUHE 2001).

According to the same report (MATUHE 2001), which is based on the observations, experiments, and analyses made by the Moroccan National Institute for Agricultural Research (INRA, Balaghi et al. 2010), among other impacts on agriculture, we may expect the following:

- A reduction in crop cycles
- A shift and reduction of the growth period
- An increase in risks of dry periods at the beginning, middle, and end of the crop cycle
- Migration of the arid zone toward the north
- Extinction of some crops in the north (such as canary grass) and reduction of the density of some forest tree species in the south, such as argan (*Argania spinosa*)
- The appearance of new diseases (e.g., the white fly of tomatoes whose infestation is favored by particular climatic conditions)

The length of the growth period for wheat has decreased in some parts of Morocco. Indeed, the analysis of the evolution of the length of the growth period between 1960 and 2000, conducted by Benaouda and Balaghi (2009), in the region of Khouribga, shows a net reduction from 180 days during the period 1960–1965 to 110 days in 1995–2000 (Figure 3.1). This change is felt by farmers in this region who are growing more cereals with shorter growth cycles, such as barley, which is better adapted to a new environment (Benaouda and Balaghi 2009).

This shift and shortening of the length of the growing period in time, if it is confirmed in this region and in other rainfed agriculture areas in Morocco, require adaptations including cultivation techniques (planting dates, varieties with shorter growing cycles, etc.). Studies on farmers' perceptions based on methods that included social science (Soleri and Cleveland 2001) have shown that farmers have their own experiences and perceptions of genetic variation and its relation to environmental variation and heritability. The study reported here aimed to (1) review the literature on climate change and its impacts on agriculture in Morocco and (2) identify and analyze strategies and agricultural practices for adapting to climate change that are being adopted by farmers in the Province of Benslimane, Morocco, notably through surveys using a participatory approach with carefully selected farmers. This study thus focuses most on Te, the technological development indicator from the equation given above from Rochdane et al. (2014), aiming to highlight the importance and possibilities of increasing efficiency with which the landscape supply is produced, combining agronomic and genetic options within a systems approach and farmers' perspectives.

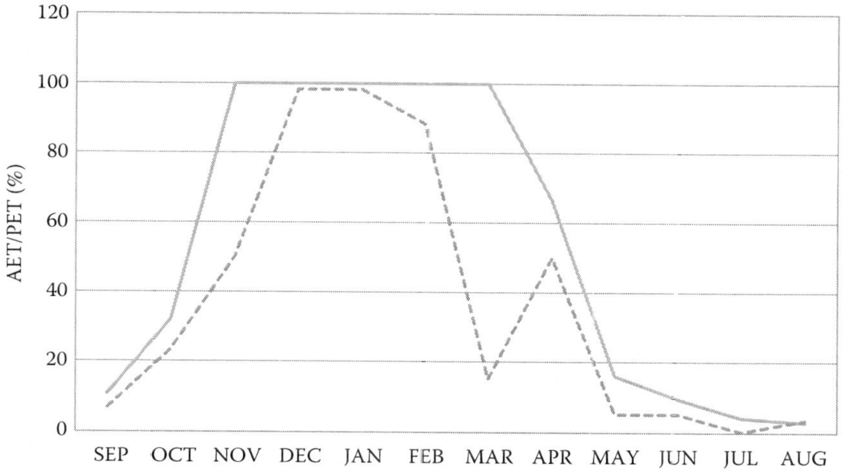

FIGURE 3.1 Length of growing period for wheat in Khouribga region, Morocco in the year 1960 (continuous line) and the year 2000 (discontinuous line). PET (potential ET) is the expected evapo-transpiration if water is not a limiting factor for given prevailing conditions of wind, temperature, and humidity. AET (actual ET) is the actual evapo-transpiration, which is the amount of water that is actually abstracted (given that water may be a limiting factor). (Modified from Benaouda, H. and Balaghi, R., Les changements climatiques: Impacts sur l'agriculture au Maroc, *Proceedings Symposium International sur L'agriculture durable en région Méditerranéenne [AGRUMED]*, Rabat, Morocco, 42–47, May 14–16, 2009.)

MATERIALS AND METHODS

Selection of the Study Area

As part of a project aimed at integrating climate-change impacts in the implementation of the Green Morocco Plan (Balaghi et al. 2010), the Ministry of Agriculture and Fisheries ranked the 16 administrative regions of Morocco according to how important would be the impact of climate change on them. Table 3.1 shows the six most vulnerable regions. The vulnerability of regions to climate change is expressed by the relative reduction in the suitability of land for agriculture in 2050 compared to the current period. The agricultural potential of the regions is expressed in Morocco by the total annual rainfall, because it is directly related to crop productivity. This classification of administrative regions is thus based on the product of these two criteria; the targeted areas are those whose products of both criteria are highest.

We have therefore chosen the Chaouia-Ouardigha region for this study because it is the most vulnerable. Within this region, we have chosen the province of Benslimane, which is divided into two zones (Benslimane and Bouznika), and within the Benslimane zone, which is itself divided into three caïdats (districts), we selected the caïdat of Fdalate. Then, because the selected area is still large, we limited the survey area to the cereal production basin, represented by the two rural municipalities of Fdalate and Moualine El Oued.

TABLE 3.1
The Six Most Vulnerable Administrative Regions Ranked according to Vulnerability to Climate Change (CC) and Agricultural Potential

Administrative Region	Vulnerability to CC (%)	Agricultural Potential (mm)	Ranking
Chaouia-Ouardigha	80	422	1
Grand Casablanca	79	408	2
Rabat-Salé-Zemmour-Zaër	37	524	3
Tadla-Azilal	33	523	4
Doukkala-Abda	51	334	5
Gharb-Chrarda-Beni Hssen	23	607	6

Source: Balaghi, R., M. Jlibene, and H. Benaouda. 2010. *Rapport de faisabilité du PICCPMV (Projet d'Intégration du Changement Climatique dans la Mise en oeuvre du Plan Maroc Vert)* Institut National de la Recherche Agronomique (INRA) and Development Finance Consultants. http://www.inra.org.ma/docs/environ/piccpmvfaisab.pdf.

Selection of Crop

The vulnerability of different crops is measured by the relative reduction in their productivity projected for 2050 compared to the current period. The priority crops are those whose vulnerability and acreages are the most important. Table 3.2 shows the ranking of vulnerable crops in the five most vulnerable regions (Table 3.1), with the exception that the second-most vulnerable region, Grand Casablanca, is not included because of its small agricultural area. In the target region of Chaouia-Ouardigha,

TABLE 3.2
The Most Vulnerable Crops in the Most Vulnerable Regions

Chaouia-Ouardigha	Rabat-Salé-Zemmour-Zaër	Tadla-Azilal	Doukkala-Abda	Gharb-Chrarda-Beni Hssen
Barley	Durum wheat	Barley	Barley	Soft wheat
Soft wheat	Barley	Soft wheat	Durum wheat	Durum wheat
Durum wheat	Durum wheat	Durum wheat	Soft wheat	Barley
Maize	Oat	Olive	Maize	Sunflower
Faba beans	Maize	Almond	Faba beans	Faba beans
Lentils	Lentils	Faba beans	Olive	Chick pea
Olive	Olive	Maize	Oat	Olive
Oat	Faba beans	Lentils	Chick pea	Maize
Chick pea	Sunflower	Vetch	Lentils	Lentils

Source: Balaghi, R., M. Jlibene, and H. Benaouda. 2010. *Rapport de faisabilité du PICCPMV (Projet d'Intégration du Changement Climatique dans la Mise en oeuvre du Plan Maroc Vert)* Institut National de la Recherche Agronomique (INRA) and Development Finance Consultants. http://www.inra.org.ma/docs/environ/piccpmvfaisab.pdf.

cereals are the most vulnerable to climate change. Therefore, our choice for this study focused on common wheat (soft wheat).

Sampling of Surveyed Farmers

The surveyed population was farmers of the Benslimane province practicing cereal production. The farms growing common wheat were classified as *large* farms, with area greater than 20 ha, *average* farms, with 5–20-ha area, and *small* farms, with area less than or equal to 5 ha. This classification allowed the inclusion of all types of farms in the data collection following a purposeful sampling strategy used successfully in sampling and conserving biodiversity (Haenn et al. 2014). The survey was conducted with a group of resource persons with a thorough knowledge of each of the municipalities studied, namely the Engineers of the Provincial Directorate of Agriculture (CCA) and the Benslimane Center for Agricultural Extension.

Data Collection

During the investigation, the information collected (from as far back in time as possible) was as follows:

- Inventory of farm equipment
- Changes over time in the yield of common wheat
- The varieties of common wheat grown
- Planting dates
- Rotations performed
- Different diseases observed
- All inputs, from soil preparation to harvest

The interviewed farmers were also asked to give opinions on changes in agricultural calendars and dates and to describe how their practices changed because of climate change.

RESULTS AND DISCUSSION

Impact of Climate Change on Farmers' Practices: The Most Prevalent Crops and Crop Rotations

Common wheat is the main crop. It is grown by most of surveyed farmers, 89% of the study sample (data not shown). It appears from the study that the main previous condition before a rotation to wheat is *tilled fallow* (Figure 3.2). About 41% of farmers practice this rotation. In second place is the use of grain legumes (chick peas, peas, faba beans, and lentils) before wheat, with 32% of farmers practicing this rotation. These practices are justified, primarily by the early release of fields for the execution of tillage and installation of cereal crops and the need to maintain soil fertility, as legumes leave a nitrogen-rich soil. Indeed, fallowing and legumes leave more nitrogen in the field for the wheat compared to any other rotation crop.

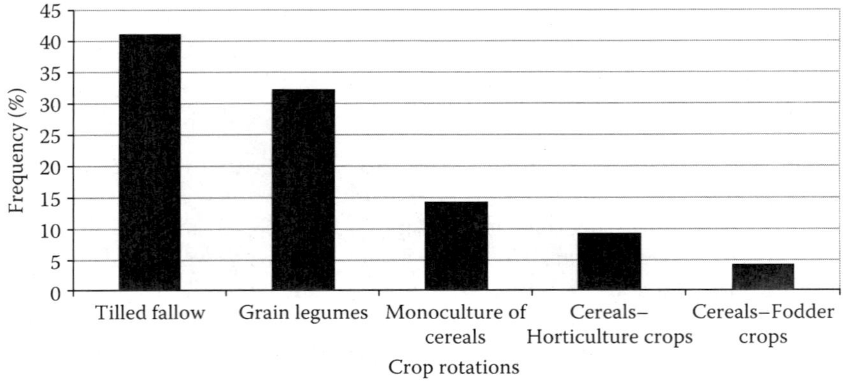

FIGURE 3.2 Crop rotation patterns and proportion of surveyed farmers using them.

Although fallowing is a necessary system in areas where rainfall is insufficient and its distribution irregular, farmers would prefer to cultivate legumes in rotation with cereals. According to surveyed farmers, it all depends on the total annual rainfall. They argued that the decrease in rainfall in recent years constrained them to opt for the tilled fallow instead of legumes. A monoculture of cereal following cereal is also practiced, but only on a small proportion of farms (14%, Figure 3.2). It is not advisable to grow wheat after another cereal (wheat, barley, oats) because monoculture always leads to yield losses. Surveyed farmers say that there has been a change in rotation practice; while in the past years, they practiced monoculture (cereal after cereal) because rain was abundant, now they are forced to consider growing cereals in rotation with tilled fallow.

The rotation cereal–tilled fallow would permit better management of time with regard to tillage and water conservation for the next growing season and would also facilitate soil tillage and sowing of the crop. The importance of fallow in crop rotations has been shown in Morocco and other countries, because it allows water and soil conservation and higher yields of the following crops.

Impact of Climate Change on Sowing Date

According to surveyed farmers, the agricultural calendar was properly respected in the region for many years. But with the climate change observed in recent years, this timetable, including planting dates, became difficult to follow. To adapt to these climate changes, there have been adjustments and adaptation in farming practices, including changing dates for planting of crops. Producers who were interviewed both individually and in groups reported (1) overall climate disruption and (2) the use of early sowing (October) of cereals as an adaptation adopted by some farmers in the region. In this region, cereal seeding often takes place after the first autumn rains, which usually occur in late October. In fact, the occurrence of the first rains reassures farmers and allows the emergence of weeds and thus their elimination during seedbed preparation. In addition, the first rain makes tillage easier, especially where tractors with low horsepower are used. Common wheat sowing is done usually between November and December, but late plantings are also made.

During our investigation, we were able to identify three sowing periods (Figure 3.3): The majority of farmers sow wheat in mid-November (44% of the sample), 25% of farmers sow wheat between late October and early November for early sowing, and 31% of farmers sow wheat from late November to early December.

Almost all farmers are aware that early sowing before the first rain (usually before November 15) is most recommended, because this allows wheat to take advantage of the rainy season and avoids water stress and high temperature at end of the growth cycle and thus increase the yields. However, the farmers do not have sufficient tractor capacity to plow soil in dry conditions, and therefore, this practice is seldom used. According to farmers, planting wheat beyond November 15 is generally accompanied by a yield loss. Fields seeded during the period from October 20 to late November recorded yields much higher than those that were seeded late; the yield gain was between 10% and 60% for wheat.

According to Debaeke et al. (2012), changing the timing of the crop cycle to fit the available resource (its quantity and distribution) can be done in different ways: (1) a marked change in the growing season can be made, (2) earlier varieties can be chosen, or (3) new planting dates can be tried to avoid drought by a shift in cycle (usually an advanced cycle). Turner (2004), in Mediterranean conditions, demonstrated that early autumn sowing (when the rains began) increases the potential yield because the stock of available water is greater, the flowering stage is earlier, and the distribution of transpiration between pre- and post-anthesis is optimized.

Low cereal production is mainly due to the combined effects of drought and attacks by Hessian fly (midge). Droughts promote the proliferation of midge populations and highlight its attacks. During dry periods, the late-planted fields are heavily attacked by the midge. Early sowing helps escape the second midge generation. Determining the best planting date with relation to the management of water resources is not easy, as many other factors are involved disease development,

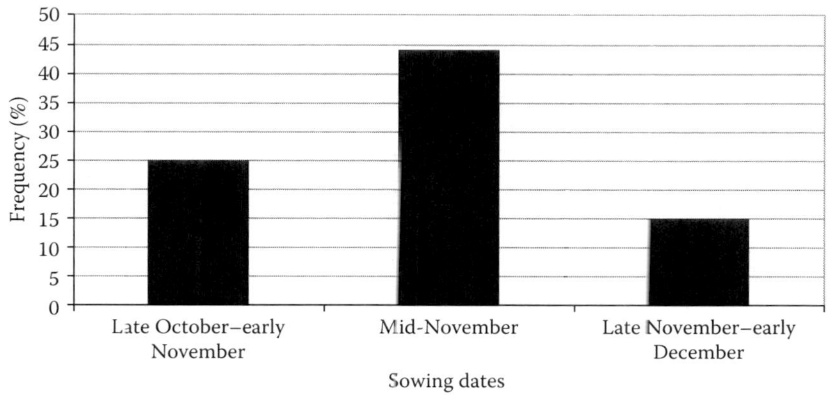

FIGURE 3.3 The most prevalent sowing dates for soft wheat as practiced by farmers in the study area.

weeds, pests, and cold sensitivity. Sowing wheat in early fall is justified in situations where the first rains can promote early vegetative growth, thus successful anticipation of the first rain allows the quick exploitation of available soil moisture (Debaeke et al. 2012).

Soft Wheat Varieties Most Used by Farmers

The survey found that adoption of new crop varieties (Figure 3.4) was a strategy of farmers, who change cultivated varieties almost annually. According to farmers, local varieties have completely disappeared; the reasons given are their long growth cycles and low yields, compared to those of new varieties. Indeed, farmers argued that short growth cycles, relatively high yield, and resistance to diseases are the main criteria for introducing new varieties into their crop cycles. The main source for identifying such new varieties is the national catalog of wheat varieties (used by 60% of respondents). Very few farmers import new common wheat varieties from abroad.

The gain in yield permitted by new common wheat varieties, compared to old varieties, has been between 10% and 50%. The gain is greater in relatively dry years (Balaghi et al. 2010). These new pest- and drought-resistant varieties allowed not only increased productivity but also meant a savings from the reduced need for pesticides, along with positive effects on the environment and quality of crops. Early varieties use less water because of their shorter growth cycle and smaller leaf-area index. In favorable years, however, these varieties are less productive. Conversely, if the biomass is reduced, the harvest index can be improved for dry conditions at the end of the wheat growth cycle (Debaeke et al. 2012).

Farmers in the small-farm category, however, have difficulty adapting to climate disruption through the adoption of new varieties. They have difficulty in accessing these varieties mainly because of their high price, but also because of an insufficient seed supply at the local level. The small farmers are losing in terms of yield potential, because they end up paying more for the common varieties of seed bought from the market or from their neighbors and because of low germination rate and low

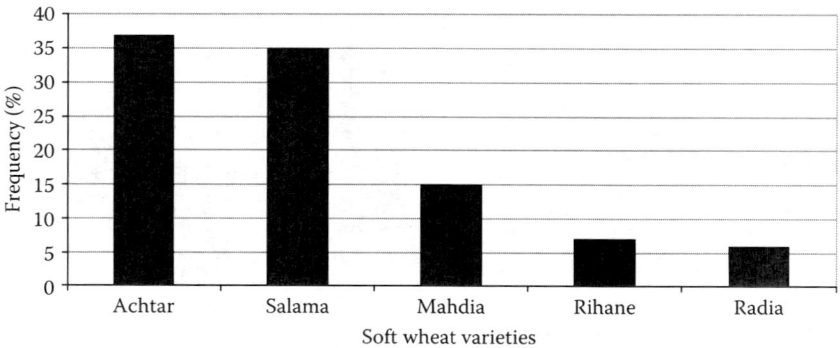

FIGURE 3.4 The soft wheat varieties most used by farmers in the study area.

vigor of these common varieties as opposed to the potential of the new varieties that they cannot afford or access.

Disruption of Traditional Agricultural Calendar

The different climate risks highlighted by this study introduce a situation of uncertainty for any decision to be taken by farmers to manage their crop operations (seedlings, fertilizer, herbicides, etc.). In fact, for more than 65% of the surveyed farmers, agricultural seasons are becoming very unstable. They claim that they are not being able to predict optimum planting dates and are fully aware that their seedlings will be at risk because of the random nature of the first and subsequent rains. Consequently, the timing and execution of the subsequent agricultural operations also become random. This affects the ultimate production at the end of the growth season.

Decrease and Fluctuation of Common Wheat Yields

Cereal yields are very low and vary among farms. All farmers who were included in our sample were unanimous that there had been yield reduction over the past 30 years. Indeed, the growth and development conditions are severely disrupted in recent years and no longer promote good wheat yields. However, the climate is not the only factor determining the performance of the crop; other determinants of performance (soil fertility, seed quality, pests, and diseases) also contribute strongly to the perceived decline in yields. This is probably due to the random nature of rainfall, which constitutes the main limiting factor for the application of appropriate management techniques that could improve yields. In addition, small holders, who constitute the majority of farmers, cannot afford to use certified seeds or apply fertilizers because of their high costs.

CONCLUSION

This study is a contribution to the knowledge of perceptions of climate change experienced by farmers in the Benslimane region (Morocco) and adaptation strategies implemented by them in response to the changes they report. The results of the survey emphasize that the impact of climate change on agriculture, and in particular on the production of common wheat, will be considerable if no action is taken to reduce the crop's vulnerability to these variations. The farmers surveyed were from two municipalities in the Benslimane region: Fdalate and Moualine El Oued. They face enormous difficulties in the exercise of their farming activities. Among these are problems caused by climate change: delay in the onset of the first rain after the long dry summer, episodes of drought during the rainy season, poor spatial distribution of rainfall, overall decrease in rainfall, violent winds, and excessive heat. These climatic changes have consequences for agriculture, for cereal production in general and wheat production in particular, that include the upheaval of the traditional farming calendar, especially the periods of sowing, the loss of crops at different growth stages, and yield reductions. Coping strategies developed and adopted by the surveyed farmers included changing sowing times, using crop varieties with shorter growing cycles and with resistance to drought and disease, and changing crop sequences (rotations).

In terms of research, we envisage exploring further with farmers the hidden potential of genetic resources to identify traits for adapting to climate change and also hope to assist in the development of innovative technologies. Genetic variation within crops will be assessed for water-use efficiency (WUE), which is generally expressed as the ratio of total dry matter production to evapotranspiration:

$$\text{WUE}\left(\text{biomass}\right)=\frac{\text{TE}}{1+\left(\text{E}_\text{S}/\text{T}\right)}$$

where:

TE is the transpiration efficiency (above-ground dry weight/transpiration)
E_S is the water lost by evaporation from the soil surface
T is water lost through transpiration by the crop (Richards et al. 2002)

This equation shows that crop WUE can be increased either by increasing its nominator TE or by decreasing the magnitude of its denominator E_S (Richards et al. 2002).

Overall genetic variation can be an important indicator of potential genotypes with high water productivity (Condon et al. 2002, Ebdon and Kopp 2004). The screening for genetic diversity for WUE traits within species provides a valuable opportunity for improving water productivity in the arid regions in light of the findings of extensive within-crop variation in relation to WUE (Richards et al. 2002), with the potential to address the negative effects of climate.

RECOMMENDATIONS

Given the impacts of climate change now and those predicted by long-term projections for 2030 and 2050 for Morocco, it is urgent for wheat cultivation to do the following:

- Encourage farmers to replace old varieties with newly created ones, especially those tolerant to drought and resistant to diseases.
- Diversify crops and crop rotations through the introduction of new species.
- Optimize planting dates and mineral fertilization.
- Adopt technologies allowing early sowing.
- Provide supplementary irrigation, where possible, to cope with droughts at the most critical periods of the wheat growth cycle.
- Maintain a portion of crop residue on the field to improve the infiltration of rainwater, reduce evaporation losses, reduce erosion, and increase soil organic matter.
- Integrate on the same farm both crop production and livestock raising, in order to decrease the impact of drought through the opportunity for livestock use of otherwise bad crop production.

REFERENCES

Achy, L. 2012a. Maroc: la Sécheresse menace la croissance et impose des réformes agricoles. *Les Echos* du March 20, 2012. http://archives.lesechos.fr/archives/cercle/2012/03/20/cercle_44774.htm.

Achy, L. 2012b. Morocco's drought threatens economic growth. *Morocco News Board*. http://www.moroccoboard.com/viewpoint-5/101-lahcen-achy/5654-morocco-s-drought-threatens-economic-growth.

Ahmed, H. and I. El Honsali. 2012. *Morocco, Grain and Feed Annual, 2012*. Global Agricultural Information Network. US Embassy, Rabat, Morocco GAIN Report MO 1203. http://gain.fas.usda.gov/Recent%20GAIN%20Publications/Grain%20and%20Feed%20Annual_Rabat_Morocco_3-12-2012.pdf (accessed October 13, 2014).

Azzam, A. and K. Sekkat. 2005. Measuring total-factor agricultural productivity under drought conditions. The case for Morocco. *J N Afr Stud* 10(1):19–31.

Balaghi, R. 2000. Suivi de l'évolution des réserves en eau sous culture de blé en région semi-aride marocaine: calibration et utilisation du modèle SOIL. In *Mémoire de Diplôme d'Etudes Approfondies*. Arlon, Belgium: Fondation universitaire luxembourgeoise.

Balaghi, R., M. Jlibene, and H. Benaouda. 2010. *Rapport de faisabilité du PICCPMV (Projet d'Intégration du Changement Climatique dans la Mise en oeuvre du Plan Maroc Vert)* Institut National de la Recherche Agronomique (INRA) and Development Finance Consultants. http://www.inra.org.ma/docs/environ/piccpmvfaisab.pdf.

Barakat, F. and A. Handoufe. 1998. Approche agro-climatique de la sécheresse agricole au Maroc. *Sécheresse* 9:201–208.

Benaouda, H. and R. Balaghi. 2009. Les changements climatiques: Impacts sur l'agriculture au Maroc. *Proceedings Symposium International sur L'agriculture durable en région Méditerranéenne (AGRUMED)*, Rabat, Morocco, May 14–16, 2009, 42–47.

Condon, A.G., R.A. Richards, G.J. Rebetzke, and G.D. Farquhar. 2002. Improving intrinsic water-use efficiency and crop yield. *Crop Sci* 42:122–131.

Debaeke, P., J.-C. Mailhol, and J.-E. Bergez. 2012. Adaptations agronomiques au risque de sécheresse, Chapitre 2.2. In *ESCo Sécheresse & Agriculture*, Amigues J.P., P. Debaeke, B. Itier, G. Lemaire, B. Seguin, F. Tardieu, A. Thomas (eds.), Sécheresse et agriculture. Adapter l'agriculture à un risque accru de manque d'eau. Rapport de l'expertise scientifique collective, INRA (France), pp 258–322.

Ebdon, J.S. and K.L. Kopp. 2004. Relationships between water use efficiency, carbon isotope discrimination, and turf performance in genotypes of Kentucky bluegrass during drought. *Crop Sci* 44:1754–1762.

Haenn, N., B. Schmook, Y. Reyes, and S. Calmé. 2014. Improving conservation outcomes with insights from local experts and bureaucracies. *Conserv Biol* 28(4):951–958.

HCP. 2003. *Le Maroc en chiffres 2003*. Rabat: Haut Commissariat au Plan, Direction de la Statistique.

MAPM. 2012. *L'agriculture marocaine en chiffres 2012*. Rabat: Ministère de l'Agriculture et Pêche Maritime. http://www.agriculture.gov.ma/sites/default/files/agriculture-en-chiffres-2012.pdf.

MATUHE. 2001. *Executive Summary. First National Communication to United Nations Framework Convention on Climate Change*, Ministère de l'Aménagement du Territoire, de l'Urbanisme, de l'Habitat et de l'Environnement, Kingdom of Morocco, October 2001.

Richards, R.A., G.J. Rebetzke, A.G. Condon, and A.F. van Herwaarden. 2002. Breeding opportunities for increasing the efficiency of water use and crop yield in temperate cereals. *Crop Sci* 42:111–121.

Rochdane, S., L. Bounoua, P. Zhang et al. 2014. Combining satellite data and models to assess vulnerability to climate change and its impact on food security in Morocco. *Sustainability* 6:1729–1746. doi:10.3390/su6041729.

Schilling, J., P.K. Freier, E. Hertig, and J. Scheffran. 2012. Climate change, vulnerability and adaptation in North Africa with focus on Morocco. *Agr Ecosyst Environ* 156:12–26.

Skees, J., S. Gober, P. Varangis, R. Lester, V. Kalavakonda, and K. Kumako. 2001. *Developing rainfall-based index insurance in Morocco*. Policy Working Paper, DECRG. Washington, DC: World Bank.

Soleri, D. and D.A. Cleveland. 2001. Farmers' genetic perceptions regarding their crop populations: An example with maize in the central valleys of Oaxaca, Mexico. *Econ Bot* 55(1):106–128.

Tabet-Aoul, M. 2008. Impacts du changement climatique sur les agricultures et les ressources hydriques au Maghreb. *Les notes d'alerte du CIHEAM*, N° 48, June 4, 2008.

Turner, N.C. 2004. Sustainable production of crops and pastures under drought in a Mediterranean environment. *Ann Appl Biol* 144:139–147.

Yessef, M. and Z. Saltani. 2009. *Mise en place d'un système d'alerte précoce à la sécheresse dans trois pays de la rive Sud de la Méditerranée: l'Algérie, le Maroc, et la Tunisie.* LIFE05TCY/TN/000150. http://www.oss-online.org/cd_envi/doc/07/03/06.pdf.

4 Assessment of the Demand–Supply Match for Agricultural Innovations in Africa

L. Woltering, L. Bachmann, T.O. Apina, B. Letty, J.A. Nyemba, and S.B. Alaoui

CONTENTS

Eighty percent of the population in Africa is estimated to be involved in agricultural production, which accounts for 30% of the continent's GDP (FAO 2013a). A rise in productivity will contribute directly to food and nutritional security and poverty reduction. However, climate-change projections indicate that agricultural productivity will decrease between 10% and 20% due to more frequent heat stress, droughts, floods, and shifts in agroecological zones (AEZs) (FAO 2013b). In addition, the population is expected to reach two billion by 2050 (UNDESA 2013). On top of that, urbanization and increased international trade demand a more market-oriented agricultural sector. This means that farmers and support services need to adapt their ways of operation to keep up with these changes.

Agricultural research centers, such as the ones of the Global Agricultural Research Partnership of the Consultative Group on International Agricultural Research (CGIAR), have the mandate to generate new and better technologies, practices, institutions, and policies for the rural poor. The World Bank (2011) stated that returns on investments in the CGIAR have generally been very good, ranging conservatively from a 175% return on investment to almost 900%, and much of the success in meeting global food needs can be attributed to these investments. In 2008, the CGIAR embarked on a major reform process to make international agricultural research more focused on farmers' needs, partnerships, and development outcomes, among other goals. Once focused on the *production of knowledge*, the international agricultural research system now puts added emphasis on ensuring that this knowledge is useful and taken up (Gildemacher and Wongtschowski 2013). The German Federal Ministry for Economic Cooperation and Development (BMZ) is a founding member of the CGIAR and has been supporting its 15 research centers (referred to as International Agricultural Research Centers [IARCs]) as well as the World Vegetable Center (AVRDC) and the International Center of Insect Physiology and Ecology (*icipe*), with about EUR 20 million per year.

The *Innovation Transfer into Agriculture–Adaptation to Climate Change (ITAACC)* program is being implemented by the Deutsche Gesellschaftfür Internationale Zusammenarbeit (GIZ) GmbH on behalf of the BMZ to support bridging the gap between research outcomes of the CGIAR, AVRDC, and *icipe* and their implementation in the agricultural sector of Africa. The program spans from 2013 to 2018, and first commissioned an assessment of the demand–supply match for agricultural innovations. Furthermore, special attention is being given to innovations that contribute to climate-change adaptation and gender equity in Africa. This chapter summarizes the findings of this assessment, which was implemented by a team of six consultants of the GFA Consulting Group GmbH (www.gfa-group.de) between September 2013 and June 2014. *Demand* is defined as the specific challenges farmers face within their farm or within the value chain they operate in, whereas *supply* refers to the innovations developed for them by the CGIAR, AVRDC, and *icipe*. Agricultural innovation is defined as the process of creating and putting into use new agricultural practices in a particular environment (Gildemacher and Wongtschowski 2013).

Given the rapid nature of the assessment, the results reported are not representative of the African agricultural sector; however, they provide some interesting insights into existing problems and allow a constructive discussion of possible solutions. The objective has been to *flag* issues, rather than making sweeping recommendations.

METHODOLOGY

Implementation Strategy

Defining the needs for agricultural innovations of almost a billion farmers in Africa, all with their specific resource base and climate, is impossible. Similarly, the 17 IARCs have produced an overwhelming amount of research findings during the past few decades. However, the innovations developed by the IARCs are evaluated regularly by a range of donors and institutes, whereas farmer needs are hardly documented. It was decided to prioritize efforts on assessing the need of farmers, thereby focusing on the kind of needs that the CGIAR, AVRDC, and *icipe* can supply. Hence, the focus is on a limited number of commodities that are cultivated by smallholder farmers, mostly staple crops. Rather than reaching out to individual farmer households, it was decided to approach stakeholders who give an aggregated demand of smallholder farmer needs. These are farmer organizations (FOs) and intermediaries, such as development organizations (GIZ, NGOs, etc.), actors along the value chain (input suppliers, seed companies, and processors), and government agencies. These actors are much easier to link up with for any follow-up activities of the ITAACC program. Individual farmer needs have been collected by the eRAILS project* in 2013 and the findings were integrated into this study. On the supply side, the regional representatives of the IARCs were asked to select up to five top innovations of their center that are ready for dissemination and that have a large positive impact on the livelihoods of smallholder farmers in their region. It was assumed that these top innovations covered the bulk of the farmer needs. Specific interviews on those innovations were conducted with the scientists heading that research, thereby reducing the number of IARC interviews considerably.

Based on a literature study and stakeholder meetings, it was found that beyond the needs for innovations, other themes such as adoption of innovations, information exchange, and extension were important to understand uptake of suitable innovations concerning climate-change adaptation. The research themes formed the framework for the development and analysis of the questionnaires for FOs, intermediaries, and IARCs. The questionnaires were designed in such a way that the questions asked on the demand side (FOs and intermediaries) related to the questions asked on the supply side (IARCs). For example, FOs and intermediaries were asked about the kind of challenges associated with a commodity, whereas IARCs were asked what key problems their top innovation addresses. This allows for a first assessment of the demand–supply match. In addition, a number of questions that characterize the farmers (FO and intermediary interviews) are compared with questions asked to the IARCs on targeting of their innovations. The questionnaires were developed, implemented, and analyzed using GrafStat Edition 2012. The questionnaires contain multiple-choice and open questions. The answers to the open questions were categorized and coded to compare the perceptions of the FOs, intermediaries, and IARCs. Three regional workshops in Africa† and two international workshops in Germany and Kenya were organized to provide a platform for actors in the innovation system to share, discuss, and validate the findings of the assessment. The workshops were set up to enable an exchange between participants by creating a healthy mix of small work groups, presentations, and networking events over two days. The workshop

* The eRails project sent technicians to 1061 households in eight African countries to ask farmers with what problems they need help. The project is funded by FARA and data were collected in 2013. http://www.erails.net/FARA/erails2/erails2/Home.

† Four regional workshops were planned, but due to the political situation in North Africa, the West and North Africa workshops were combined into one, held in Cotonou, Benin, on March 19 and 20, 2014.

was used to collect and interpret data. The results of the workshop were captured in separate workshop reports and their key findings are integrated in the results section of this report. All reports (inception, workshop, and final) and all analyzed data can be found on the ITAACC website (http://www.icipe.org/itaacc).

Data Collection

Data collection took place between November 2013 and April 2014, and it was done based on the following three steps:

1. Countries were selected by ranking all countries in East, Southern, West, and North Africa according to the number of IARCs and BMZ-funded research and development projects present (Table 4.1).
2. The FAO data network (FAO 2013a) was used to target the most important commodities for the country's dominant (>10% land cover) AEZs. The major commodities provided an entry point for identification of important stakeholders to consult for data collection in the countries to be visited. The commodities were categorized into crops, livestock, and (agro-) forestry in the questionnaires. By categorizing according to the AEZ, it was intended to increase the relevance of findings across Africa because each zone has a similar combination of constraints and potentials for land use.
3. Identification of stakeholders on the demand and supply sides: The primary stakeholders to target were mid-sized FOs that represent smallholder farmers of a target commodity. One hundred and fifty-two interviews were conducted with FOs, of which about 25% were conducted with organizations with fewer than 100 members (Figure 4.1a). Intermediaries, such as NGOs, government, and private sector actors, were the second important stakeholders to identify, as they not only articulate the needs of farmers but also promote innovations. One hundred and forty-one interviews were conducted with intermediaries, most of them with NGOs (Figure 4.1b). On the supply side, 24 interviews were conducted with regional representatives of the IARCs who selected the top innovations, for which 94 separate interviews were conducted with scientists. Figure 4.1c shows that most interviews were conducted at The International Council for Research in Agroforestry (ICRAF), International

TABLE 4.1
Seventeen Countries Were Targeted for the Assessment

East Africa	Southern Africa	West Africa	North Africa
Kenya	South Africa	Cameroon	Morocco
Ethiopia	Zimbabwe	Mali	Tunisia
Uganda	Zambia	Benin	
Tanzania	Mozambique	Ghana	
Rwanda	Malawi	Burkina Faso	
		Niger	

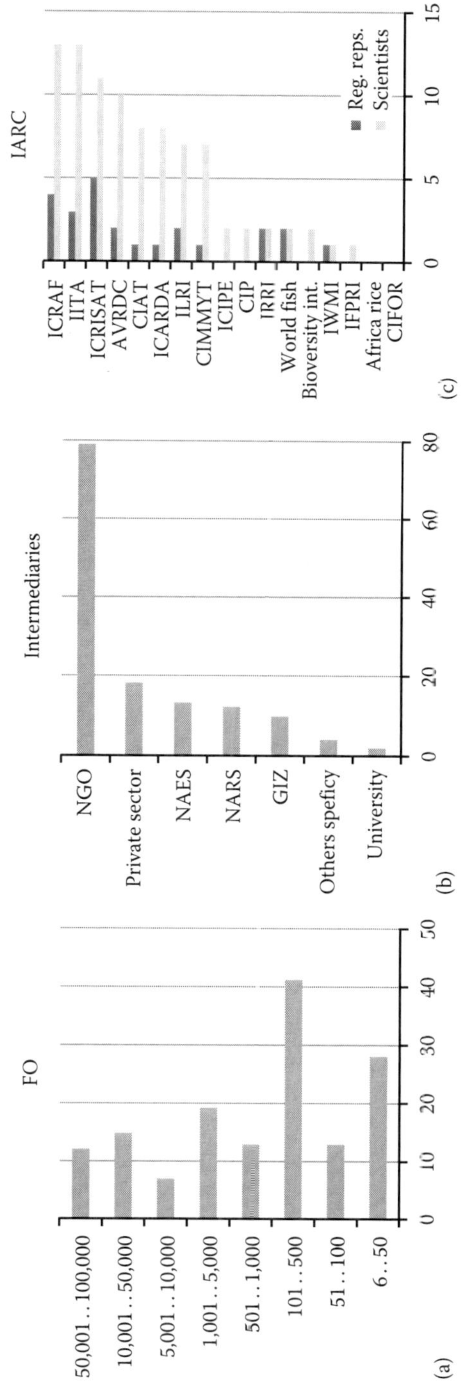

FIGURE 1.1 (a) Number of FO interviews per membership size (y-axis), (b) number and type of intermediary organization interviewed, and (c) number of interviews per IARC.

Institute of Tropical Agriculture (IITA), and International Crops Research Institute for the Semi-Arid Tropics (ICRISAT). Not all centers could be reached, and not all scientists working on selected top innovations could be interviewed. About 90% of the interviews were conducted face to face, and the remaining via email.

DEMAND–SUPPLY MATCH FOR AGRICULTURAL INNOVATIONS

Demand for Innovations

More than 400 interviews were conducted on a range of crops, livestock, and agroforestry species. Most of them were on staple crops such as maize, cassava, and sorghum, or on cattle, indicating that smallholder farmers are organized around these commodities. Yet, the data of the eRAILS project show that individual farming households face the most challenges with vegetables, livestock (poultry), and fruit cultivation. This shows that smallholder farmers are organized around those commodities in which most IARCs are active as well. Only AVRDC focuses on vegetables, which are largely cultivated by smallholder farmers at the household level.

The results of the needs of farmers and top innovations that exist for maize are presented here, because most interviews were conducted on maize, of which 41 were with FOs, 40 with intermediaries, and 17 with IARCs. This is a reflection that maize is a staple crop in most of the countries included in the study. Table 4.2 shows the coded

TABLE 4.2
Major Challenges for Farmers according to FOs ($n = 199$) and Intermediaries ($n = 119$), and Key Problems or Top Innovations of IARCs That Address Maize in Africa ($n = 42$)

Main Challenges: Maize	FO (%)	Intermediary (%)	IARC (%)
Access to (quality) seeds	16	14	5
Marketing	10	8	5
Access to credit or finance, high production costs	9	5	5
Post-harvest processing or storage	8	8	5
Tools, machinery, and irrigation equipment	8	3	2
Access to, quality and use of fertilizer and other inputs	7	7	7
Institutional, regulatory, and policy issues	6	12	2
Drought, flood, any climate-related problem	6	8	5
Pest and diseases (including rodents, animals)	5	6	14
Training, knowledge, knowhow, awareness	5	8	7
Cultivation practices, weeding, and harvesting	5	3	7
Social issues: theft, adulteration, crime, and conflict	4	5	0
Natural resources: land and water availability	4	4	2
Transport and infrastructure	4	1	0
Labor (including shortage, price)	3	1	2
Poor soil fertility	1	3	10
Low yields and poor quality	1	3	14

challenges for maize farmers according to the open answers of FOs and intermediaries. The last column shows what challenges are being addressed by the selected top innovations of the IARCs according to scientists. The bottom row, nominations, shows the number of coded answers extracted from the responses. The challenge most frequently raised by FOs was access to (quality) seeds (16%). This was also mentioned most frequently by intermediaries (14%), but for IARCs, this problem ranked only at a medium position (5%). Limited seed multiplication capacity and poor distribution of quality seeds deprive farmers from accessing good-quality seeds. Furthermore, many complaints were collected on sales of adulterated seeds. The FOs seemed to shy away from hybrid seeds, preferring to (re-)grow open-pollinated varieties (OPVs) themselves. *Marketing*, the second most important problem for the FOs, was ranked only medium by intermediaries and IARCs in the interviews. This also had to do with poor policies on import and export regulations for maize, as it is a highly traded commodity. The FOs raise the issue that food aid and input-support programs too often lead to market distortions. The top innovations of the IARCs seem to address the issue of pests and diseases more (14% of responses) than it is mentioned as an important problem by FOs (5%) and intermediaries (6%). Pests and diseases are a very big problem for farmers in East Africa, especially maize lethal necrosis, but since Table 4.3 gives the overall picture of Africa, this is evened out. Surprisingly, FOs did not mention *Striga* as a major problem, though IARCs offer very elaborate packages on *Striga* control. Lack of access to and availability of drought-resistant maize varieties were found to be problems for farmers, leading to promotion of crop diversification to more drought-resistant crops (sorghum, millet, cassava, sweet potato, etc.). A very different perception was recorded on *poor soil fertility* and *low yields* that were raised by only 1% of FOs, while this ranked top for researchers (10%–14%). Even though poor soil fertility *is* a major limitation to improved production, farmers perceive other problems (e.g., access to inputs and marketing) as much more pressing. The dataset for other rainfed crops, such as cassava, potato, beans, and sorghum, show a better balance between the challenge of farmers to access quality seeds and the top innovations that address that (data not shown). Nevertheless, also for these crops, farmers are more worried about adding value and marketing their crops than to increase yields.

TABLE 4.3
Climate-Change Coping Strategies Promoted by FOs ($n = 266$) and Intermediaries ($n = 400$)

Strategy	FO (%)	Intermediary (%)
Improve water usage (irrigation and water harvesting)	22	13
Improved cultivation practices (GAP, conservation agriculture, storage, etc.)	18	22
Improving current variety	16	27
Diversification of production	14	10
Other (early warning, insurance, information systems, etc.)	12	13
No coping strategy even though climate change is a threat	9	2
Planting trees (agroforestry approach)	7	7

Coping with Climate Change

The Intergovernmental Panel on Climate Change (IPCC 2013) states that the warming of the climate system is unequivocal and is likely to triple in the next 40 years. The large majority of stakeholders consulted during this study confirm that climate change entails serious problems for farmers. Others state that climate variability has always been there (often mentioned in the Sahel), or that climate change is not among the most pressing problems that farmers are facing. Increased drought, erratic rainfall, increased temperature, and vulnerability are seen as major challenges in Africa. Eighty-four percent of the FOs and 98% of the intermediaries claim that farmers experience a loss in productivity due to climate change. Shorter growing periods (44%–61%) and increased pests and diseases (43%–49%) are mentioned as the second and third most common consequences of climate change. Table 4.3 shows the climate-change coping strategies being promoted by FOs and intermediaries among the farmers they represent. FOs are mostly promoting ways of using water more efficiently at a watershed scale (water harvesting) and the farm scale (e.g., drip irrigation). Intermediaries, on the other hand, rather promote the use of varieties that are better adapted to the consequences of climate change. This even ranks higher than the promotion of different crops that might be more suitable to the changing conditions. Especially in East Africa, the FOs recommend a shift from cows to camels and from maize to sorghum. Almost 10% of the FOs admitted that they were not promoting any climate-change coping strategy among their members, largely because they did not know what a suitable strategy would be. It is remarkable that only 7% of the responses indicated that agroforestry and planting trees could be a strategy to promote. The latter was analyzed in work groups at the final workshop, and it was concluded that there was still a need for extension and awareness-raising activities, particularly traditional practices that include the integration of trees. In addition, improved access to suitable tree species for beneficial tree–crop interactions, in particular fertilizer trees, and the promotion of the multiple benefits (fuel, fodder, fuel wood, fertilizer, fruits, and timber) from agroforestry systems would improve integration of trees in farming systems. During the workshop, off-farm income generation was also highlighted as another important coping strategy, but this was not considered in the analysis.

Supply of Top Innovations

The top innovations selected by the regional representatives of the IARCs are categorized according to the type of innovation (Table 4.4). The majority of top innovations aimed to improve varieties, enhance drought and pest resistance, and improve the nutritional value, for example, by biofortification. This is not surprising considering the history, infrastructure, and the competitive advantage that IARCs possess to do that kind of work. Technologies not directly related to varieties, such as repellents or biopesticides (*icipe*), vaccines (ILRI), processing techniques for cassava (IITA), or metal silos to vacuum-store seeds (CIMMYT), were often cited (17%). Many of the innovations on improved cultivation practices revolved around integrated soil

TABLE 4.4
Types of Top Innovations Supplied by IARCs ($n = 109$)

Types of Top Innovations	(%)
Development of improved varieties	26
Development of technologies (not varieties)	17
Cultivation methods (conservation agriculture, agroforestry, irrigation)	13
Promotion of using specific crop/variety	10
Information tools/equipment (analysis)	6
Improve policies/institutions	6
Value addition (processing and storage)	5
Pest and disease management	5
Improved service to farmers	4
Access to inputs and markets	4
Knowledge systems	3
Innovation platforms and public–private partnerships	3

fertility management, micro-dosing, and conservation agriculture. Innovations that aim to improve access to inputs, services, and markets for farmers are not often mentioned among the top innovations of the IARCs, even though this was found to be a major challenge for maize farmers. Twenty-eight percent of the top innovations selected by the IARCs could be linked, to a variable extent, to past funding from the BMZ (Kasten 2014), which is quite significant considering the large number of donors supporting IARCs.

Suitability of Innovations

Yet, innovations should not only address the challenges, they should also fit into the socioeconomic context in which farmers live. A number of characteristics of the top innovations and of the farmers were evaluated to assess the suitability of innovations, and thereby its likelihood for sustainable adoption. The findings are summarized as follows:

- Even though arid AEZs are gaining importance for food security due to population growth and climate change, only 1% of the top innovations have been specifically designed for arid areas.
- According to the FOs, around 65% of farmers used no or very little chemical pesticides and fertilizer, and about half of them use no or very little animal manure or compost for crop production. Especially, the relatively low use of manure and compost is surprising, indicating that more efforts are needed to promote crop–livestock systems. While this low use of inputs needs to be recognized, still between 25% and 30% farmers use moderate levels of inputs.
- Similarly, 23% of farmers have farm sizes of less than 0.4 hectares and 17% exceed 3 hectares.

- One-third of the scientists could not provide any information on the investment cost, and only a minority had independent information on the benefit–cost ratio of the innovation. Nevertheless, it was estimated that only one-third of the current top innovations are affordable to (very) resource-poor farmers. The need for innovations is high for very resource-poor farmers, but the number of suitable IARC innovations for them is rather limited.
- Even though all actors recognized the important role women play along the agricultural value chain, gender could be considered more in the design of innovations or dissemination programs. For example, by re-engineering technologies in production, processing, and marketing to account for the capacity of women to handle certain tools and equipment, or by recognizing the complementary role of men and women, on and off the farm.
- The scientists claimed that for 11% of the top innovations, farmers had full control of the innovation development process, for example, in farmer-led breeding programs. On the other hand, 27% of the top innovations were fully controlled by researchers.

Clearly, there is variability in the resource base of farmers, and innovations should cater to that. Information on the suitability of innovations to certain AEZ or socio-economic contexts is critical for dissemination partners. The extent to which farmers can play a role as partners, or leaders, in the development of innovations rather than being recipients of the end product should be further explored. Greater involvement of farmers in the design and testing of innovations is likely to influence the level of adoption positively.

LINKAGES BETWEEN DEMAND AND SUPPLY

Information Exchange

There was a consensus among FOs, intermediaries, and IARCs that face-to-face interaction (training, demonstration, and farmer exchanges) was the best way to share information with farmers. Sharing information through media (radio, television, telephone, etc.) ranked second, and through written material (leaflets, newspaper, etc.) ranked last. Interestingly, only 7% of scientists use mobile phone applications to disseminate their innovation, whereas 14%–21% of FOs and intermediaries, respectively, thought that mobile phone applications would be useful for dissemination. Most farmers have mobile phones but they mainly use them for voice communication and money transfer and not as a source to obtain agricultural information. This is a new and potential avenue for disseminating agricultural innovations if well utilized. Overall, poor information exchange is a major bottleneck for extension of innovations. This is confirmed by the eRAILS project, where it was found that for 54% of the problems mentioned by farmers, the solution already existed for more than 20 years. This shows that a lot of knowledge on farmer problems already exists, but that it often does not reach the intended beneficiary. This also becomes clear when looking at the collaboration and complementarities between the different actors in the innovation system.

Quality of Extension Services to Farmers

FOs and intermediaries were asked to appraise the quality of extension services to farmers by the National Agricultural Extension Services (NAES), National Agricultural Research Services (NARS), NGOs, and the private sector. Table 4.5 shows that both FOs and intermediaries are least satisfied with the services of the NAES. Moreover, about 15% of the responses collected contain a reference that staff of the NAES is not competent or motivated to support the farmer. Many interviewees assign this to lack of resources to reach the farmers, forcing farmers or development partners to pay for transport and other fees. Yet, a considerable number of farmers appreciate the support of the NAES, especially for staple crops, and they are found to be much more accessible than NARS, NGOs, and the private sector. Twenty percent of the FOs mention that interventions of NGOs are limited to duration and location, which raises questions by FOs on the capacity of NGOs to find a healthy balance between providing support and creating dependency with project beneficiaries. NGOs are very much appreciated for their ability to provide useful training and exposure to innovations.

More than 20% of FOs and intermediaries mention cases where the private sector is exploiting farmers, for example, the sales of fake inputs, market monopolies, and extortion. This has resulted in increased mistrust toward the private sector. Services are concentrated around urban areas and are mostly reserved for the better-off farmers.

TABLE 4.5
Coded Responses of FOs and Intermediaries When Asked about the Quality of Services to Farmers by NARS, NAES, Private Sector (PS), and NGOs

	View of Intermediaries On			View of FOs On		
Response	**NARS (%)**	**NAES (%)**	**PS (%)**	**NGO (%)**	**NAES (%)**	**PS (%)**
Quality of services high	16	5	25	16	8	16
Quality of services low	1	12	9	1	26	4
Lack adequate resources to provide services (e.g., poor logistics)	11	28	1	6	15	0
Inaccessible or poor presence	17	8	13	20	7	22
Staff competent and motivated	2	3	1	6	1	1
Staff not competent or motivated	6	15	4	5	15	3
Exploiting farmers	0	0	20	0	0	23
Providing only specific services (e.g., for single product only)	9	3	13	14	6	18
Provision of capacity building	5	3	2	16	4	1
Provide services indirectly through partnerships and collaborations	4	3	2	4	2	3
Extension services are not their responsibility	20	0	1	0	0	0
Other comments	9	21	11	8	17	9
$n =$	179	190	179	159	147	170

But, numerous positive examples of good after-sale services are also given, thereby praising the sensible approach of the private sector to extension and good training capacities. However, the private sector still tends to focus on one product only, thereby jeopardizing the optimization of resources for the entire farming system. There is also recognition that government interventions, such as subsidized inputs and fixed sales of produce, are making it very risky for the private sector to operate.

Sixteen percent of intermediaries claim that the NARS are doing good work, which responds to the needs of farmers. Others claim that the NARS are inaccessible and lack resources (especially logistics) to provide services to the farmers. Intermediaries call for an urgent need to strengthen the soft skills of the NARS to improve collaboration with agencies working on dissemination of innovations.

It is important to note that during the presentation of the results at the final ITAACC workshop in Nairobi (May 8, 2014), there were mixed reactions on the results. Many participants were happy that criticism or praise were finally quantified and shared, but other participants found it very confronting, raising the question if the results can be presented without doing in-depth research on the origins of such comments.

Quality of Services of IARCs to Farmers

FOs and intermediaries were asked with which IARC they have collaborated so far, and what were the strengths and weaknesses of this collaboration. Out of the 152 FOs that were interviewed, 102 FOs (67%) stated that they had never worked with IARCs. This is not surprising since IARCs usually approach farmers through intermediary organizations, which reduces their visibility. Then again, 62 out of 141 intermediaries (43%) claim never to have undertaken activities with IARCs, suggesting that there is still scope for building a bigger partnership base.

FOs and intermediaries mention the approach of the IARCs toward farmers more often as a weakness (32%) than as a strength (25%) (Table 4.6). It was mentioned that the IARCs focus too much on the research methodology than on the impact, or focus rather on the need of the donor than that of the farmer. FOs appreciate the training opportunities and high-quality standards of IARCs, and intermediaries appreciate the openness to share and test germplasm, and the opportunities provided for networking and fund acquisition through IARCs. The FOs liked to be seen as equal partners and desired equal share in the funding for activities, and claim that IARCs can do much more to connect them to interesting partners or to help them gain knowledge from abroad. The intermediaries wanted the IARCs to be more accessible and, for example, have more than one open day per year. They also feel that they have little influence on project designs and budget sharing, and that ownership of results tends to be an issue when working with some IARCs.

Links of IARCs with Intermediaries

In return, the IARCs were asked about the strengths and weaknesses of working with the NAES, NARS, NGOs, and the private sector as a partner. The NARS are

TABLE 4.6
Frequency of Coded Responses by FOs and Intermediaries on the Strengths and Weaknesses of IARCs

	Strengths	Weaknesses
FOs	25% approach meets farmers needs	32% approach does not meet farmer needs
	19% offer good training opportunities	14% poor/no funding to FOs
	13% high professional standards	14% poor facilitator of partnerships
Intermediaries	17% provide good technologies	24% poorly accessible
	13% high professional standards	19% approach does not meet farmer needs
	10% approach meets farmers needs	16% poor partner (poor mutual respect)
	10% good sharing of germplasm	
	9% opportunities for networking/ acquisition	

generally seen as a good partner with whom germplasm is exchanged and field days are organized. The NARS also have an important mandate to test and multiply, for example, seeds from IARCs before they are officially released. Still 16% of responses hint that they are a poor partner, highlighting that the NARS rely too much on funds of IARCs to collaborate (Table 4.7). Lack of adequate resources for field visits or experimentation is the most frequently mentioned reason hampering the collaboration with the NARS and NAES. Many of the regional representatives of IARCs characterize the NAES as adhering to poor-quality standards and working with outdated approaches to extension. The IARCs call for a revision of the way the NAES operate, starting at the universities where, for example, in Kenya, the *poor*

TABLE 4.7
Frequency of Coded Responses by IARCs on Their Collaboration with Key Partners: NAES (n = 56), NARS (n = 41), NGOs (n = 36), and the Private Sector (PS) (n = 37)

Response	NARS (%)	NAES (%)	NGO (%)	PS (%)
Good partner	25	12	31	35
Poor partner	16	7	17	17
Lack adequate resources	27	22	17	3
Poor financial/HR management	12	15	3	0
Approach does not meet farmer needs	2	20	6	16
Approach meets farmer needs	2	0	8	8
High-quality standards and competencies	14	0	6	5
Poor-quality standards and competencies	0	17	3	5
Good information on products	2	0	3	3
Poor information on products	0	0	6	0
Hardly any cooperation	0	7	3	8

students go to the *extension* department, and the *better performing* students can do research. Developing methods for extension is part of social science research and the NAES should do that much more in collaboration with the NARS. One of the regional directors of IARC claims that mutual respect is the basis for good collaboration and that the IARC staff should invest much more in person-to-person contact with the government staff.

Both the NGOs and the private sector are generally regarded as good partners. IARCs mention that individual NGOs often lack resources, or flexible-enough funds, to accompany an entire research-development cycle. Scientists claim that some NGOs lack in-depth knowledge about the innovations and the way to upscale them adequately, taking the research results out of context and thereby failing to provide feedback to the researchers on the dissemination process. The IARCs regard the private sector as an important player to scale up innovations, but complain that they leave out remote rural areas and are not interested in OPV seeds. Then again, the private sector is more interested in tradable commodities, whereas the CGIAR works mostly on staple crops for food security.

According to the views of FOs and intermediaries, it can be concluded that farmers are not getting the services they need to improve their farming systems. The analysis of the perceptions that the various stakeholder groups have in regard to one another shows that there is considerable room to improve trust, understanding, and true partnerships to support smallholder farmers.

Upscaling Innovations

The regional representatives of the IARCs found the NARS (36%) and NGOs (34%) most instrumental when it comes to upscaling of their innovations ($n = 80$). IARCs do not seem to rely much on the NAES (selected by only 19%) and the private sector (9%) for upscaling. Almost two-thirds of the top innovations were considered at a stage of upscaling. According to scientists, half of the top innovations had a big, or very big diffusion potential, being suitable for potentially all farmers in the country. However, only 28% of respondents state that more than 20,000 households have adopted their innovation, and 58% of the innovations reached out to less than 5,000 household ($n = 64$). Thus, the interpretation of *big diffusion* potential is relative when considering the low number of households that adopted a specific top innovation. The regional representatives of the IARCs are remarkably self-conscious when asked about the major bottlenecks for adoption of innovations. The most frequent answer was that the approach of IARCs to upscaling and collaboration with farmers and partners is poor and the poor design of the IARC innovations ranks fifth. A lack of understanding of the public sector delivery system and a lack of support for farmer innovations were mentioned by the regional representatives as weaknesses of IARCs. This may be explained by the fact that the performance of the scientists is largely evaluated based on the number of peer-reviewed articles that are published (91%), and hardly at the level of engagement with other actors, in particular with farmers (13%) (Table 4.8). The second most important bottleneck is the lack of policies supporting dissemination, especially when it comes to crops that are important for food security.

TABLE 4.8
Most Important Criteria for Performance Evaluation of the IARC Scientists ($n = 89$)

Evaluation Criterion	(%)
Peer-reviewed publications	91
Acquisition of funds	70
Quality of scientific work delivered	65
Perceived impact of research	57
Novelty of research	44
Engagement with other actors of the value chain	30
Engagement with farmers	13

CONCLUSIONS AND RECOMMENDATIONS

The findings of the study helped to create a better understanding of the reasons underlying the apparent gap between available agricultural research results within the IARCs and the needs of practical farmers in Africa. It provides an independent assessment of matches and links according to the perceptions of the key actors in the innovation system. Even though perception may not always reflect reality, it is an important element to consider in design of innovations.

The results show that even though a large number of research outputs match with farmer needs, the most important challenges are not met by them. Poor infrastructure to access input and output markets, lack of financial and advisory services, and inadequate law enforcement (e.g., seed quality) are some of the challenges faced by the average African farmer on a daily basis. These challenges are primarily outside the mandate of the IARCs. Most innovations focus on improving traits of crop varieties and only a few innovations aim at improving the marketability of commodities. Farmers are poorly informed about existing innovations and they receive little support to experiment with innovations on their farms. Farmers are not satisfied with the services provided by the NAES, and their most frequent comment about the private sector is that it exploits them. FOs and intermediaries mention the approach of the IARCs toward farmers more often as a weakness than as a strength. The IARCs rely mostly on the NARS and NGOs to disseminate their innovations. Farmers want themselves to be seen more as partners, or leaders, in the development of innovations, rather than as recipients of the end product. To summarize, there is considerable room to improve trust, understanding, and true partnerships among key actors in the innovation system, so that farmers are better able to adopt innovations.

The ITAACC project is advocating for a shift from linear technology transfer toward innovation system thinking (GIZ and CGIAR 2014). It follows the principle that taking a new *practice* from one place to the next requires by default the re-creation of the innovation process, to assure local fit and the re-ordering of actor relations required for its success (Gildemacher and Wongtschowski 2013). It is therefore important for IARCs to involve target beneficiaries and partners for early upscaling in the innovation development process. If the IARCs want to reach the intended development impact (CGIAR 2011), they should be provoked to revise the evaluation criteria by which scientists are measured to reward bridging, brokering, and catalyzing innovation besides scientific relevance. Furthermore, IARCs should also make changes at an institutional level, so that positions for communication, partnerships, and technology brokers and catalysts can work side by side with researchers. However, development partners need to play a more effective role in supporting farmers to innovate their farming practices. Much can be gained by aligning research and development projects. This already starts at the project design, for example, by having long-term projects that can be phased to balance the influence of research and development. Room for failure and flexibility in budgets are a prerequisite for successful projects. Research and development should go hand in hand to bridge the gap between the existing *solutions* and the challenges faced by smallholder farmers in Africa.

ACKNOWLEDGMENTS

This work was supported by the Deutsche GesellschaftfürInternationale Zusammenarbeit (GIZ) GmbH on behalf of the German Federal Ministry for Economic Cooperation and Development (BMZ). We are grateful to the respondents of the interviews and the participants of the regional workshops for their contribution to the findings of this study. Special thanks go to Jörg Lohmann, program manager for ITAACC at GIZ, and Michel Bernhardt (GIZ-BEAF) for their continuous support during the implementation of this work.

REFERENCES

CGIAR. 2011. *A strategy and results framework for the CGIAR*. For submission to the CGIAR Funders Forum. http://library.cgiar.org/bitstream/handle/10947/2608/Strategy_and_Results_Framework.pdf?sequence=4.

FAO. 2013a. *CountrySTAT. Food and agriculture data network*. United Nations Food and Agriculture Organization. http://www.fao.org/economic/ess/ess-capacity/country stathome/en/.

FAO. 2013b. *Women, agriculture and rural development: A synthesis report of the Africa region*. United Nations Food and Agriculture Organization Programme of Assistance in Support of Rural Women in Preparation for the Fourth World Conference on Women. http://www.fao.org/docrep/x0250e/x0250e03.htm.

Gildemacher, P.R. and M. Wongtschowski. 2013. Catalysing innovation: From theory to action. Keynote paper for the GIZ Workshop "Demand driven agricultural research: Bridging the gap." November 19–22, 2013, Feldafing, Germany. http://www.kit.nl/sed/wp-content/uploads/sites/2/2015/06/WPS1_2015_online.pdf.

GIZ and CGIAR. 2014. *The Feldafing principles for enhancing agricultural innovation systems*. November 22, 2013, Feldafing, Germany. Deutsche Gesellschaftfür Internationale Zusammenarbeit (GIZ) GmbH and Consultative Group on International Agricultural Research (CGIAR). Editorial committee: Ann Waters-Bayer, Michel Bernhardt, Piers Bocock, Patrick Dugan, Joerg Lohmann, and Sidi Sanyang. http://www.icipe.org/itaacc/index.php/component/docman/doc_download/111-feldafing-principles-update?Itemid=.

IPCC. 2013. *Climate change 2013: The physical science basis. Contribution of Working Group I to the Fifth Assessment Report of the Intergovernmental Panel on Climate Change*, eds. T.F. Stocker, D. Qin, G.-K. Plattner, M. Tignor, S.K. Allen, J. Boschung, A. Nauels, Y. Xia, V. Bex, and P.M. Midgley. Cambridge: Cambridge University Press. doi:10.1017/CBO9781107415324. https://www.ipcc.ch/report/ar5/.

Kasten, W. 2014. International Agricultural Research List of BMZ Funded Projects. Deutsche Gesellschaftfür Internationale Zusammenarbeit (GIZ). http://www.giz.de/expertise/downloads/giz2014-09-en-list-of-bmz-funded-projects.pdf.

UNDESA. 2013. *World population ageing 2013*. United Nations Department of Economic and Social Affairs (UNDESA), Population Division. ST/ESA/SERA/348. http://www.un.org/en/development/desa/population/publications/pdf/ageing/WorldPopulationAgeing2013.pdf.

World Bank. 2011. *Forty findings on the impacts of CGIAR research, 1971–2011*. CGIAR Fund Office, World Bank, Washington, DC. http://www.cgiar.org/www-archive/www.cgiar.org/pdf/Forty-findings-CGIAR%20_March2011.pdf.

Section II

Potential of Using Genetic Resources and Biodiversity to Adapt to and Mitigate Climate Change

5 Exploitation of Genetic Resources to Sustain Agriculture in the Face of Climate Change with Special Reference to Wheat

A.B. Damania, A. Morgounov, and C.O. Qualset

CONTENTS

The global agricultural community has now acknowledged that climate change is a fact and is here to stay. Climate change could become the biggest driver of biodiversity loss and thus poses a new challenge in the conservation and management of plant genetic resources. The average global surface temperature has increased by 0.2°C per decade in the past 40 years and is further predicted to increase by 6°C by 2050 (Rana and Sharma 2009). The time has come when the vast quantities of genetic resources conserved in genebanks around the world need to be evaluated specifically to address the need for genes to combat or mitigate the effects of climate change. Some of the collections may not have adequate representation of the global resource for these genes, and those gaps will have to be filled by additional collections to cover the full geographic distribution of a crop species, especially from areas that are at the ecogeographic extremes of the distribution, where novel genes for tolerance to abiotic stresses may exist. As wild progenitors are the ones most exposed to climate change, consolidation of those collections from various genebanks for evaluation of adaptation to climate change should be given high priority.

However, there appears to be a large disparity in agricultural vulnerability to climate change. Although temperature change in lower latitudes, where developing countries are located, tends to be less than the global average, modeled yield changes are primarily negative there, in contrast to predominantly positive yield changes in the middle and higher latitudes, where many developed countries are located (Rosenzweig and Parry 1994). This phenomenon occurs after taking into account the fact that global simulation of agricultural production of major grain crop declines under climate-change conditions. Climate change accompanied by increase in CO_2 emissions has had a profound effect on crop yields and groundwater use. Scientific evidence now tilts toward the fact that, over the last century or so, human activity has begun to have a discernible influence on the world's climate causing it to warm up (Watson and Zinyowera 1998). However, despite technological advances such as bioengineering, genetic modification, improved cultivars, and sophisticated irrigation systems, weather and climate, including rainfall and its temporal and spatial distribution, are still, and will remain, a key factor in agricultural production for years to come.

Global warming and climate change affect food security and agricultural production systems worldwide. Prediction of increase in temperature and decrease in rainfall are still very imprecise; therefore, adaptation to climate change could be an ill-defined crop-breeding objective at present and is unlikely to be uniform for the same crop from region to region. Therefore, the need for genetic adaptation seems to be at odds with the tendency toward genetic uniformity of modern plant breeding, consequently decreasing agrobiodiversity. A suite of plant traits are known to play an important role in the adaptation of crops to high temperature and drought. However, despite our ability to manipulate them through genetic improvement, because of the uncertainty of what exactly to breed for, the emphasis should not be so much on which trait to breed for, but rather to adopt a breeding strategy that allows a highly flexible and efficient suite of traits for variety deployment to farmers' fields.

During the past two decades, international agricultural commodity trade has grown substantially, and it is key to the food supply to importing nations, thus providing considerable income to the food-producing nations. For example, a weak monsoon season in South Asia results in shortfalls in crop production of grains and food legumes in India, Bangladesh, and Pakistan, which contributes to the importation of food grains, whereas India and Pakistan are normally exporters. There is continuous decline in food production in all of Africa during the past 20 years caused in part by drought and in part by low production potential. However, the population of Africa, despite all odds, including war and sectarian violence, is increasing and international relief efforts are coming continuously to prevent widespread famine. These examples underline the close links between agricultural production and climate change and the need to examine the global impact of the change in weather and rainfall patterns in the entire world. For example, California, has suffered from drought for the last four consecutive years, resulting in losses not only to cattle and sheep ranchers but also to farmers who need to totally rethink about the crops they will be planting. Walnuts and almonds require water throughout the year, but wheat, maize, sorghum, barley, and forage legumes do not require. Farmers are installing bore-wells in their farmlands, but the water table is descending rapidly. In certain coastal regions, the water table is so low that seawater is pumped instead of freshwater. These regions in California are turning to treatment of sewage water, which is costly; but in the majority of cases, there are insufficient inhabitants to generate the quantity of sewage water needed for treatment and use for irrigation. In April 2015, California Governor Brown introduced 35% water-use restrictions across the board. Water available for irrigation has become so expensive that farmers make more profit from sale of their allocated surface water or water from bore-wells than they would by planting crops. Farmers are also adjusting their cropping systems to reuse irrigation water. The adjustments must accommodate higher salinity of the reused water. Desalinization of seawater has application for municipal water uses, but is prohibitively expensive for agriculture.

Climate change can also introduce new pests and pathogens into farming systems, and varieties that were not previously exposed are susceptible. For example, in Imperial county of California, farmers are dealing with a new race of downy mildew in spinach (caused by *Peronospotafarimosa* f. sp. *spinaciae*). It is believed that the disease has become a problem due to warming climate (Warnert 2014). The citrus disease huanglongbing, caused by a bacterial pathogen spread by the Asian citrus psyllid (*Diaphorina citri*), has moved from Mexico into southern California and, as the climate warmed up, moved northward into central California. No genetic resistance or chemical controls are available for this disease at the present time, but entomologists and pathologists are working on it at the University of California, Riverside (White 2014).

YIELD STAGNATION: WHEAT EXAMPLE

It is well known that wheat is one of the most important food crops of the world, second perhaps only to rice. Approximately 1.5–2.5 billion people in the developing countries rely on wheat for their daily sustenance as bread or other foods. While its demand is increasing, effects of climate change are likely to reduce wheat production by 20%–30% during the next three decades (WHEAT 2011). Yields of winter wheat have been stagnating for a number of years: Global estimates show only 0.8% increase in wheat yields per annum (Fischer et al. 2014). For 27% of the wheat-growing area of the United States, 64% in Turkey, 56% in China, and the same percentage in the whole of western and eastern Europe, climate change is one of the contributing factors for this stagnation of yields, despite the introduction of new cultivars (Ray et al. 2012). Genetic progress is difficult when available moisture is limited. Today, most of the exploitable genetic variation available in wheat genetic resources has been tapped, and future prospects for enhanced genetic potential for grain yield of wheat in the face of climate change will depend on more sophisticated breeding tools and environmental assessments (Graybosch et al. 2014).

Morgounov et al. (2013), while studying the effects of climate change at winter-wheat-growing sites in central Asia, eastern Europe, and the United States, concluded that changing climate patterns as expressed through rising temperatures during the crucial stages of winter-wheat development most probably contributed to the stagnation of winter-wheat yields, especially in eastern Europe. However, there are some gains to be made too, such as less investment needed in adding traits associated with winter survival.

Rising temperatures in the month of June are detrimental to grain development and grain filling, and so heat tolerance would require a higher priority in wheat-breeding programs. This is where genebanks, which store wheat genetic resources, come into the picture because there are some accessions conserved that may harbor genes for heat tolerance as seen in the case of two samples, discussed in Section "Genetic Resources Available", collected in the 1930s from Afghanistan (Damania 2008).

TEMPERATURE CHANGE

Braun et al. (2010) summarized the effect of climate change on wheat production across mega-environments. They concluded that climate change in the coming years is expected to result in warmer winters, thereby facilitating the cultivation of spring wheat. However, rising temperatures exacerbate water deficits and affect crop development before anthesis, as well as during grain filling (Morgounov et al. 2014). The use of practices promoted in the conservation agriculture movement mitigates the effects of rising temperatures, weather extremes, and variation in precipitation (Hobbs and Govaerts 2010). The modification of common agronomic practices (such as planting dates, fertilizer application, crop protection, crop residue management, and rotation) to match the diverse patterns of crop development is another approach to mitigating the effects of climate change.

On the other hand, the effect of climate change on grain quality in wheat, especially as related to protein content, is much less understood and is associated with a trade-off in yield (Pleijel and Udding 2012). Hence, there is a need for breeders to develop wheat varieties with a broad adaptation band combined with resistance to the major diseases as well as biotic and abiotic stresses prevailing in the target area. Genetic gains thereby obtained will need to be sustained and coupled with traits that possess resilience to changes in climate from year to year.

An agronomic interpretation of the effect of rising temperatures on winter wheat can indicate to breeders that they should focus on traits that will provide better adaptation to a changing climate. Warmer autumns could provide favorable conditions for winter-wheat germination and establishment, if rainfall is sufficient. It would become necessary to study the response of different genotypes to temperature changes during germination. For example, lack of moisture associated with higher temperatures in autumn would require deeper planting and longer coleoptiles (Liatukas and Ruzgas 2011), although, as Bhatt and Qualset (1976) have shown, coleoptile length is decreased in warmer soil conditions.

GENETIC RESOURCES AVAILABLE

Annicchiarico et al. (1995) gave information on relationships between climatic features at collection sites and morpho-physiological variability of genetic resources that could enhance our understanding of evolving adaptation patterns. In a study involving durum wheat landraces from Ethiopia, Morocco, Syria, and Turkey that were evaluated in Syria, several relationships were found that mainly concerned drought and high-temperature stress among climatic collecting variables and earliness of heading and maturity among agronomic traits. Higher drought tolerance and heat stress were associated with lower yields in moderately favorable environments. Drought tolerance and earlier heading were found in Ethiopian landraces. Landraces from Turkey showed that temperature influenced positively both yield and earliness. Spike length was positively correlated with milder drought and heat stress as well as lower protein content in the Syrian gene pool. Overall morpho-physiological variation as summarized by the first axis of a principal component analysis was associated with climatic variables influencing drought and heat stress in the landraces collected from Ethiopia and Syria and to maximum temperatures in the Turkish germplasm (Annicchiarico et al. 1995).

Working with freshly collected wheat landraces of *Triticum turgidum* L. (*durum* group) from Turkey, Damania et al. (1996) came to the conclusion that distinct differences among clusters of provinces are such that population samples collected from one area in Turkey do not adequately represent samples from another area. In the light of climate change, it would be worthwhile to survey provinces in the principal durum wheat-growing areas of the Fertile Crescent that have not been sampled before to identify and collect genetic resources with unique gene combinations. Also, Turkey, which forms part of the center of diversity for wheat, has three distinct wheat-growing areas: the coastal areas, the middle plains, and the high-elevation mountainous region. The areas that can yield the best landrace genetic resources are those that can be found in small farmers' fields where there is low soil fertility and inputs. These are fields in which the modern varieties generally do not grow well. However, this was less evident in the durum wheat landraces with solid stems collected from Morocco (Damania et al. 1987). Recent collections of landrace wheat samples from farmers' fields throughout Turkey have revealed a wealth of genetic diversity that may be exploited for breeding for environmental stresses (Kan et al. 2015).

Jack Harlan enjoyed telling of a nondescript wheat sample that he collected in eastern Turkey whose value to wheat improvement he could never have guessed (Harlan 1975). He maintained that the potential value of a collection cannot be assessed in the field at the time of collection. To make his point, Harlan gives an example of a sample of wheat he collected in a remote part of eastern Turkey in 1948, which was subsequently maintained by the U.S. National Plant Germplasm System as PI 178383. It was a miserable-looking wheat: tall statured, thin stemmed, susceptible to lodging, susceptible to leaf rust, lacking winter hardiness but difficult to vernalize, and with poor baking qualities. Understandably, no one paid any attention to it for some 15 years. Suddenly, after stripe rust became a serious problem in the northwestern United States, it was discovered that PI 178383 was resistant to 4 races of stripe rust, 35 races of common bunt, 10 races of dwarf bunt, and also possessed good tolerance to flag smut and snow mold. The improved cultivars developed based on PI 178383 germplasm reduced losses to the U.S. wheat farmers by millions of dollars (USD) per year (Harlan 1975). Hence, one of the most important principles learned from Harlan is that no matter how insignificant a sample may appear in the field, its full potential should never be discounted until it is evaluated for important traits. In other words, no sample should be rejected at the collection site on the basis of its appearance. For instance, in 1986, while the senior author was evaluating wheat in the vicinity of the saline Lake Jabboul in northeastern Syria, four accessions stood out in contrast to the rest of 5000 samples evaluated at this highly saline and heat-prone site. They alone were still green at harvest time and produced spikes with viable seeds. It was found that two of these accessions had originated from Afghanistan: one had been collected in Kabul by Wilbur Harlan (PI 134116) in 1939, and the other (PI 125367) in Baghlan by Walter Koelz in 1937. Jack Harlan, in a personal communication in the 1980s, confirmed that if we were looking for germplasm with drought, salinity, and heat tolerance, areas in Afghanistan and Iran were more likely to yield the desired germplasm than anywhere else in the world (Damania 2008). The implications for growing wheat in the face of global warming are obvious.

WILD CROP RELATIVES TO THE RESCUE

Variability in wild wheat progenitors, such as wild emmer, *T. turgidum* L. ssp. *dicoccoides*, makes them indispensable as a source for genes to combat abiotic stresses due to climate change, such as heat, soil salinity, and aridity. Several authors have conducted evaluation studies on wild emmer whose genetic structure is intimately associated with ecological background heterogeneity affected primarily through climatic factors (Nevo 1998). For example, in contrast to the relatively uniform samples collected from western Israel, Kato et al. (1998) observed that the variation was much larger in the southeastern area where wild emmer is found to grow under severe aridity stress. They concluded that ecogeographical heterogeneity, or instability of growing conditions, is an important indicator for possession of genetic variability in a population for adaptive traits to climate change. Therefore, these areas in Israel should be further explored for collecting germplasm specifically targeted to capture genes useful for mitigating the effects of global warming on wheat. However, when each of the chromosomes of a single *T. turgidum dicoccoides* accession was introduced to the common wheat variety Bethlehem, the results were disappointing. None of the introduced chromosomes alone enhanced productivity of Bethlehem (Millet et al. 2013). This further indicates that populations of wild crop relatives must be studied to discover useful traits to counter the effects of climate change.

GLOBAL FOOD IMPACTS

The United States produces 41% of the world's corn and 38% of the world's soybeans. These crops comprise two of the four largest sources of caloric energy produced and are thus critical for world food supply. Schlenker and Roberts (2009) paired a panel of county-level yields for these two crops, plus cotton (a warmer weather crop), with a new fine-scale weather dataset that incorporates the whole distribution of temperatures within each day and across all days in the growing season. They found that yields increased with temperature up to 29°C for corn, 30°C for soybeans, and 32°C for cotton, but that temperatures above these thresholds are very harmful and yields dropped dramatically. The slope of the decline above the optimum is significantly steeper than the incline below it. The same nonlinear and asymmetric relationship is found when they isolated either time-series or cross-sectional variations in temperatures and yields. This suggests limited historical adaptation of seed varieties or management practices to warmer temperatures because the cross section includes farmers' adaptations to warmer climates but the time series does not. Holding current growing regions fixed, area-weighted average yields are predicted to decrease by 30%–46% before the end of the century under the slowest (B1) warming scenario and decrease by 63%–82% under the most rapid warming scenario (A1FI) under the Hadley III model (Schlenker and Roberts 2009). The impact of reduced crop yields in the United States could have a major impact on prices of these commodities in the global markets.

There have been signs that the climate is changing significantly in West Africa. In the northern part of the Sahel, the rainfall in the 1980s was one-half of what was received in 1950s (OFEDI and GRAIN 2009). There have been changes in the weather patterns, especially the rainy season. For example, in 2008, torrential rains led to floods that devastated agricultural lands and also resulted in loss of human lives in Togo and Ghana. The *harmattan*, the dry cold, northeasterly trade wind that blows along the coast of West Africa, has weakened in recent years, particularly in Benin and Ivory Coast.

Biodiversity is essential for humankind, because it supplies the raw material and the genes that give rise to new plant varieties on which farmers and pastoralists depend. Biodiversity in flora and fauna increases resilience to stresses and to changes in the environmental conditions through adaptation. Climate change represents a threat not only to the existence of individual plant species but also to the genetic diversity hidden within them. DNA studies have revealed that traditional species, as defined by taxonomists, contain a vast amount of *cryptic* diversity, such as different lineages, or even species within species. Bálint et al. (2011) attempted to understand how global warming might affect this cryptic diversity, and found that the lost evolutionary potential could hinder the ability of a species to adapt to change. Furthermore, that study showed how global climate change may lead to the loss of significant amounts of hidden diversity, even if some of the traditionally defined species will persist. What is not clear is how to extend this approach to other species with different powers of migration or dispersal. In Europe, the most genetically diverse regions are also the most endangered. The study further predicted that losses of genetic diversity will be the greatest in the Mediterranean region, which is also the region with the greatest genetic diversity.

A historical look at grain growth in North America, particularly wheat, shows that past generations of farmers have coped well with significant climate changes. Olmstead and Rhode (2010) showed that, for wheat, farmers have adapted their practices to suit the climate in past centuries. From 1839 to 2007, a time over which median annual rainfall dropped by one-half and average annual temperatures dropped by 3.7°C in North America, wheat production expanded into areas once thought to be too arid, variable, or harsh for cultivation. During this period, wheat output rose by 26-fold in the United States and by more than 270-fold in Canada. Farmers managed to increase their output through innovations such as the use of hardier wheat varieties and mechanized equipment, with most of the distributional shift in wheat production having occurred by 1929, long before the green revolution of the 1960s (Olmstead and Rhode 2010). In Mexico, wheat harvests were six times greater per acre after the green revolution years of the 1960s than in 1950s (Werblow 2015). Wheat yields in South Asia tripled in a short period of time averting starvation.

However, despite these successes, the green revolution bypassed millions of people in Africa. Degraded soil structure, misguided policies, and poor infrastructure (inputs could not reach the farmers in time and their produce did not reach the markets) were the main reasons for this. The challenges that farmers will face in the future are comparable to those faced by farmers in the past, emphasizing that development of new varieties adapted to future climates require accessing genetic resources of crop plants conserved in genebanks.

SOLUTIONS

The challenge of climate change expressed through higher temperatures and more frequent droughts affecting wheat production requires international and global efforts in genetic resources characterization and utilization. There are several major initiatives focusing on enhancement of wheat genetic resources to combat the climate change. The Seeds of Discovery project (http://seedsofdiscovery.org/en) implemented by the International Maize and Wheat Improvement Center (CIMMYT) aims to genotype and phenotype all wheat genetic resources maintained at its genebank (more than 120,000 accessions). The first priority was wheat landraces and more than 30,000 accessions have been genotyped already using the genotyping by sequencing (GBS) technology. All the landraces are also being evaluated for agronomic traits and abiotic stresses tolerance. Eventually, a wheat genetic resources database will be publicly available merging the genomic and phenotypic information, thus, making their utilization much more efficient. The Heat and Drought Wheat Improvement Consortium (HEDWIC) aims at identification of valuable wheat genetic resources and their use to enhance drought and heat tolerance using modern genomic and physiological tools. HEDWIC encourages public–private partnerships and builds its research agenda on the latest scientific discoveries. More than 70 research ideas on how to make wheat more tolerant to moisture and heat stress were selected in the first round of consultations and future project directions will be formulated in 2015. On the other hand, International Wheat Yield Partnership (http://iwyp.org) aims to increase the crop yield by as much as 20%–25% for the areas where environments and production technologies allow full expression of genetic potential. Screening genetic resources to identify the genes and alleles contributing to higher biomass and its better partitioning is one of the important thrusts of this partnership. The international donor community is very supportive of this initiative and the first 10–15 projects will be awarded up to $1 million in 2015. These three global initiatives exemplify how the advances in modern science, especially in genomic and phenotyping tools, can be synergistically assembled into global networks with the main focus on genetic resources and utilization of genetic diversity.

Prebreeding refers to all activities designed to identify desirable traits or genes from materials that are not adapted to crop-production conditions and then introgress them into intermediate materials that can be used directly in breeding populations for producing new varieties for farmers. For wheat, prebreeding gains more importance as wild relatives are included into crossing programs to enhance tolerance to abiotic stresses caused by climate change. There is substantial evidence summarized by Ogbonnaya et al. (2013) that synthetic hexaploid wheat represents an enormous resource of genes and alleles for resistance to diseases and abiotic stresses. Current commonly used synthetics were developed in the 1990s at CIMMYT by crossing tetraploid wheats (*Triticum turgidum*, genomes AABB) with the progenitor of the D genome (the diploid *Aegilops tauschii*). Following crossing and backcrossing with modern varieties, new D-genome variation contributed to a number of physiological traits useful for drought tolerance. Several varieties having synthetic wheat in their pedigree have been released in developing countries (China, Iran, India, and Pakistan) and provide stable yield combined with disease resistance. However, prebreeding is an expensive, state-of-the-art activity not easily applied in developing countries. Strong wheat-prebreeding programs are located in Europe and the United States in both public and private programs, for example, the Wheat Improvement Strategic Programme in the United Kingdom (Moore 2015). The outcomes of prebreeding are not always accessible to developing-country breeders, and the new genetic variation may be incorporated into backgrounds adapted primarily for production regions in Europe or the United States. International public assistance to wheat breeding and research programs in South Asia, central and West Asia, and Africa is important if these countries are to keep pace with the genetic gains required to overcome the detrimental effects of climate change.

CONCLUSION

Continuing population and consumption growth will mean that the global demand for food will continue to increase. Growing competition for land water, and energy, in addition to the overexploitation of fisheries, will affect our ability to produce food, as will the urgent requirement to reduce the impact of the food production systems on the environment. The effects of climate change are a further threat. But, the world can produce more food, and can ensure that it is used more efficiently and equitably. A multifaceted and linked global strategy is needed to ensure sustainable and equitable food security, different components of which are explored here. Today, the methods, applicability, and affordability of omics technologies are improving at a fast pace, giving us the tools that we need to uncover the molecular secrets behind the complex set of genes that control plant response to climate change.

Many of our major crops, which have received the attention of plant breeders for over 100 years, have reached a yield plateau that may be impossible to breach with conventional breeding efforts. Several plant scientists now think that genetic engineering can take a leading role in increasing yields, especially to focus on potential impacts of climate change. Nothing short of an international effort to dig systematically into the databases of the world's genebanks using modern tools such as mathematics and omics will be needed to reveal the genes required to alter the genetic composition of our principle food crops and enable them to overcome the adverse effects of climate change.

Some of the ways of doing this is through greater use of genetic resource collections for (1) genetic enhancement and utilization, (2) focused identification of germplasm strategy (FIGS, ICARDA 2013), (3) bioprospecting for novel/unique biomolecules and genes, (4) greater use of bioinformatics, and (5) setting up a phenomics and genomic resource center. Emphasis should be given to increasing yields through improved crop management systems, efficient use of available water resources, and improved soil–nutrient management. Intensive rotations and making most of the growing season by planting a short-duration crop between two major crops could also be some of the ways to boost food output under the threat of a climate change.

REFERENCES

Annicchiarico, P., L. Pecetti, and A. Damania. 1995. Relationships between phenotypic variation and climatic factors at collecting sites in durum wheat landraces. *Hereditas* 122:163–167.

Bálint, M., S. Domisch, C.H.M. Engelhardt et al. 2011. Cryptic biodiversity loss linked to global climate change. *Nat Climate Change* 1:313–318. doi:10.1038/nclimate1191.

Bhatt, G.M. and C.O. Qualset. 1976. Genotype-environment interactions in wheat: Effects of temperature on coleoptile length. *Exp Agr* 12:17–22.

Braun, H.-J., G. Atlin, and T. Payne. 2010. Multi-location testing as a tool to identify plant response to global climate change. In *Climate change and crop production*, ed. M. Reynolds, 115–138. Wallingford, CT: CABI.

Damania, A.B. 2008. History, achievements, and current status of genetic resources conservation. *Agron J* 100:9–21.

Damania, A.B., S. Miyagawa, T. Kuwabara, M. Furusho, F. Llabas, and L. Ali. 1987. Cereal germplasm collecting mission to Morocco. *Rachis* 6(2):50–51.

Damania, A.B., L. Pecetti, C.O. Qualset, and B.O. Humeid. 1996. Diversity and geographic distribution of adaptive traits in *Triticum turgidum* L. (*durum* group) wheat landraces from Turkey. *Genet Resour Crop Ev* 43:409–422.

Fischer, T., D. Byerlee, and G. Edmeades. 2014. *Crop yields and global food security: Will yield increase continue to feed the world?* ACIAR Monograph No. 158. Canberra: Australian Centre for International Research.

Graybosch, R., H.E. Bockelman, K.A. Garland-Campbell, D.F. Garvin, and T. Regassa. 2014. Wheat. In *Yield gains in major U.S. field crops*, CSSA Special Publication 33, eds. S. Smith, B. Diers, J. Specht, and B. Carver, 459–487. Madison: Crop Science Society of America.

Harlan, J.R. 1975. Seed crops. In *Crop genetic resources for today and tomorrow*, eds. O.H. Frankel and J.G. Hawkes, 111–115. Cambridge: Cambridge University Press.

Hobbs, P. and B. Govaerts. 2010. How conservation agriculture can combine buffering climate change. In *Climate change and crop production*, ed. M. Reynolds, 177–200. Wallingford, CT: CABI.

ICARDA. 2013. *A new approach to mining agricultural gene banks—To speed the pace of research innovation for food security.* Research to Action 3. Beirut, Lebanon: ICARDA. https://apps.icarda.org/wsInternet/wsInternet.asmx/DownloadFileToLocal?filePath=Research_to_Action/Research_to_action-3.pdf&fileName=Research_to_action-3.pdf.

Kan, M., M. Küçükçongar, M. Keser, A. Morgounov, H. Muminjanov, F. Özdemir, and C.O. Qualset. 2015. *Wheat landrace inventory of Turkey.* Konya, Turkey: Bahri Dagdas International Agricultural Research Institute.

Kato, K., C. Tanizoe, A. Beiles, and E. Nevo. 1998. Geographical variation in heading traits in wild emmer wheat, *Triticum dicoccoides*. II. Variation in heading date and adaptation to diverse eco-geographical conditions. *Hereditas* 128:33–39.

Liatukas, Z. and V. Ruzgas. 2011. Relationship of coleoptile length and plant height in winter wheat accessions. *Pakistan J Bot* 43:1535–1540.

Millet, E., J.K. Rong, C.O. Qualset et al. 2013. Grain yield and grain protein percentage of common wheat lines with emmer chromosome-arm substitutions. *Euphytica* 195:69–81.

Moore, G. 2015. Strategic pre-breeding for wheat improvement. *Nat Plants* 1:15018. doi:10.1038/nplants.2015.18.

Morgounov, A., A. Abugalieva, and S. Martynov. 2014. Effect of climate change and variety on long-term variation of grain yield and quality in winter wheat in Kazakhstan. *Cereal Res Commun* 42:1–10.

Morgounov, A., S. Haun, L. Lang, S. Matynov, and K. Sonder. 2013. Climate change at winter wheat breeding sites in central Asia, eastern Europe, and USA, and implications for breeding. *Euphytica* 194:277–292.

Nevo, E. 1998. Genetic diversity in wild cereals: Regional and local studies and their bearing on conservation ex situ and in situ. *Genet Resour Crop Ev* 45:355–370.

OFEDI and GRAIN. 2009. Climate change in West Africa—The risk to food security and biodiversity. *Seedling* October:35–37.

Ogbonnaya, F., O. Abdalla, A. Mujeeb-Kazi et al. 2013. Synthetic hexaploids: Harnessing species of the primary gene pool for wheat improvement. *Plant Breed Rev* 37:35–122.

Olmstead, A.L. and P.W. Rhode. 2010. Adapting North American wheat production to climatic challenges. *Proc Natl Acad Sci USA* 108(2):480–485.

Pleijel, H. and J. Udding. 2012. Yield vs. quality trade-off for wheat in response to carbon dioxide and ozone. *Global Change Biol* 18:596–605.

Rana, J.C. and S.K. Sharma. 2009. Plant genetic resources management under emerging climate change. *Indian J Genet Pl Br* 69(4):267–283.

Ray, D.K., N. Ramankutty, N.D. Mueller et al. 2012. Recent patterns of crop yield growth and stagnation. *Nat Commun* 3:1293. doi:1038/ncomms2296.

Rosenzweig, C. and M.L. Parry. 1994. Potential impact of climate change on world food supply. *Nature* 367:133–138.

Schlenker, W. and M.J. Roberts. 2009. Nonlinear temperature effects indicate severe damages to U.S. crop yields under climate change. *Proc Natl Acad Sci USA* 106:15594–15598.

Warnert, J. 2014. Unwelcome arrivals. *California Agr* 68(4):102–104.

Watson, R.T. and M.C. Zinyowera. 1998. The regional impacts of climate change: An assessment of vulnerability. In *A special report of IPCC Working Group II*, ed. R.H. Moss, 201–204. Cambridge: Cambridge University Press.

Werblow, S. 2015. Green revolution—Hunger fighters survey a broader battlefield. *Furrow* 120:10–13.

WHEAT. 2011. *Global alliance for improving food security and the livelihoods of the resource-poor in the developing world*. Mexico City: CIMMYT.

White, H. 2014. A look at EIPD strategic initiative projects. *California Agr* 68(4):105–108.

6 Utilizing the Diversity of Wild Soybeans in China for Accelerating Soybean Breeding in the Genome Era

Y. Li, R. Chang, and L. Qiu

CONTENTS

Soybean (*Glycine max* [L.] Merr.) and its wild progenitor (*G. soja* Sieb. and Zucc.) belong to the genus *Glycine*, subgenus *soja*. Both *G. soja* and *G. max* are annual and predominantly self-pollinating (Hymowitz 2004). *G. soja* is endemic to East Asia including China, far eastern Russia, Japan, and Korea. It is believed that *G. max* was domesticated from *G. soja* about 4,500 years ago in China (Qiu et al. 2011). The domesticated soybean was introduced from China to Europe in 1737, to North America in 1765, and to South America in 1882 (Chang 1989). At present, modern soybean is an important economical crop for food, feed, and fuel, grown worldwide. In addition, although the nomenclature and classification of *G. gracilis* Skvortzow, a unique type of soybean bearing morphologically intermediate characteristics between typical *G. soja* and *G. max* accessions, was always controversial, many scientists believe that these semi-wild soybeans may serve as important materials for understanding the evolutionary history of soybean (Skvortzow 1927, Hymowitz 1970, Broich and Palmer 1980, Wang and Li 2011).

Artificial selection has without a doubt played a significant role in the course of soybean domestication and genetic improvement, which resulted in the gradual displacement of local traditional landraces and wild soybeans for agricultural production. Moreover, soybean breeders tended to use modern cultivars as direct parents in breeding programs. The analysis of 1,019 cultivars bred and released during 1923–2005 in

China revealed that most of them derived from a small number of ancestors, and 71.78% of cultivars were derived from crosses involving modern cultivars or lines as direct parents (Xiong et al. 2008). These agricultural practices led to a dramatic narrowing of the genetic base and a reduction in the genetic diversity within modern cultivars, producing eventually a dead end for genetic improvement (Hyten et al. 2006, Li et al. 2013). As a means of effectively responding to adverse environmental conditions resulting from global climate changes and associated pests and diseases, a considerable level of genetic variation should be identified and introduced into modern cultivars for increasing and sustaining soybean production. After domestication and genetic improvement, modern soybeans exhibit substantial morphological and physiological differences from *G. soja*. In addition, genetic diversity in modern cultivars as compared to *G. soja*, especially rare alleles, has declined (Hyten et al. 2006), suggesting *G. soja* as a promising natural source of novel genes for soybean breeding. However, only a few genes have been discovered in *G. soja* despite numerous examples of plant genetic resources providing genes and traits to improve crops (Cooper et al. 2001). Capturing the majority of genetic diversity present within soybean germplasm is urgently needed for improving soybean.

The diverse soybean genetic resources are collected and conserved in the ex-situ gene banks, including undomesticated *G. soja*, landraces, modern cultivars, and provide a rich source of genetic variation for modern soybean improvement (McCouch et al. 2013). As a center of diversification of wild and cultivated soybeans, China is the location for most of the world's soybean genetic resources. Among more than 45,000 unique soybean collections worldwide, two-thirds are from China, including 23,587 *G. max* and around 7,000 *G. soja* accessions in the Chinese National Soybean Genebank (CNSGB) (Qiu et al. 2011). Most (93%) of the conserved *G. max* accessions are landraces; modern cultivars or lines accounted for only 7%. These Chinese soybean accessions have great phenotypic and genetic diversity, and they have played an important role in developing soybean cultivars worldwide (Carter et al. 2004, Qiu et al. 2011). Completely resolving genetic diversity at the whole-genome level in Chinese soybean germplasm, especially *G. soja*, will be helpful for enhancing soybean improvement in the near future, not only in China but also worldwide.

The development of high-throughput genotyping and next-generation sequencing (NGS) technologies and publication of a soybean reference genome (cultivar Williams 82, *GmaxW82*) (Schmutz et al. 2010) provided opportunity for Chinese scientists to comprehensively analyze genetic diversity and identify genome-wide genetic variation controlling agronomic traits in *G. soja* in an efficient manner. Here, we discuss recent advancements in this area in China.

ASSEMBLING DRAFT GENOMES FOR THE PROGENITOR OF CULTIVATED SOYBEAN (*G. SOJA*)

In 2010, the chromosome-scale draft sequence of cultivated soybean (cultivar Williams 82, Glyma 1.0) using a whole-genome shotgun (WGS) approach was published (Schmutz et al. 2010). The assembled GmaxW82 genome was 950 Mb in size and showed a highly duplicated pattern resulting from two major genome duplications at around 59 and 13 million years ago (Schmutz et al. 2010). Although *G. soja* is important as a natural source for novel genes for broadening the genetic base of modern soybean cultivars, its genome sequence was not known until the draft genomes of *G. soja* were published (Table 6.1). In order to maximize the discovery of sequence diversity in wild soybean (*G. soja*), seven accessions (referred to as GsojaA through GsojaG for brevity) representing China (including northeastern, northern, Huanghuai, and southern regions), Japan, Korea, and Russia gene pools were de novo sequenced (Li et al. 2014c). Seven wild soybean genomes were assembled using a total of 779.2 Gbp data at an average depth of 111.9× (ranging from 83.7× to 136.3×), generated from short-insert paired-end (~180 and ~500 bp) and long-insert mate-pair (~2000 bp) sequencing libraries. The contig N50 sizes ranged from 8 to 27kbp, and scaffold N50 sizes ranged from 17 to 65.1kbp. Moreover, a wild soybean accession (W05) from central China (Henan Province), which is highly tolerant of salt stress, was also de novo sequenced using a combination of insert size libraries, that is, 180 bp, 260 bp, 326 bp, 317 bp, 2 kbp, 2.3 kbp, 6 kbp, and 10 kbp. The 868-Mb genome of W05 was assembled with a contig N50 of 24.2 kbp and a scaffold N50 of 401.3 kbp (Qi et al. 2014). In addition, the draft genome of a wild accession Lanxi 1, collected from southern China, was constructed based on a 55-fold genome coverage of de novo sequence generated from two libraries with different insertion sizes: 500 bp and 2 kbp. The assembled Lanxi 1 genome was 929.9 Mb with a contig N50 of 21.7 and a scaffold N50 of 51.0 kb (Qiu et al. 2014). Wild soybean protein-coding genes were annotated and the gene number estimates ranged from 49,560 (W05) to 57,631 (GsojaD), with an average of 54,983 (Table 6.1). Around 65% of annotated genes were expressed, as inferred from RNA-Seq data of transcriptomes generated from tissue mixtures (Li et al. 2014c, Qi et al. 2014). Although, larger scaffold N50s, ~50 kbp, would have been preferred, the sizes obtained in these analyses are sufficient for comparative and evolutionary analyses. Moreover, the genome sizes of *G. soja* were estimated to range from 889 (GsojaG) to 1170 Mbp (W05), which were larger than the assembled genomes (813–985 Mbp). If needed, the genomes of *G. soja* can later be improved with additional sequencing of large mate-pair insert libraries.

Based on the seven genome assemblies, the first pan-genome of annual wild soybean was established, which contained 59,080 gene families and had an overall size of 986.3 Mbp (Li et al. 2014c). Among the 59,080 annotated gene families, nearly half (48.6%) are conserved across all seven *G. soja* genomes as core genomic units, which were enriched in biological processes such as growth and immune system processes, reflecting the biological characteristics of wild soybean species. Approximately half of the gene families (51.4%) were present in more than one, but not all, of the seven *G. soja* genomes, and represent the dispensable genome, perhaps reflecting a role in adaptation to various abiotic and biotic stresses.

CLARIFICATION OF WHOLE GENOME-WIDE GENETIC VARIATION IN SOYBEANS USING BOTH RESEQUENCING AND DE NOVO SEQUENCING APPROACHES

Subsequent to sequencing the soybean reference genome (GmaxW82), several resequencing analyses of soybean germplasm (mainly of Chinese origin) had been conducted and reported. Lam et al. (2010) first resequenced 31 soybean accessions (including

TABLE 6.1
Sequencing, Assembly, and Annotation Information for Nine Published *G. soja* Genomes

ID	Origin	No. of Libraries (Insertion Size, bp)	Fold Sequencing Depth (x)	Estimated Genome Size (Mbp)	Assembled Genome Size (Mbp)	Contig N50 (kbp)	Scaffold N50 (kbp)	Annotated Gene Number	Reference
GsojaA	Zhejiang, China	3 (180, 500, and 2000)	118	981	813	9.0	18.3	55,061	Li et al. (2014c)
GsojaB	Ibaraki, Japan	3 (180, 500, and 2000)	116	1001	895	22.2	57.2	54,256	Li et al. (2014c)
GsojaC	Chungcheong Puk, Korea	3 (180, 500, and 2000)	123	1054	841	8.0	17.0	56,542	Li et al. (2014c)
GsojaD	Shandong, China	3 (180, 500, and 2000)	136	1118	985	11.0	48.7	57,631	Li et al. (2014c)
GsojaE	Shanxi, China	3 (180, 500, and 2000)	103	956	920	27.0	65.1	55,901	Li et al. (2014c)
GsojaF	Heilongjiang, China	3 (180, 500, and 2000)	84	993	886	24.3	52.4	54,805	Li et al. (2014c)
GsojaG	Khabarovsk, Russia	3 (180, 500, and 2000)	104	889	878	19.2	44.9	54,797	Li et al. (2014c)
W05	Henan, China	8 (180, 260, 326, 817, 2,000, 2,300, 6,000, and 10,000)	80	1170	868	24.2	401.3	49,560	Qi et al. (2014)
Lanxi1	Zhejiang, China	2 (500 and 2000)	55		930	21.7	51.0	56,298	Qui et al. (2014)

17 *G. soja* and 14 *G. max* genomes) with an average depth of approximately 5× coverage. All of 17 *G. soja* accessions were from central and northeastern China. Of the 14 *G. max* accessions, 12 were from southern to northeastern China, and the other two were from the United States and Brazil, respectively. By comparison of the 32 soybean genomes with the reference genome of GmaxW82 (Glyma 1.0), a total of 6,318,109 single nucleotide polymorphisms (SNPs) were identified. In order to increase the representation of the germplasm analyzed, Li et al. (2013) resequenced 25 soybean accessions, including eight *G. soja* accessions, eight landraces, and nine elite cultivars. These accessions were chosen from the Chinese soybean core collection (Qiu et al. 2009, 2013), mainly based on (1) their genetic diversity revealed by molecular markers (e.g., simple sequence repeat [SSRs] and SNPs) in a previous study (Li et al. 2010) and (2) geographic distribution. These samples well represent the major gene pools of the whole germplasm collection deposited in CNSGB. For these 25 soybean accessions, a total of 1356 million high-quality paired-end reads consisting of 93.55 Gbp of sequences was generated, with an average of 3.4Gbp coverage per accession. In order to improve the quality of SNP calling and secure the reliability of selection region, the dataset of Lam et al. (2010) was also incorporated into the current dataset. There were 5,102,244 SNPs and 707,969 small insertion/deletions (indels, ≤5bp) identified across 55 soybean genomes.

Although millions of SNPs and small indels among wild and cultivated soybeans have been cataloged using the genome of GmaxW82 as a reference, variants that may be present only in wild accessions and/or landraces, especially structural variations (SVs) including *G. soja*-specific present–absent variation (PAV) and copy-number variation (CNV) derived from large indels remain unidentified. Since the genome sequences of *G. soja* and/or original landraces are still not completely resolved, these SVs cannot be effectively identified using the resequencing approach alone. Moreover, the resequencing approach alone was unable to discover genetic variation in divergent regions with high evolutionary rates or in highly repetitive regions, where short sequencing reads were difficult to map. Therefore, the assembly-based method was widely used for clarifying the genetic variation in Chinese soybean germplasm, and greater numbers and types of genetic variation have been identified. By comparison of the assembled W05 genome with the GmaxW82 genome, 186,177 indels were discovered. Although more than 50% of them were small (≤5bp), there were 5,592 large indels (>500 bp), including 4444 insertions in the W05 genome and 1,148 insertions in the GmaxW82 genome (Lam et al. 2010). After integrating and comparing the WGS data of Chinese landrace NN 1138-2 (91× coverage) and modern cultivar KFNo.1 (13× coverage) with resequencing data generated from 31 soybean accessions (Lam et al. 2010), 33,127 PAVs (>50 but <2600 bp) were observed (Wang et al. 2014). By comparative analysis of the *G. soja* pan-genome and the cultivated GmaxW82 genome, 1978 genes were affected by CNV, and 354 genes containing species-specific PAV were identified using a three-filter criteria: (1) the size of PAV sequence was larger than 100 bp, (2) the identity of PAV sequence was lower than 95%, and (3) more than 50% of the length of the CDS contained PAV sequence (Li et al. 2014c). Limited to the current size of contigs and scaffolds, few large-range structural variants have been discovered, except for 60 inversion and 579 translocation events (160 intra-translocation and 419 inter-chromosomal translocation events) observed by screening semi-wild accessions (Qiu et al. 2014).

Until now, more than 10 million genetic variations with detailed information have been identified, which provide a valuable resource for facilitating the development of molecular biology, genetics, and molecular breeding in soybean. For example, the development of high-throughput genotyping techniques made the published soybean SNP dataset a promising resource for classical quantitative trait locus (QTL) mapping. To solve the issue of low polymorphism between *ZhongHuang13* and *ZhongPin03-5373*, two parental lines of one large recombinant inbred line population, the whole-genome-wide SNP and indels were identified based on the information of the resequencing of *ZhongHuang13* and *ZhongPin03-5373* (Li et al. 2014a,b). Since resequencing analysis could provide an opportunity to take a closer look at the landscape of SNP and indels between the two parental lines, researchers could select the polymorphism based on their objectives. Using a customized SNP array containing 384 SNPs detected between *ZhongHuang13* and *ZhongPin03-5373*, researchers constructed a soybean genetic linkage map, clarified the landscape of recombination, and identified several QTLs controlling plant height and seed weight (Liu et al. 2013, Li et al. 2014b). Furthermore, with biparental resequencing, additional indel markers were developed. Adding these indel markers to the previously used SNP and SSR markers facilitated the discovery of further recombination events, allowing the fine-mapping of a QTL to a 0.5 Mbp region (Li et al. 2014a). This suggested that the combination of SNPs and the indel dataset produced by whole-genome resequencing, high-throughput genotyping, and QTL genetic mapping is an effective approach to reveal novel genomic information for the improvement of soybean modern cultivars.

UNDERSTANDING EVOLUTIONARY HISTORY OF SOYBEAN

G. max and *G. soja* form a species complex that is around five million years old (Lavin et al. 2005). However, the analysis of deep resequencing data of a single *G. soja* accession suggested that *G. soja* was domesticated and diverged ~0.27 mya (Kim et al. 2010). To explain this inconsistency, researchers had suggested that soybean was domesticated from the *G. soja/G. max* complex, which diverged from a common ancestor. The evolutionary analysis of the *G. soja* pan-genome somewhat supported this scenario (Li et al. 2014c). However, the estimated divergence time (~0.8 mya) was approximately three times more than the older estimate (0.27 mya). This difference might not be caused simply by the different methodologies adopted to estimate the divergence time and different samples. One other possible reason is that the cultivated soybean is a subfamily of one branch in the wild soybean family. Another explanation could be the closest *G. soja* is actually a hybridization of *G. max* and ancient *G. soja*. Further analysis is required to solve this puzzle.

In contrast to the estimated divergence time between *G. soja* and *G. max*, phylogenetic analyses, Bayesian clustering, and PCA analysis revealed that *G. max* is genetically distinct from *G. soja* and supports the hypothesis of a single event in soybean domestication in contrast to multiple domestication events in legumes (Li et al. 2013). Moreover, different levels of introgression between the cultivated and the wild groups have been discovered. The finding of some original cultivated soybeans with at least one of the early evolutionary traits and wild accessions carrying at least one of the typical cultivated traits underlined that substantial gene flow had occurred between wild and cultivated soybean (Lam et al. 2010, Li et al. 2013). Serving as an evolutionary bridge between *G. soja* and *G. max*, these accessions with admixture genomes tended to mix with typical wild or cultivated accessions, respectively, rather than forming an independent phylogenetic branch (Qiu et al. 2014).

EVALUATING THE POPULATION BOTTLENECK AND TRACKING THE FOOTPRINTS OF SELECTION

For mining the genetic diversity in *G. soja*, genetic variation was compared between 55 genome sequences of wild and cultivated soybeans. Of the 6,318,109 SNPs identified in 17 *G. soja* and 14 *G. max*, 2,148,585 were polymorphic in the *G. soja* population but fixed in the *G. max* population (Lam et al. 2010). Of the 5,102,244 high-quality SNPs identified in 55 soybean accessions (including the 30 soybean accessions mentioned above), 1,661,945 were specific to 25 *G. soja* accessions (Li et al. 2013). By comparing 10 semi-wild and 19 wild soybean genomes, the number of SNPs in semi-wild soybean populations (910,373) only accounted for 55.9% of SNPs identified in wild soybean populations (1,628,253) (Qiu et al. 2014). Although *G. max*-specific SNPs were also identified, the number was much lower than that of *G. soja*-specific SNPs. Only 463,409 and 529,724 SNPs were *G. max*-specific, which accounted for around 10.4% and 7.3% of total SNPs identified in the 17 and 25 wild soybean genomes, respectively (Lam et al. 2010, Li et al. 2013). Further analyzing PAV between the *G. soja* pan-genome and the GmaxW82 genome revealed 338 *G. soja*-specific and 16 *G. max*-specific PAV genes (Li et al. 2014c). In addition, 1978 genes were identified to be affected by CNV, including 1179 loss-CNV with consistently lower copy number, 726 gain-CNV with consistently higher copy number, and 73 loss/gain-CNV with consistently lower/higher copy number in *G. soja* populations than in the GmaxW82 genome (Li et al. 2014c). These analyses of the pattern of allelic diversity between *G. soja* and *G. max* suggested that soybean largely lost genetic diversity during domestication. Further analysis indicates this set of genetic variation/genes may contribute to variation of agronomic traits such as pathogen resistance, seed composition, flowering and maturity time, organ size, and final biomass. For example, *G. soja* contained more R (Resistant)-gene domain architectures than *G. max*, possibly reflecting adaptation to varied biotic stresses (Li et al. 2014c). Qi et al. (2014) identified a candidate causal gene (*GmCHX1*) for salt tolerance by the comparison of the W05 and GmaxW82 genomes, QTL mapping, and functional analysis. In *GmCHX1*, a ~3.4 kb retrotransposon PAV was involved in the salt tolerance; it was present in the salt-sensitive cultivated accessions (GmaxW82 and C08), but absent in the salt-tolerant wild accession (W05) (Qi et al. 2014). Li et al. (2014c) observed an 8-kbp *G. soja*-specific PAV, present in seven *G. soja* genomes, but absent in the GmaxW82 genome, which contained three annotated genes. Homologous analysis revealed that they related to the tolerance for biotic and abiotic stress or plant development, respectively, and perhaps reflecting a role in the biological characteristics of wild soybean species, such as adaptation to various abiotic and biotic stresses. These analyses indicated that the reservoir of genetic variation in *G. soja* can be utilized to broaden the genetic base of modern soybean cultivars that can adapt to global changes.

SELECTION SIGNALS AND DOMESTICATION SYNDROME

From the wild soybeans with prostrate growth habit and small black seeds, the shift has been to modern soybeans with erect plants and large seeds, which could be considered as a domestication syndrome. The formation of cultivated traits was probably a slow process, which took a few thousand years or more, as the result of human-mediated selection of desired agronomic traits. Based on resequencing and comparing multiple *G. soja* and *G. max* genomes, a set of genomic regions with a significantly high degree of inter-species differentiation identified using a population genetic approach were defined as regions significantly impacted by selection, which accounted for 1.47% or 5% of the soybean genome in different analyses (Lam et al. 2010, Li et al. 2013). These candidate genomic regions may contain loci responsible for traits underlying the soybean domestication. Li et al. (2013, 2014c) examined changes in the genetic architecture of key domestication traits and found candidate genes with SV and selection signals, which are, in many cases, supported by previously mapped QTLs. Many domestication syndrome traits are controlled by a relatively small number of genes (major QTLs or Mendelian loci), but because of hitchhiking effects, the selection of these traits can lead to genome-wide reduction of genetic diversity and long linkage disequilibrium, which led to difficulties in finding causal genes through QTL analyses. Further work, involving analyses of unique genetic resources (such as wild, semi-wild, or semi-cultivated) or segregating populations (such as interspecific populations) using the integration of multiple analysis methods (such as linkage mapping, association mapping, and function analysis) is necessary to identify the causal genes for important domestication syndrome traits.

MEETING FUTURE FOOD NEEDS BY USING WILD GENETIC RESOURCES

In order to meet the food and energy demands of the increasing human population with decreasing arable land area, under changing climatic conditions and with associated new plant diseases, agricultural production should be increased significantly. However, the increase of crop yields by depending only on the resources within domesticated crops cannot catch up with the increasing demands. The understanding and use of wild genetic resources is needed to meet future food needs (McCouch et al. 2013). Genomics can help do that more efficiently. The wild soybean (*G. soja*) is a valuable source of new genetic variation, which can be used to replenish the genetic diversity lost or largely changed during natural and artificial selection. Not only will the introduction of these variations from the gene pool of wild soybean to soybean cultivars broaden the extremely narrow genetic variation found in the cultivars used in current soybean production, but they will also play an important role in facilitating yield increase of modern cultivars and enable them adapt to a broad range of changes in natural environments. However, the identified variation still needs to be associated with specific traits and the associations need to be tested in crosses in order to deeply understand the basis of domestication syndromes and the diverse adaptability of wild soybeans left behind after natural and artificial selection. The utilization of wild soybean needs to be promoted.

REFERENCES

Broich, S.L. and R.G. Palmer. 1980. A cluster analysis of wild and domesticated soybean phenotypes. *Euphytica* 29:23–32.

Carter, T.E., R.L. Nelson, C.H. Sneller, and Z. Cui. 2004. *Genetic diversity in soybean*. Madison, WI: ASA, CSSA, and SSSA.

Chang, R.Z. 1989. The utilization of Chinese soybean genetic resources in foreign countries. *World Agr* 20–21, 55.

Cooper, H.D., C. Spillane, and T. Hodgkin. 2001. Broadening the genetic base of crops: An overview. In *Broadening the genetic base of crop production*, eds. H.D. Cooper, C. Spillane, and T. Hodgkin, 1–23. New York: CABI Publishing.

Hymowitz, T. 1970. On the domestication of the soybean. *Econ Bot* 24:408–421.

Hymowitz, T. 2004. Speciation and cytogenetics. In *Soybeans: Improvement, production and uses*. 3rd edition, eds. H.R. Boerma and J.E. Specht. Madison, WI: ASA, CSSA, and SSSA.

Hyten, D.L., Q. Song, Y. Zhu et al. 2006. Impacts of genetic bottlenecks on soybean genome diversity. *Proc Natl Acad Sci USA* 103:16666–16671.

Kim, M.Y., S. Lee, K. Van et al. 2010. Whole-genome sequencing and intensive analysis of the undomesticated soybean (*Glycine soja* Sieb. and Zucc.) genome. *Proc Natl Acad Sci USA* 107:22032–22037.

Lam, H.M., X. Xu, X. Liu et al. 2010. Resequencing of 31 wild and cultivated soybean genomes identifies patterns of genetic diversity and selection. *Nat Genet* 42:1053–1059.

Lavin, M., P.S. Herendeen, and M.F. Wojciechowski. 2005. Evolutionary rates analysis of Leguminosae implicates a rapid diversification of lineages during the tertiary. *Syst Biol* 54:575–594.

Li, Y.H., W. Li, C. Zhang et al. 2010. Genetic diversity in domesticated soybean (*Glycine max*) and its wild progenitor (*Glycine soja*) for simple sequence repeat and single-nucleotide polymorphism loci. *New Phytol* 188:242–253.

Li, Y.H., B. Liu, J.C. Reif et al. 2014a. Development of insertion and deletion markers based on biparental resequencing for fine mapping seed weight in soybean. *Plant Genome* 7(3). doi:10.3835/plantgenome2014.04.0014.

Li, Y.H., Y.L. Liu, Z.X. Liu et al. 2014b. Biparental resequencing coupled with SNP genotyping of the segregating population offers new insights into the landscape of recombination and identical by state regions in soybean. *G3* 4(4):553–560. doi:10.1534/g3.113.009589.

Li, Y.H., S.C. Zhao, J.X. Ma et al. 2013. Molecular footprints of domestication and improvement in soybean revealed by whole genome re-sequencing. *BMC Genomics* 14:579.

Li, Y.H., G.Y. Zhou, J.X. Ma et al. 2014c. De novo assembly of soybean wild relatives for pan-genome analysis of diversity and agronomic traits. *Nat Biotechnol* 32:1045–1052.

Liu, Y.L., Y.H. Li, J.C. Reif et al. 2013. Identification of quantitative trait loci underlying plant height and seed weight in soybean. *Plant Genome* 6(3). doi:10.3835/plantgenome2013.3803.0006.

McCouch, S., G.J. Baute, J. Bradeen et al. 2013. Agriculture: Feeding the future. *Nature* 499:23–24.

Qi, X., M.W. Li, M. Xie et al. 2014. Identification of a novel salt tolerance gene in wild soybean by whole-genome sequencing. *Nat Commun* 5:4340.

Qiu, J., Y. Wang, S. Wu et al. 2014. Genome re-sequencing of semi-wild soybean reveals a complex *Soja* population structure and deep introgression. *PLoS ONE* 9:e108479.

Qiu, L.J., P.Y. Chen, Z.X. Liu et al. 2011. The worldwide utilization of the Chinese soybean germplasm collection. *Plant Genet Resour* 9:109–122.

Qiu, L.J., L.L. Xing, Y. Guo, J. Wang, S.A. Jackson, and R.-Z. Chang. 2013. A platform for soybean molecular breeding: The utilization of core collections for food security. *Plant Mol Biol* 83:41–50.

Qiu, L.J., Y.H. Li, R.X. Guan, L.X. Wang, and R.Z. Chang. 2009. Establishment, representative testing and research progress of soybean core collection and mini core collection. *Acta Agron Sin* 35:571–579.

Schmutz, J., S.B. Cannon, J. Schlueter et al. 2010. Genome sequence of the palaeopolyploid soybean. *Nature* 463:178–183.

Skvortzow, B.V. 1927. The soybean-wild and cultivated in Eastern Asia. *Proc Manchurian Res Soc Publ Ser A, Natural History, History Sect* 22:1–8.

Wang, K.-J. and X.-H. Li. 2011. Genetic differentiation and diversity of phenotypic characters in Chinese wild soybean (*Glycine soja* Sieb. et Zucc.) revealed by nuclear SSR markers and the implication for intraspecies phylogenic relationship of characters. *Genet Resour Crop Ev* 58:209–223.

Wang, Y., J. Lu, S. Chen et al. 2014. Exploration of presence/absence variation and corresponding polymorphic markers in soybean genome. *J Integr Plant Biol* 56:1009–1019.

Xiong, D.J., T.J. Zhao, and J.Y. Gai. 2008. Parental analysis of soybean cultivars released in China. *Sci Agr Sinica* 41:2589–2598.

7 Adaptation of Potato (*Solanum tuberosum*) and Tomato (*S. lycopersicum*) to Climate Change

R. Schafleitner

CONTENTS

The Solanaceae, or nightshade family, is composed of more than 3000 species and encompasses a number of economically important food crops, ornamental plants, and medicinal herbs (Mueller et al. 2005). Potato (*S. tuberosum* L.) is probably the most economically important solanaceous crop, with 368 million tons harvested on 19.5 million hectares of land (FAOSTAT 2014). Tomato (*S. lycopersicum* L.) is the most widely grown horticultural crop with annual worldwide production of more than 160 million tons in 2012 (FAOSTAT 2014), and an economic value of US$ 1.9 billion in the United States alone (USDA 2013). Although potato and tomato originate from different ecosystems and differ in respect to their sensitivity to climate stress, climate change is expected to impact the cultivation of both of these important crops.

Climate change is assumed to bring about strong negative effects for agriculture (Rosenzweig et al. 2014). Higher temperatures and changing rainfall patterns will likely affect agricultural production in many regions of the world in the twenty-first century. Elevated temperatures lead to greater plant transpiration, increasing the water requirement of plants and augmenting the risk of drought stress for crops. Adapting planting calendars, shifting production zones toward more accommodating regions, and replacing heat-sensitive varieties with more climate-resilient cultivars will contribute to maintain agricultural production under climate-change scenarios.

POTATO

Potato evolved in cool zones and heat affects its yield in many ways. Optimum temperatures for potato development and yield range between 14°C and 22°C (Marinus and Bodlaender 1975). Elevated temperatures generally cause increased stem growth, smaller leaves, and reduced tuber formation (Ben Khedher and Ewing 1985, Menzel 1985, Levy 1986, Prange et al. 1990, Lafta and Lorenzen 1995). Tuber yield in potato is determined by the dynamics of many organs involving a range of physiological mechanisms, some of which are heat sensitive (reviewed by Struik 2007). However, hormonal effects restricting tuberization under hot conditions seem to affect yield under heat stress more than photosynthetic limitations (Reynolds et al. 1990). Tuber initiation can become delayed at temperatures above 22°C and can be blocked at higher temperatures, especially at elevated night temperatures. Tuber growth is temperature sensitive and is affected at moderate temperatures, while other plant functions such as photosynthesis do not show any stress symptoms (Hancock et al. 2014).

Due to these growth requirements, potato cultivation is restricted to the temperate zones of the Northern and Southern Hemispheres and to tropical highlands (Hijmans 2001). The effect of an average global temperature increase (excluding Antarctica) between 2.1°C and 3.2°C on potato harvests in current potato-growing regions was modeled, resulting in an estimated yield decline of up to 32% (Hijmans 2003). Schafleitner et al. (2011) estimated changes in climate suitability for potato production using statistically downscaled outputs of selected global circulation models from the Intergovernmental Panel on Climate Change (IPCC) for 2010 through 2039 and emission scenarios from the IPCC Special Report on Emission Scenarios (SRES-A2). This research suggested that precipitation changes in the current potato-growing regions are likely to remain modest. Temperature increases, however, are likely to cause many current potato-cultivation zones to become less suitable for potato production. The most significant decreases in suitability for potato cultivation were predicted to occur in tropical highlands and in southern Africa. Yet, with rising temperatures and longer frost-free periods due to climate change, vast regions at high latitudes and also subtropical highland areas will become suitable for potato production, resulting in an expected increase of average global suitability of 1.3% (Schafleitner et al. 2011). It appears that a shift of potato production away from heat-prone areas to zones that will likely benefit from higher temperatures in the future may even lead to an increase in global potato area and yield.

Adaptation strategies such as genetic improvement of abiotic stress tolerance of potato may mitigate negative effects of climate change. Improved heat-stress tolerance of potato would benefit some 7.7 million ha (more than 60%) of the current cultivation area and would allow expansion of potato cropping to 15.5 million new ha, while improvements of water-stress tolerance might have much less impact (Schafleitner et al. 2011). Projected rises of the CO_2 concentration could bring about additional benefits for potato production in the form of a substantial net increase of photosynthesis, resulting in yield increases in some regions (Miglietta et al. 2000).

TOMATO

Tomato is grown worldwide in temperate, subtropical, and tropical climates. Constraints to tomato production imposed by climate change may include increased heat stress, drought, and flooding. Estimates of tomato losses (and gains) from climate change have not been addressed by global modeling approaches, as most of the fresh market tomato production is on relatively small plots, often under irrigated and protected conditions. Although climate change might also affect irrigated and protected agriculture, most impact is expected for tomato production in the open field, such as processing tomato cultivation in the Xinjiang region of China, in parts of California, and in the Mediterranean. Jackson et al. (2011) investigated climate-change effects on different crops grown in Yolo County, California, and the data suggested hardly any effect of climate change on tomato in this specific region. For southern Italy, in contrast, climate change was predicted to reduce yield and increase the irrigation requirement for tomato (Lovelli et al. 2012, Ventrella et al. 2012). Wang et al. (2014) suggested further investment in improved irrigation methods to prevent negative effects of climate change in tomato-growing regions of Xinjiang. In general, climate change is thought to aggravate the abiotic and biotic stresses that already limit tomato production in warm regions, while in cooler regions, vegetable production in general might benefit from temperature rises, which may extend the growing season for tomato as long as sufficient water remains available to sustain the crop.

The vegetative organs of tomato adapt relatively well to heat stress, but the reproductive organs, especially pollen, are strongly affected by high temperatures, resulting in large yield drops. Anther and pollen development is altered by elevated mean daily temperatures. Negative effects on pollen development already have been observed at quite moderate increases of daily average temperatures ranging from 25°C to 29°C (Harel et al. 2014). Day/night temperatures of 32°C/26°C reduced pollen viability from 90% at 28/22°C to 30%, and only 3% of the flowers set fruit, while fruit set at more permissible temperatures was 13% (Sato et al. 2006). The comparison of viable pollen and fruit set between summer and autumn tomato cultivation at AVRDC—The World Vegetable Center headquarters in Shanhua, Taiwan, is shown in Figure 7.1. The average day/night temperatures during sampling were 32.5°C/26°C during summer (August) and 27.7°C/19.6°C in autumn (November). Pollen viability of the heat-tolerant genotype CLN1621L was equivalent in summer and autumn, while the heat-sensitive variety CA4 showed a sharp decrease in the proportion of viable pollen during the hot summer. Additionally, the total pollen count under both heat stress and normal conditions was much higher in CLN1621L than in CA4 (data not shown). As a result, CLN1621L still yielded tomatoes during the hot summer, in contrast to CA4, which did not set fruit.

Pollen development in tomato is particularly sensitive to heat 8–13 days before anthesis, corresponding to the shift from pollen mother cells to unicellular microspores. Heat at this stage produces pollen grains with morphological changes of the tapetum layer and the vacuole; such grains are unable to germinate (Peet et al. 1998). On the molecular level, heat-stress transcription factors and heat-shock proteins are

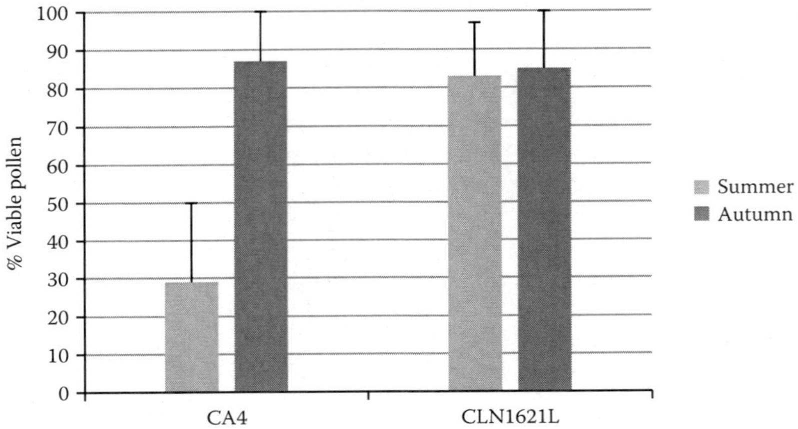

FIGURE 7.1 Pollen viability (in %) of heat-sensitive tomato variety CA4 and the heat-tolerant line CLN1621L during summer (August) and autumn (November) tested at AVRDC-The World Vegetable Center headquarters in Shanhua, Taiwan. The error bars indicate the standard deviation of three measurements at different time points among five replicated plants.

activated in response to heat in both vegetative tissues and developing pollen (Scharf et al. 2012). These factors are also likely to play a role in pollen development in the absence of stress (reviewed by Giorno et al. 2013). In addition to impairing pollen development, extreme heat stress can alter flower morphology in tomato, leading to sterility and yield loss.

AVRDC has put considerable effort into developing heat-tolerant tomato. Recombinant inbred lines (RILs) derived from CLN1621L, a heat-tolerant small-fruited variety, have greater plant vigor and, most importantly, much larger counts of viable pollen grains compared to heat-sensitive varieties, which allows for better fruit set and higher yields under heat stress (Figure 7.2). Some RILs also show resistance against flooding.

Increased drought risks are predicted under climate-change scenarios for tomato-production areas in the Mediterranean. Drought and reduced irrigation water availability may influence tomato yield directly, and additionally through unfavorable tomato–weed interactions whose relative effect on yield may become more important under water-limiting conditions (Valerio et al. 2013).

CONCLUSION

Climate-change predictions and crop-yield modeling in potato suggest yield reductions in the tropics and gains in high latitudes and altitudes. Reducing temperature sensitivity of potato would have the largest impact on the crop's adaptation to climate change. Although heat might affect the tomato crop, lack of irrigation water is more likely to impact yield in current production zones.

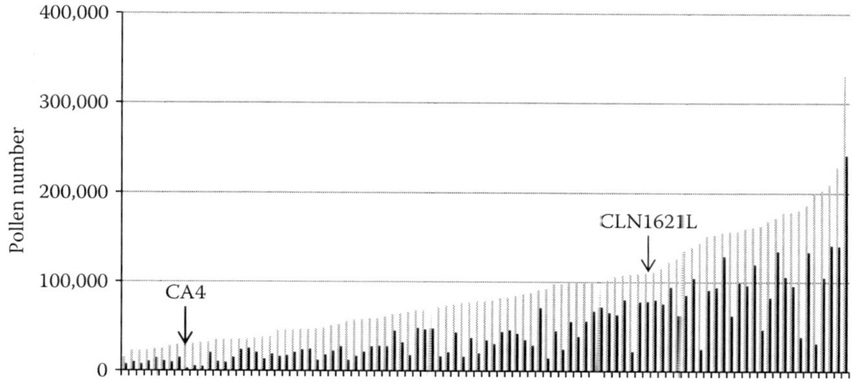

FIGURE 7.2 Number of total (gray bar) and viable (black bar) pollen in F6 families derived from the cross CLN1621L (heat tolerant) × CA4 (heat sensitive) exposed to natural heat-stress conditions during the summer season 2013 at AVRDC-The World Vegetable Center headquarters in Shanhua, Taiwan (Schafleitner et al. 2014). The count of total and viable pollen has been determined as described in Heslop-Harrison et al. (1984). The bars for the parental lines of the population are labeled.

Increased temperatures might, on the other hand, provide opportunities to expand the tomato crop to new regions.

REFERENCES

Ben Khedher, M. and E.E. Ewing. 1985. Growth analysis of eleven potato cultivars grown in the greenhouse under long photoperiods with and without heat stress. *Am Potato J* 62:537–554.

FAOSTAT. Statistics Division. 2014. http://faostat3.fao.org//faostat-gateway/go/to/download/Q/*/E (accessed October 2, 2014).

Giorno, F., M. Wolters-Arts, C. Mariani, and I. Rieu. 2013. Ensuring reproduction at high temperatures: The heat stress response during anther and pollen development. *Plants* 2:489–506.

Hancock, R.D., W.L. Morris, L.J. Ducreux et al. 2014. Physiological, biochemical and molecular responses of the potato (*Solanum tuberosum* L.) plant to moderately elevated temperature. *Plant Cell Environ* 37(2):439–450. doi:10.1111/pce.12168.

Harel, D., H. Fadida, A. Slepoy, S. Gantz, and K. Shilo. 2014. The effect of mean daily temperature and relative humidity on pollen, fruit set and yield of tomato grown in commercial protected cultivation. *Agronomy* 4:167–177.

Heslop-Harrison, J., Y. Heslop-Harrison, and K.R. Shivanna. 1984. The evaluation of pollen quality, and a further appraisal of the fluorochromatic (FCR) test procedure. *Theor Appl Genet* 67:367–375.

Hijmans, R.J. 2001. Global distribution of the potato crop. *Am J Potato Res* 78:403–412.

Hijmans, R.J. 2003. The effect of climate change on global potato production. *Am J Potato Res* 80:271–279.

Jackson, L.E., S.M. Wheeler, A.D. Hollander et al. 2011. Case study on potential agricultural responses to climate change in a California landscape. *Climatic Change* 109:407–427.

Lafta, A.M. and J.H. Lorenzen. 1995. Effect of high temperature on plant growth and carbohydrate metabolism in potato. *Plant Physiol* 109:637–643.

Levy, D. 1986. Genotypic variation in the response of potatoes (*Solanum tuberosum* L.) to high ambient temperatures and water deficit. *Field Crop Res* 15:85–96.

Lovelli, S., M. Perniola, E. Scalcione, A. Troccoli, and L.H. Ziska. 2012. Future climate change in the Mediterranean area: Implications for water use and weed management. *Ital J Agron* 7:e7.

Marinus, J. and K.B.A. Bodlaender. 1975. Response of some potato varieties to temperature. *Potato Res* 18:189–204.

Menzel, C.M. 1985. Tuberization in potato at high temperatures: Interaction between temperature and irradiance. *Ann Bot* 55:35–39.

Miglietta, F., M. Bindi, F.P. Vaccari, A.H.M.C. Schapendonk, J. Wolf, and R.E. Butterfield. 2000. Crop ecosystem responses to climatic change: Root and tuberous crops. In *Climate change and global crop productivity*, eds. K.R. Reddy and H.F. Hodges, 189–212. New York: CAB International.

Mueller, L.A., T.H. Solow, N. Taylor et al. 2005. The SOL genomics network: A comparative resource for *Solanaceae* biology and beyond. *Plant Physiol* 138:1310–1317.

Peet, M.M., S. Sato, and R.G. Gardner. 1998. Comparing heat stress effects on male-fertile and male-sterile tomatoes. *Plant Cell Environ* 21:225–231.

Prange, R.K., K.B. McRae, D.J. Midmore, and R. Deng. 1990. Reduction in potato growth at high temperature: Role of photosynthesis and dark respiration. *Am Potato J* 67:357–369.

Reynolds, M.P., E.E. Ewing, and T.G. Owens. 1990. Photosynthesis at high temperature in tuber-bearing *Solanum* species. *Plant Physiol* 93:791–797.

Rosenzweig, C., J. Elliott, D. Deryng et al. 2014. Assessing agricultural risks of climate change in the 21st century in a global gridded crop model intercomparison. *Proc Natl Acad Sci USA* 111:3268–3273.

Sato, S., M. Kamiyama, T. Iwata et al. 2006. Moderate increase of mean daily temperature adversely affects fruit set of *Lycopersicon esculentum* by disrupting specific physiological processes in male reproductive development. *Ann Bot* 97:731–738.

Schafleitner, R., P. Kadirvel, and P. Hanson. 2014. Heat stress tolerance in tomato. *Poster presented at Plant and Animal Genome XXII Conference*, San Diego, CA, January 10–15.

Schafleitner, R., J. Ramirez, A. Jarvis, D. Evers, R. Gutierrez, and M. Scurrah. 2011. Adaptation of the potato crop to changing climates. In *Crop adaptation to climate change*, eds. S.S. Yadav, R.J. Redden, and J.L. Hatfield, 287–297. Chichester: Wiley-Blackwell.

Scharf, K.D., T. Berberich, I. Ebersberger, and L. Nover. 2012. The plant heat stress transcription factor (Hsf) family: Structure, function and evolution. *Biochim Biophys Acta* 1819:104–119.

Struik, P.C. 2007. Above-ground and below-ground plant development. In *Potato biology and biotechnology*, ed. D. Vreugdenhil, 219. Oxford: Elsevier.

USDA. 2013. National Agricultural Statistics Service, Crop Values 2012 Summary. http://usda.mannlib.cornell.edu/MannUsda/viewDocumentInfo.do?documentID=1050 (accessed October 2, 2014).

Valerio, M., S. Lovelli, M. Perniola, T. Di Tommaso, and L. Ziska. 2013. The role of water availability on weed-crop interactions in processing tomato for southern Italy. *Acta Agr Scand B-S P* 63:62–68.

Ventrella, D., L. Giglio, M. Charfeddine et al. 2012. Climate change impact on crop rotations of winter durum wheat and tomato in southern Italy: Yield analysis and soil fertility. *Ital J Agron* 7:e15.

Wang, J., X. Lv, J.L. Wang, and H.R. Lin. 2014. Spatiotemporal variations of reference crop evapotranspiration in northern Xinjiang, China. *Sci World J.* doi:10.1155/2014/931515.

8 Barley Genetic Resources for Climate-Change Adaptation

Searching for Heat-Tolerant Traits through Rapid Evaluation of Subsets

A. Jilal, H. Ouabbou, and M. Maatougui

CONTENTS

Barley (*Hordeum vulgare* L.) has been cultivated over millennia, making it one of the oldest domesticated crops (Salamini et al. 2002). Ranking fourth in global cereal-crop production, it is used for animal feed, brewing malts, and human consumption (von Bothmer et al. 2003). Given barley's long history as a crop involving migration and selection leading to adaptation to different environments and agroecology gradients, leading to a plethora of uses, it has the potential of being an excellent model for further elucidation of agricultural responses to changing and evolving climates (Dawson et al. 2015). Barley landraces are genetically heterogeneous populations comprising inbreeding lines and hybrid segregates generated by a low level of random outcrossing in each generation (Nevo 1992). The genetic structure of landraces may be considered as an evolutionary approach to survival and performance under arid and semiarid conditions (Schulze 1988). They are composed of several genotypes reported for both cultivated and wild barley (Brown 1978, 1979, Asfaw 1989). Natural selection accompanied by human selection during centuries of cultivation resulted in landraces that are genetically variable for qualitative and quantitative characters, have good adaptation to specific environmental conditions, and give dependable yields (Harlan 1992). Barley landraces have developed abundant patterns of variation and represent a largely untapped reservoir of useful genes for adaptation to biotic and abiotic stresses (Brush 1995, Tarekegn and Weibull 2011) to contribute to the improvement of modern cultivars (Hadjichristodoulou 1995).

The escalating size of ex-situ collections of plant genetic resources (e.g., 31,000 barley accessions held in ICARDA genebanks) and the limited funds available to assess data were earlier identified as limitations to the use of germplasm collections (Frankel and Brown 1984). Exploiting accessions for useful gene discovery, the focused identification of germplasm strategy (FIGS) has been developed by Mackay and Street (2004) to improve the use of genebank collections by increasing the chances of finding useful traits in relatively small subsets of germplasm reported for stem rust resistance in a FIGS wheat subset (Bari et al. 2012). Reducing time, space, and money consumption and assessing a large genetic resource are considered real challenges for genetic resource managers and breeders.

To exploit genetic resources collections for useful genes, partitioning techniques were proposed, among which is the core-collection strategy, which aims to capture diversity and be amenable to the FIGS trait-based approach. The development of the core-collection concept and theory was by Frankel and Brown (1984, Brown 1989a,b). A core collection consists of a representative sample of a base collection and accounts for the genetic variability of a crop and its related species, with minimum redundancy (Frankel and Brown 1984). The common practice to develop a core is based on setting up a hierarchical structure that will capture the alleles present in the base collection, within a conservation perspective. The core should capture most alleles that contribute to adaptation to specific environments (Brown 1989a, Allard 1992) and the accessions are selected based on stratification. The two main factors to be considered in the stratified random sampling procedure are geographic distribution and genotypic composition (Brown 1989a, Crossa et al. 1994, van Hintum 2000). This chapter addresses the issue of space allocated for collection evaluation and the reduction of collection size based on partitioning using both the core concept and trait-based concepts.

METHODOLOGY

The approach was a combination of a partitioning strategy and a field-evaluation strategy to reduce, on the one hand, the number of accessions to be screened and, on the other hand, the area where evaluation was conducted (Figure 8.1).

Partitioning Using Prior Information

Partitioning was conducted using environmental data (*a priori* information) to develop one subset that is likely to contain climate-change–related traits and another subset representative of the different environments where the barley accessions were originally sampled across Morocco. Each subset contains 100 accessions, of which 30 accessions were selected at random (Figure 8.2). Both subsets were grown in the same field for comparison based on *a posteriori* evaluation. The environmental

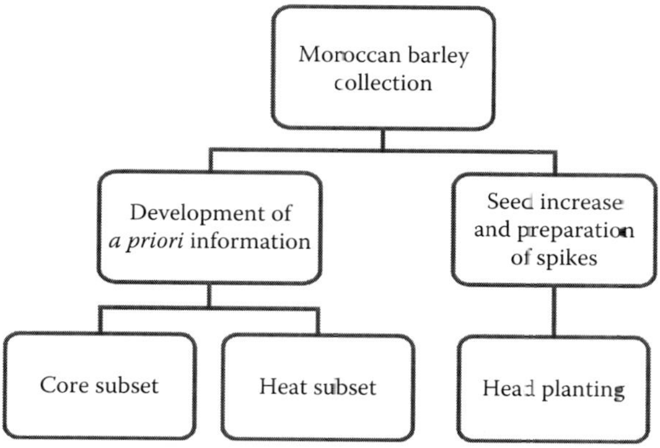

FIGURE 8.1 Partitioning and evaluation scheme.

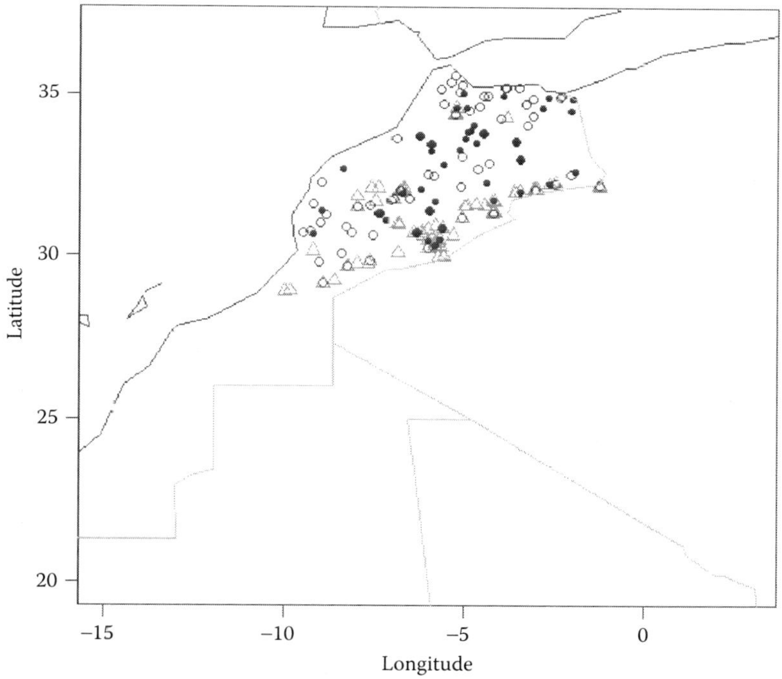

FIGURE 8.2 Barley landraces site distribution map for the heat subset in triangles and the representative subset in circles; the random set in dots is part of each of the two subsets as controls.

data used in the partitioning consisted of 19 climate variables extracted from world climate data (Table 8.1).

Climate data for each of the variables in Table 8.1 were extracted from bioclim surface data, which is used in the development of Hutchinson's environmental niche model as an *n-dimensional hyperspace* of climatic variables where a species can

TABLE 8.1
Climate Data Used in the Development of Subsets

Variable	Description
BIOCLIM1	Annual mean temperature
BIOCLIM2	Mean diurnal range (Mean of monthly [max temp–min temp])
BIOCLIM3	Isothermality (BIO2/BIO7 × 100)
BIOCLIM4	Temperature seasonality (standard deviation × 100)
BIOCLIM5	Maximum temperature of warmest month
BIOCLIM6	Minimum temperature of coldest month
BIOCLIM7	Temperature annual range (BIO5–BIO6)
BIOCLIM8	Mean temperature of wettest quarter
BIOCLIM9	Mean temperature of driest quarter
BIOCLIM10	Mean temperature of warmest quarter
BIOCLIM11	Mean temperature of coldest quarter
BIOCLIM12	Annual precipitation
BIOCLIM13	Precipitation of wettest month
BIOCLIM14	Precipitation of driest month
BIOCLIM15	Precipitation seasonality (coefficient of variation)
BIOCLIM16	Precipitation of wettest quarter
BIOCLIM17	Precipitation of driest quarter
BIOCLIM18	Precipitation of warmest quarter
BIOCLIM19	Precipitation of coldest quarter

Source: World climate data at http://www.worldclim.org/.

occur (Booth et al. 2014). This encompasses other important elements supporting the contention that variations in climate exert a strong influence on the distribution of a species (Booth et al. 2014). These bio-climatic variables (bioclim data) were derived from monthly surface data as synthetic variables to represent biologically meaningful variables coded as BIOCLIM1 through BIOCLIM19. They represent trends, seasonality, and extreme or limiting environmental factors that are detrimental to plant development (phenology) (Hijmans et al. 2005). Monthly data (*wtmin*, *wtmax*, and *wprec*) were also extracted and used in the analysis of data, especially to superimpose these data with the crop-growing period as per the crop-growing day requirements in terms of temperature. Under heat stress, these requirements will be expected to increase inversely proportional to the range for the crop stages.

This part of the partitioning process is to identify traits that breeders have long sought in order to combine an optimized grain-filling period with maturity rather than with earliness alone. As depicted in Figure 8.2, partitioning has been carried out for both the core subset and the trait-based subsets, in particular, to assess genetic resources of barley for heat traits based on agro-morphological data and climate data.

A POSTERIORI EVALUATION

To compare and validate the partitioning based on climate data, the accessions were compared based on their evaluation attributes. Both subsets were grown among 697 entries at Merchouch experimental station (33°.6049N; 06°.71600W; 410 masl) in one 5-m row for each accession. Observations were taken on several agro-morphological and physiological traits. After harvesting, all the accessions were analyzed with near-infrared spectroscopy (Infraneo machine) and the absorbance data (in the range from 850 to 1048 nm) and protein content have been determined.

Further evaluation was carried out based on a honeycomb design of hill plots with the aim to also reduce the area needed for evaluation (Figure 8.3). Subsequent evaluation was also carried out focusing mostly on the two subsets and heat-related traits, among which were canopy temperature and phenology traits, specifically grain-filling period.

RESULTS

The subsets were classified based on their environment and their evaluation scores. Based on the monthly average maximum temperature (*wtmax*) of March, the successful heat-subset accessions are all from areas where temperatures are high, while those of the core or representative samples are more from areas with mild to low temperatures. When the classification is based on April *wtmax*, the distinction in origin between the heat subset and the random and core subsets is even more apparent (Figure 8.4).

Similarly, when the partitioning is based on evaluation data (*a posteriori*), there is also a difference among the subsets, although it is not as clear as when environmental data is used. For example, the histograms of Figure 8.5 for evaluation of accessions with absorption spectroscopy and for protein content show that the heat subset

FIGURE 8.3 Barley head planting using the honeycomb planting design.

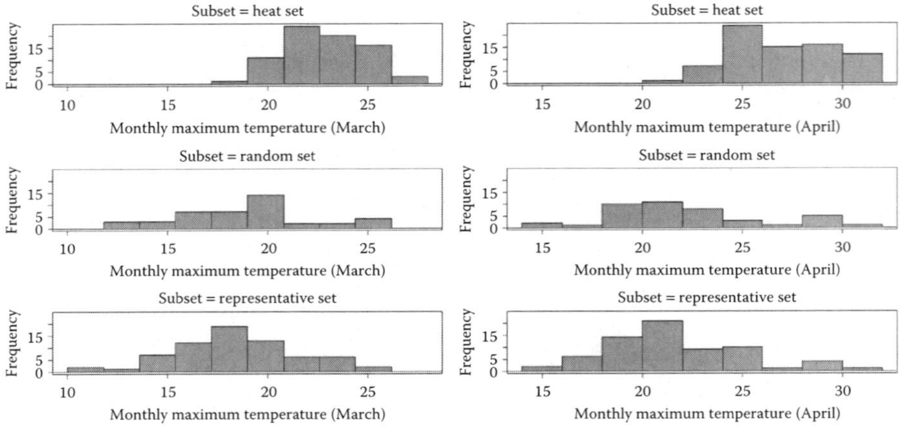

FIGURE 8.4 Distribution of accessions per climate data (monthly average maximum temperatures) for March (*left*) and April (*right*).

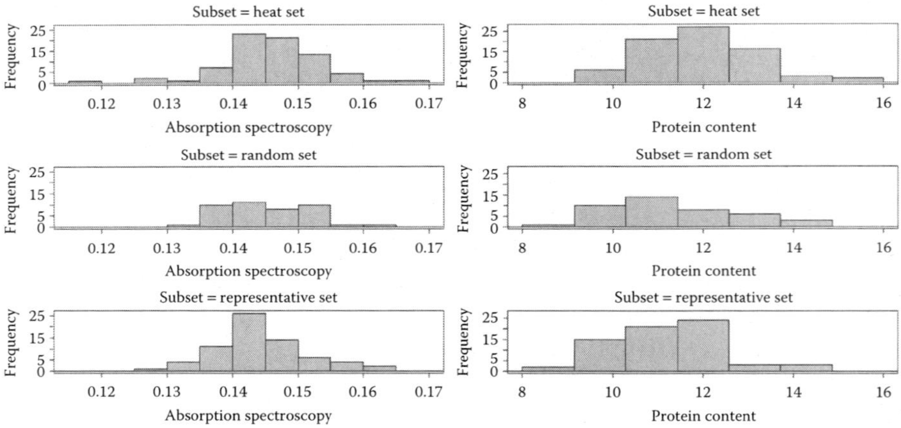

FIGURE 8.5 Distribution of accessions per two examples of evaluation data: absorption spectroscopy (*left*) and protein content (*right*).

may have traits that are different when compared to the core or random subsets, with the heat subset more likely to yield heat-related traits.

The subsequent evaluation involving heat-related traits such as days to maturity (a measure of phenology) and canopy temperature, especially in the grain-filling period, has shown that the two subsets are different (Figure 8.6). The heat subsets are most likely to yield heat tolerance when compared to the representative and random subsets.

These results indicate that there is potential not only to identify useful traits but also to combine traits, such as the desired grain-filling period with the optimum time of reaching maturity. The identification of a genotype with a grain-filling period that

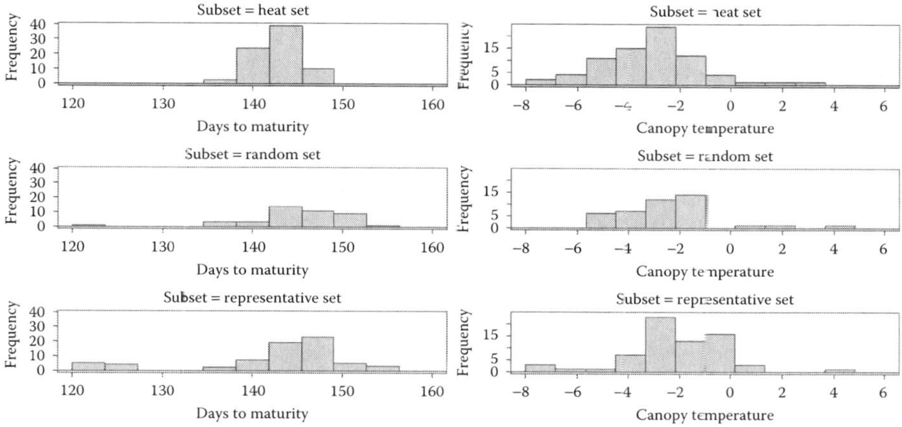

FIGURE 8.6 Distribution of accessions based on two climate-related traits: days to maturity (*left*) and canopy temperature (*right*); the range of days to maturity for most accessions in the heat subset fall within a tight range when compared to the random and representative sets (*left*).

allows the plant to achieve optimum grain filling while adjusting its time to maturity when conditions are not favorable could be more interesting than a genotype with simply increased earliness.

DISCUSSION

The rejuvenation of barley accessions in ex-situ conservation is time, space, and money consuming, in addition to the constraint of limited screening capacity for traits of interest of the massive genebank collection. The barley spike represents a good source of information on the accession since it gives feedback on the grain and spike characters (spike density, spike width, row type, spike length, spike shape, number of spikelet/spike, grain weight/spike, grain plumpness, etc.). It has been used traditionally for seed storing as a bouquet for a long time. Head planting is used for increasing pre-basic seed in some countries such as Algeria and for testing heat tolerance during off-season for cereals. Combining this important source of information on the accession with an adequate experimental design such as the honeycomb design, which compares the middle plant to the encircled ones and perform the evaluation in the absence of interplant (Fasoulas and Fasoula 1995), could be an effective way to screen, increase, and disseminate genetic accessions throughout the world (Figure 8.1).

Targeted partitioning is likely to yield the desired traits including climate-change-related traits, such as heat tolerance. The environmental data can be used as surrogate information when evaluating collections. Crops such as barley, which has been grown over different environmental gradients, may harbor genetic variation with the potential to adapt to changing climate conditions. Recent research by Dawson et al. (2015) showed that barley, in particular, can be an excellent model to elucidate the adaptation of crops to climate change. Barley genetic resources are thus of great relevance to the future as stated at the 2014 Lillehammer International Conference on *Genetic Resources for Food and Agriculture in a Changing Climate*, emphasizing the value of genetic resources for farming in the future (Præbel and Groeneveld 2014).

Genetic resources overall have helped in providing traits for crop improvement. A combination of partitioning strategies along with careful seed increase, taking into account optimal times for evaluation and space availability can help to address the limited screening capacity for traits in the face of the large size of genebank collections.

This new procedure has many advantages, such as the following:

- Large numbers of accession can be evaluated (10,000 accession/ha).
- Intraplant competition is possible.
- Artificial inoculation is facilitated for disease and pest screening.
- Tolerance for abiotic stresses (drought, salinity, and cold) can be easily screened.
- Biotic and abiotic stresses are easily scored.
- Samples and sample integrity can be tracked.
- Regeneration of accessions takes place in a reduced area.
- Fragile seeds, such as naked barley, can be easily handled, avoiding embryo damage during threshing.
- Duplication of accessions is facilitated, since they are stored as spikes.
- Reduced threshing.
- Easy planting and harvesting.
- Seed loss while threshing is avoided.
- Overall, time, money, and space are conserved.

Lack of evaluation of genetic resources has been reported by many authors as the main factor preventing the utilization of genetic resources. To evaluate all the genetic resources of barley, for example, for climate-change-related traits, would require large field areas, assessing large number of accessions, a considerable outlay of funds, and a great amount of time. The research reported in this chapter can help to address both the timing and the space requirement for evaluation by proposing new ways of planting and assessing the large collections that are presently conserved at genebanks. The scheme in Figure 8.7 outlines our proposal; however, the issue of accession integrity, given that a genebank landrace accession is usually a mixture, has yet to be amply addressed. In order to screen high numbers of accessions in a small space (40,000 accessions/ha) within two years, from the few seed per accession of barley in the genebanks, we can conduct a hill-plot trial, in order to evaluate the agro-morphological traits (by development of core and FIGS subsets) and to increase seed and spikes (developing information on the stored germplasm). The honeycomb design using head planting in multilocation tests will allow an efficient screening of germplasm for traits of interests (even physiological traits), always in a small space. After two years of screening, useful traits will be identified and subsets will be validated in addition to the regeneration of the accessions for future study and storage (Figure 8.7).

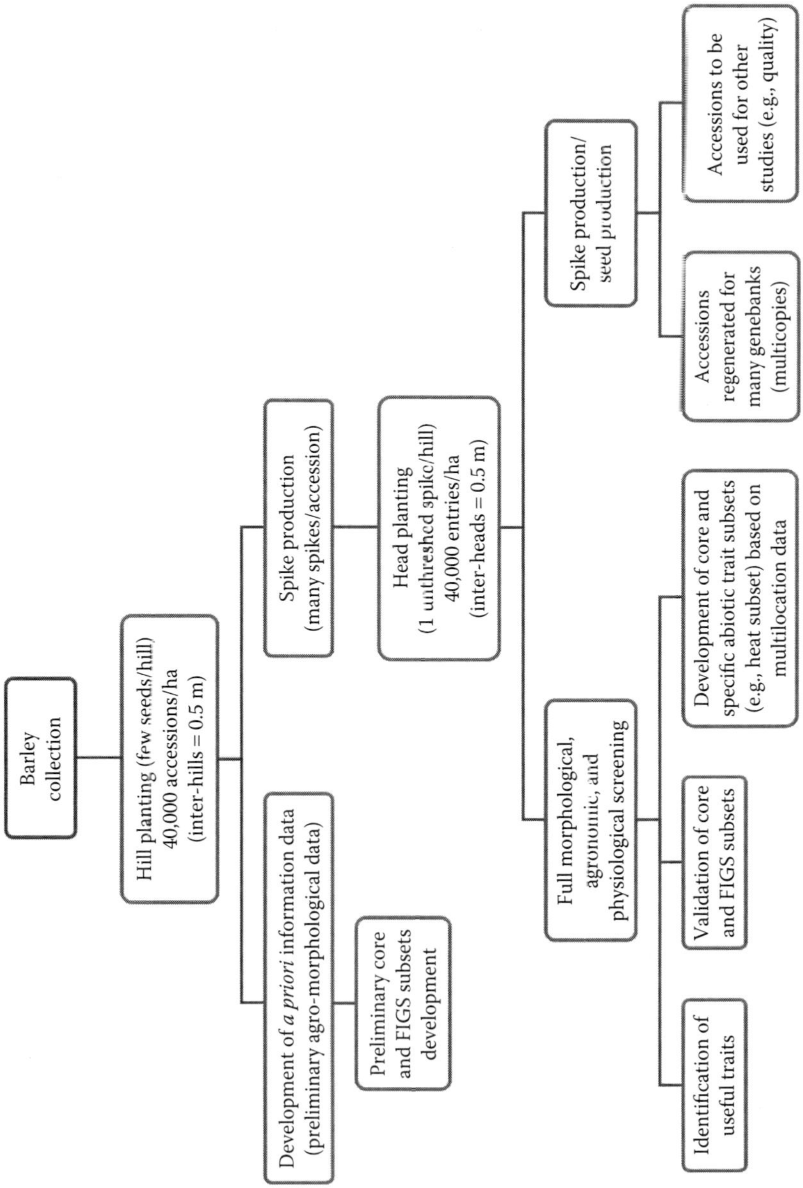

FIGURE 8.7 Proposal for partitioning collections of barley genetic resources while managing accession evaluation.

FUTURE WORK

The search for climate-related traits continues with further evaluation, along with analysis of evaluation data to compare morphological and physiological data with results at three other stations. The data will also include absorbance data (linked to all molecules with C, H, O, and N atoms, e.g., NDVI, $\Delta^{13}C$, $\Delta^{18}O$, photosynthesis, and quality parameters).

Other comparison will be also conducted using other partitioning tools. In terms of core collections, we will be using PowerMarker to select accessions from the Moroccan barley collection based on the accumulated data from the three stations (three contrasting environments). For trait-based partitions, different mathematical modeling techniques will be carried out using both linear and nonlinear approaches. These analyses can be performed with Unscrambler (as well as with MATLAB® or R program). The analysis of the data may also involve a moving grid adjustment, which is a spatial method to adjust for environmental variation in field trials common in unreplicated plant breeding field trials (Technow 2011).

REFERENCES

Allard, R.W. 1992. Reproductive systems and dynamic management of genetic resources. In *Reproductive biology and plant breeding*, eds. Y. Dattee, C. Dumas, and A. Gallais, 325–334. Berlin, Germany: Springer-Verlag.

Asfaw, Z. 1989. Variation in hordein polypeptide pattern within Ethiopian barley *Hordeum vulgare* L. (Poaceae). *Hereditas* 110:185–191.

Bari, A., K. Street, M. Mackay, D.T.F. Endresen, E. De Pauw, and A. Amri. 2012. Focused identification of germplasm strategy (FIGS) detects wheat stem rust resistance linked to environmental variables. *Genet Resour Crop Ev* 59:1465–1481.

Booth, T.H., H.A. Nix, J.R. Busby, and M.F. Hutchinson. 2014. Bioclim: The first species distribution modelling package, its early applications and relevance to most current MaxEnt studies. *Diversity Distrib* 20:1–9. doi:10.1111/ddi.12144.

Brown, A.H.D. 1978. Isozymes, plant population genetic structure, and genetic conservation. *Theor Appl Genet* 52:145–157.

Brown, A.H.D. 1979. Enzyme polymorphism in plant populations. *Theor Popul Biol* 15:1–42.

Brown, A.H.D. 1989a. The case for core collections. In *The use of plant genetic resources*, eds. A.H.D. Brown, O.H. Frankel, D.R. Marshall, and J.T. Williams, 136–156. New York: Cambridge University Press.

Brown, A.H.D. 1989b. Core collections: A practical approach to genetic resources management. *Genome* 31:818–824.

Brush, S.B. 1995. In situ conservation of landraces in centers of crop diversity. *Crop Sci* 35:346–354.

Crossa, J., S. Taba, S.A. Eberhart et al. 1994. Practical considerations for maintaining germplasm in maize. *Theor Appl Genet* 89:89–95.

Dawson, I.K., J. Russell, W. Powell, B. Steffenson, W.T. Thomas, and R. Waugh. 2015. Barley: A translational model for adaptation to climate change. *New Phytol* 206(3):913–931.

Fasoulas, A.C. and V.A. Fasoula. 1995. Honeycomb selection designs. *Plant Breed Rev* 13:87–139.

Frankel, O.H. and A.H.D. Brown. 1984. Current plant genetic resources: A critical appraisal. In *Genetics: New frontiers, Vol. 4. Applied genetics Proceedings of the International Congress of Genetics, 15th,* December 12–21 1983, eds. V.L. Chopra, B.C. Joshi, R.P. Sharma, and H.C. Bansal, 1–11. New Delhi, India: Oxford and IBH.

Hadjichristodoulou, A. 1995. Evaluation of barley landraces and selections from natural outcrosses of *H. vulgare* ssp. *spontaneum* with ssp. *vulgare* for breeding in semi-arid areas. *Genet Resour Crop Ev* 42:83–89.

Harlan, J.R. 1992. *Crops and man*. Madison, WI: American Society of Agronomy.

Hijmans, R.J., S.E. Cameron, J.L. Parra, P.G. Jones, and A. Jarvis. 2005. Very high resolution interpolated climate surfaces for global land areas. *Int J Climatol* 25:1965–1978.

Mackay, M. and K. Street. 2004. Focused identification of germplasm strategy—FIGS. In *Proceedings of the 54th Australian Cereal Chemistry Conference and the 11th Wheat Breeders' Assembly*, September 21–24, Canberra, ACT, Australia, eds. C.K. Black, J.F. Panozzo, and G.J. Rebetzke, 138–141. Melbourne, Victoria: Cereal Chemistry Division, Royal Australian Chemical Institute (RACI).

Nevo, E. 1992. Origin, evolution, population genetics and resources of wild barley, *Hordeum spontaneum*, in the fertile crescent. In *Barley: Genetics, biochemistry, molecular biology and biotechnology*, ed. P.R. Shewry, 19–43. Wallingford, CT: CABI.

Præbel, A. and L.F. Groeneveld (eds.). 2014. *Book of abstracts: Genetic resources for food and agriculture in a changing climate*, January 27–29, Lillehammer, Norway. Ås, Norway: The Nordic Genetic Resource Center.

Salamini, F., H. Ozkan, A. Brandolini, R. Schäfer-Pregl, and W. Martin. 2002. Genetics and geography of wild cereal domestication in the near east. *Nat Rev Genet* 3:429–441.

Schulze, E.D. 1988. Adaptation mechanisms of noncultivated arid-zone plants: Useful lesson for agriculture? In *Drought research priorities for the dryland tropics*, eds. E.R. Bidinger and C. Johansen, 159–177. Patancheru, India: ICRISAT.

Tarekegn, K. and J. Weibull. 2011. *Adaptation of Ethiopian barley landraces to drought stress conditions*. Saarbrücken, Germany: Lambert Academic Publishing.

Technow, F. 2011. R Package mvngGrAd: Moving grid adjustment in plant breeding field trials. R package version 0.1. University of Hohenheim, Institute of Plant Breeding, Seed Science and Population Genetics, Stuttgart, Germany.

van Hintum, Th.J.L., A.H.D. Brown, C. Spillane, and T. Hodgkin. 2000. *Core collections of plant genetic resources*. IPGRI Technical Bulletin No. 3. Rome, Italy: IPGRI.

von Bothmer, R., Th.J.L. van Hintum, H. Knüpffer, and K. Sato (eds.). 2003. *Diversity in barley* (Hordeum vulgare). (*Developments in plant genetics and breeding, Vol. 7.*) Amsterdam, The Netherlands: Elsevier Science.

9 Fruit Genetic Resources Facing Increasing Climate Uncertainty

O. Saddoud Debbabi, S. Mnasri, S. Ben Abedelaali, and M. Mars

CONTENTS

Climate changes often adversely affect farming systems and biodiversity (Halewood et al. 2013). Morton (2007) has shown that these impacts are hard to predict because of their complexity and specificity, and they are mainly affecting small farmers in developing countries. These factors may be diverse including emergence of new pests and diseases, change of flowering and pollination period, and timing of some agriculture activities. Plant genetic resources are affected by the climate change and can respond to adverse effects by adapting crops and varieties. Farm conservation is a very efficient solution to adapt to climate change because it can maintain the evolutionary process of crops and their agricultural systems (Bellon 2009). Furthermore, field genebank, seed genebank, and community genebank can make plant materials available for adaptation to diverse environment conditions. Custodian farmers can play an important role to conserve plant genetic resources and to offer adapted germplasm to climate change. Adapted germplasm can be evaluated and improved in collaboration with farmers and research institutes. Performed accessions can be distributed to other regions, particularly vulnerable ones (Ramirez-Villegas et al. 2013, Vermeulen et al. 2013). The ability of farmers, plant breeders, and natural resource managers to identify and access such germplasm is becoming increasingly important as climates continue to change. Tunisia is one of the countries most vulnerable to climate change in the region, given its limited natural resources and the dominance of an arid and variable climate. Scenarios of climate change agree on an increase in temperature of about 2°C, more pronounced in the south of the Mediterranean (GIZ 2013). The country is particularly vulnerable to accelerated desertification and degradation of its coastline facing the rising sea levels. This natural vulnerability and increased human pressure on ecosystems and natural resources are more and more increased by the need of socioeconomic development. Tunisia has already developed policies for sustainable management of natural resources in line with its economic development goals. But limited resources and the effects of climate change are still threatening the equilibrium between socioeconomic development and resources (GIZ 2013).

Fruit tree genetic resources in Tunisia have been sampled for a diverse array of species. The study of climate-change effects on agriculture in Tunisia highlighted the vulnerability of the majority of the agroecosystems that support this fruit tree diversity. Extending the area of annual crops to the limit of suitable areas increases the vulnerability of agriculture to climate change (reduced rainfall). At the same time, the extension of olive and almond trees in the areas suitable for this type of planting threatens whole swathes of fruit tree growing in Tunisia (Sghaie and Ouessar 2011). One of the most adapted fruit species is fig (*Ficus carica* L.). It is a typical fruit tree of the Mediterranean area and is recognized as one of the oldest domesticated fruit trees. Fig tree is very adapted plant species to drought conditions. Such kind of thermophilic trees may be useful to adapt to the consequences of climate change and global warming, thus allowing the cultivation of fig trees on hot and dry areas, where other species may not survive (Sugiura et al. 2007). The work reported in this chapter focuses on describing vulnerability of fruit tree genetic resources and agrobiodiversity to climate changes in Tunisia.

TUNISIA AND CLIMATE CHANGE

Many countries have established specific policies and breeding strategies for climate-change mitigation and adaptation. Policy design and implementation occurs through a range of legislation, strategies, plans, programs, and projects. It occurs at local, regional, national, and international levels, and is enacted and financed by a range of actors. The links between climate change and biodiversity mean that there is great potential for policies to be implemented that achieve multiple objectives. There is also the potential risk of activities striving for one objective to have unintended impacts on other objectives (Mant et al. 2014).

Aware of these issues, Tunisia was among the first countries to ratify the UN Convention Framework on Climate Change (UNFCCC) in 1993 and the Kyoto Protocol in 2002. Many initiatives have been undertaken by the Tunisian authorities with the support of international cooperation. Moreover, these initiatives address both mitigation of and adaptation to climate change by several sectors (tourism, agriculture, health, coastal, energy, biodiversity, etc.) and crosscutting issues (capacity building and positioning of Tunisia in negotiations that are international, institutional, legal, or regulatory). However, these approaches are varied and diverse and require alignment and, consequently, an establishment of a National Strategy on Climate Change (SNCC). To ensure its operability, this approach should be developed by an inter-sector approach while ensuring its integration into national plans for economic and social development.

Furthermore, although not contributing to global emissions of greenhouse gases, Tunisia is expected to contribute to international efforts to mitigate the effects of climate change in a collective perspective of limiting global warming to 2°C. The numerous 2030 uncertainties about the availability and future prices of fossil fuels will also lead Tunisia to adopt a proactive policy of energy conservation (and mitigation) without compromising its development goals.

EFFECTS OF CLIMATE CHANGES

Oases

Studies have shown that southern Tunisia will be most affected by climate change: increasing temperatures and decreasing precipitation. In oases areas, an average warming of 1.9°C by 2050 and a decrease in rainfall of 9% by 2030 and 17% by 2050 are anticipated. The oases will be severely affected by climate change as follows:

- Increased water needs of crops because of the worsening situation of water resources with a continuous decline in static level drilling, increasing the salinity of the water, and increasing the cost of pumping.
- The gradual rise in sea level favoring the intrusion of seawater into groundwater in coastal oases, making it more difficult for the natural flow of drainage water.
- The risk of higher temperatures for tree species requiring a chilling period, resulting in a drop in production.
- Dryness of dates resulting from successive days of high heat.
- A higher incidence of attacks of the palm mite.
- Negative effects of high temperature on blooming dynamics, flower structure, fruit set, and fruit quality.
- Changes in macroclimate may affect variety adaptation, especially for the date palm that has particular needs for fruit maturity development.

Vulnerability of Islands

Elevated islands (Galite Zembra and Zembretta) will be unaffected by accelerated elevation of sea level. Most likely, the low-lying islands, especially those that are inhabited, are particularly vulnerable (Kerkennah, Jerba, and Kuriates Kneiss). For the Kerkennah islands, the danger of erosion is likely to become increasingly important and will accelerate. Salinization continues to gain ground. Some 70 km of coastal swamps and sebkhas (salt flats) will be lost, about one-third of its total current area. Fragmentation of the archipelago into more islands is expected (about 30% of the total area is exposed to marine erosion, Ministry of Environment of Tunisia 2005).

For Djerba Island, the consequences of accelerated rise in sea level will be extreme on beaches and tourist facilities. A long coastline of 25 km that includes hotels may be left with no natural beaches. Coastal sebkhas evolve into environments more often invaded by the waters of the lagoons. More than 3400 ha of wetlands on the island are threatened by erosion in the marine areas of Rmal and Ras-el-Bin ouedien (Ministry of Environment of Tunisia 2005).

FRUIT TREE GENETIC RESOURCES IN TUNISIA AND THEIR VULNERABILITY

More than 36 cultivated species adapted to different climates and environmental conditions have been recorded in Tunisia. More than 80% of Tunisia's area consists of arid and semiarid regions (most vulnerable to climate changes). While old varieties have been conserved, biotic and abiotic stresses are threatening these fruit tree resources. Moreover, the extension of commercial olive and almond plantations over inappropriate areas is causing extinction of some valuable fruit tree material. For example, 60% of almond trees perished due to drought in the late 1990s and early 2000s. Similarly, there is risk for olive-planted areas that are currently at the limit of traditional olive-growing areas. Climate change may make such plantations unsustainable.

FIG TREE (*F. CARICA* L.) AS A SPECIES ADAPTED TO CLIMATE CHANGE

The fig tree ($2n = 26$) belongs to the Moraceae family, genus *Ficus*. It has been grown since 11,400–11,200 years ago as a gynodioecious and insect-pollinated species (Kislev et al. 2006). It propagates vegetatively. Two fig tree types are cultivated: one called common figs (unisexual female trees) and the other called caprifigs (bisexual with functional male flowers: pollinator). Both occur in similar frequencies in wild populations (Valdeyron and Lloyd 1979). This fruit crop is widespread in Mediterranean basin countries. It is well adapted in Tunisia to diverse conditions, soils, and climate change. In Tunisia, fig tree germplasm consists of numerous landraces mainly selected by farmers for their fruit qualities and maintained in orchards. Hodgson (1931) reported the prevalence of the *Smyrna* (crossbreeding) ecotypes in southern Tunisia, while in northern Tunisia, the *common* (parthenocarpic) and *Smyrna* ones are equally distributed. Thus, a wide phenotypic diversity characterizes the large number of ecotypes distinguishable by taste, color, and flavor of fruits (Rhouma 1996). In addition, fig trees represent the principal component of several agroecosystems in the southern areas such as the Jessours region (Matmata, Beni Khédache, and Douiret) and constitute the second largest fruit crop in the Tunisian oases (Mars 1995, 2003). Saddoud et al. (2007) have surveyed the genetic polymorphism in Tunisian fig trees and produced an ecotype identification key based on simple sequence repeat (SSR) data. Seventy-two Tunisian fig ecotypes conserved in situ and ex situ were analyzed using six microsatellite loci. A total of 58 alleles and 124 genotypes were revealed and yielded evidence of a high degree of genetic diversity, maintained at the intra-group level. Cluster analysis based on genetic distances proved that a typical continuous genetic diversity characterizes the local germplasm. In addition, microsatellite multilocus genotyping unambiguously distinguished 70 well-defined ecotypes (resolving power of 97.22%). These results demonstrated the utility of SSR markers to evaluate the conformity of plant material and can be a tool to rationally manage the conservation of this crop. Both SSR markers and morphological characters were used to characterize Tunisian fig cultivars (Saddoud et al. 2011). Morphological traits suggested a high level of variation in the germplasm. Principal component analysis differentiated the studied cultivars. In the derived dendrogram, the cultivars clustered independently of their geographical origin and sex of trees. The Mantel test confirmed a disparity between morphological variation and genetic polymorphism of these cultivars (the correlation was negative and not significant: -0.031, $p = .7$). In line with this, the dendrogram for cultivars based on morphology differed from the one based on SSR genetic distance. Morphological characters are highly influenced by environmental conditions. Microsatellites (SSRs) are environment neutral; they clustered the cultivars based on divergence between genotypes. Morphological features and molecular markers were used to assess the diversity of Tunisian fig cultivars sampled from north to south. The two types of markers contribute differently to diversity evaluation. Fig germplasm is fairly straightforward to characterize morphologically, but microsatellite markers allow precise genotyping and give a picture of genetic diversity independent of geographic origin. Both morphological characters and SSR markers provide important information about the diversity of fig germplasm (Saddoud et al. 2011). Together they provide a powerful tool for future agricultural and conservation tasks.

CONCLUSION

Climate change effects have been observed in agroecosystems and for species and landraces. Several projections have shown that Tunisia is located among the regions vulnerable to the effects of climate change. Thus, it is imperative to plan more research related to climate-change effects on genetic resources and to enhance studies on plant adaptation, especially for fruit tree species. Generally, governments focus too much on top–down approaches to climate change adaptation, without considering the local needs of rural communities. It is important to develop participatory planning tools to allow communities to be involved in decision-making processes and share their valuable indigenous knowledge and solutions to climate change. Working together, local communities and scientists can identify feasible solutions to the negative impacts of climate change in Tunisia.

REFERENCES

Bellon, M.R. 2009. Do we need crop landraces for the future? Realizing the global option value of *in situ* conservation. In *Agrobiodiversity and Economic Development*, eds. A. Kontoleon, U. Pascual, and M. Smale, 51–61. London and New York: Routledge.

GIZ. 2013. Mise en place d'un système de suivi évaluation de l'adaptation au changement climatique: cas de l'agriculture. Deutsche Gesellschaft für Internationale Zusammenarbeit (GIZ) Gmb.

Halewood, M., P.N. Mathur, C. Fadda, and G. Otieno. 2013. *Using crop diversity to adapt to climate change: Highlighting the importance of the Plant Treaty's policy support.* Rome, Italy: Bioversity International. http://hdl.handle.net/10568/33831.

Hodgson, R.W. 1931. *La culture fruitière en Tunisie, son état actuel, ses possibilités et son amélioration: Rapport de mission d'études fruitières en Tunisie.* Tunis, Tunisia: Société Anonyme de l'Imprimerie Rapide de Tunis.

Jarvis, A., H. Upadhaya, C.L.L. Gowda, P.K. Aggarwal, S. Fujisaka, and B. Anderson. 2010. Climate change and its effect on conservation and use of plant genetic resources for food and agriculture and associated biodiversity for food security. Thematic Background Study. In *The Second Report on the State of the World's Plant Genetic Resources for Food and Agriculture.* Food and Agriculture Organization of the United Nations, Rome, Italy.

Kislev, M.E., A. Hartmann, and O. Bar-Yosef. 2006. Early domesticated fig in the Jordan Valley. *Science* 312:1372–1374.

Mant, R., E. Perry, M. Heath et al. 2014. Addressing climate change: Why biodiversity matters. Cambridge: UNEP World Conservation Monitoring Centre. http://www.unep-wcmc.org/system/dataset_file_fields/files/000/000/221/original/IKI_report_2_accessible_version_20140530.pdf?1401884844.

Mars, M. 1995. La culture du grenadier (*Prunica granatum* L.) et du figuier (*Ficus carica* L.) en Tunisie. *Cah Options Medit* 13:85–95.

Mars, M. 2003. Conservation of fig (*Ficus carica* L.) and pomegranate (*Prunica granatum* L.) varieties in Tunisia. In *Conserving biodiversity in arid regions*, eds. J. Lemons, R. Victor, and D. Schaffer, 433–442. Dordrecht, The Netherlands: Kluwer Academic Publishers.

Ministry of Environment of Tunisia. 2005. National report, CGE hands-on training workshop on vulnerability and adaptation for the Africa region Maputo, Mozambique, April 18–22, 2005. United Nations Framework Convention on Climate Change.

Morton, J.F. 2007. The impact of climate change on smallholder and subsistence agriculture. *Proc Natl Acad Sci USA* 104:19680–19685.

Ortiz, R. 2011. Agrobiodiversity management and climate change. In *Agrobiodiversity management for food security: a critical review*, eds. J.M. Lenné and D. Wood, 189–211. Wallingford, CT: CABI.

Ramirez-Villegas, J., A. Jarvis, S. Fujisaka, S. Fujisaka, J. Hanson, and C. Leibing. 2013. Crop and forage genetic resources: International interdependence in the face of climate change. In *Crop genetic resources as a global commons: Challenges in international law and governance*, eds. M. Halewood, I. López Noriega, and S. Louafi, 78–98. London: Routledge.

Rhouma, A. 1996. Les ressources phytogénétiques oasiennes: le figuier (*Ficus carica* L.). In *Proc. 3èmes Journées Nationales sur les Acquis de la Recherche Agronomique, Vétérinaire et Halieutique*, Nabeul, Tunisia.

Saddoud, O., G. Baraket, K. Chatti, M. Mars, M. Marrakchi, and M. Trifi. 2011. Using morphological characters and simple sequence repeat (SSR) markers to characterize Tunisian fig (*Ficus carica* L.) cultivars. *Acta Biol Cracov S Bot* 53(2):7–14. doi:10.2478/v10182-011-0019-y.

Saddoud, O., K. Chatti, A. Salhi-Hannachi et al. 2007. Genetic diversity of Tunisian figs (*Ficus carica* L.) as revealed by nuclear microsatellites. *Hereditas* 144:149–157.

Sghaie, M. and M. Ouessar. 2011. *L'oliveraie tunisienne face au changement climatique. Méthode d'analyse et étude de cas pour le gouvernorat de Médenine.* Deutsche Gesellschaft für Internationale Zusammenarbeit (GIZ) GmbH, Institut des Régions Arides, Tunisia, and German Federal Ministry for Economic Cooperation and Development (BMZ). http://www.environnement.gov.tn/PICC/wp-content/uploads/Loliveraie-tunisienne-face-au-changement-climatique.pdf.

Sugiura, T., H. Kuroda, and H. Sugiura. 2007. Influence of the current state of global warming on fruit tree growth in Japan. *Hort Res* 6:257–263.

Valdeyron, G. and D.G. Lloyd. 1979. Sex differences and flowering phenology in the common fig (*Ficus carica* L.). *Evolution* 33:673–685.

Vermeulen, S.J., A.J. Challinor, P.K. Thornton et al. 2013. Addressing uncertainty in adaptation planning for agriculture. *Proc Natl Acad Sci USA* 110(21):8357–8362.

Section III

Applied Mathematics (Unlocking the Potential of Mathematical Conceptual Frameworks)

10 Applied Mathematics in Genetic Resources

Toward a Synergistic Approach Combining Innovations with Theoretical Aspects

A. Bari, Y.P. Chaubey, M.J. Sillanpää, F.L. Stoddard, A.B. Damania, S.B. Alaoui, and M.C. Mackay

CONTENTS

> Instead of allowing the flexibility of our computer tools to continue to overwhelm us with a surplus of riches, we should make use of theory to help focus these powerful resources upon the task at hand.
>
> **Nilsson, N**
> *The Interplay Between Experimental and Theoretical Methods in Artificial Intelligence,* Cognition and Brain Theory, *January 1981*

Genetic resources consist of genes and genotypes with frequencies and patterns generated over space and time that may significantly enhance their potential for adaptive evolution to changing conditions (Darwin 1859, Harlan 1992). Capturing these patterns requires digging into large and complex datasets, including people's knowledge associated with the resources. These data consist mostly of nonreplicated records or observations with limited information on a number of variables. Analyzing such complex and large datasets with limited information requires the elaboration of new mathematical conceptual frameworks and new approaches for a cost-effective and timely utilization of these resources. The lack of availability of *ex ante* evaluation of genetic resources for indicative traits has also been highlighted as the most prevalent and long-standing impediment to their effective use in plant improvement (Koo and Wright 2000, FAO 2010). There is also a lack of methodologies or more elaborated approaches specific to mining genetic resources data, restricting their effective deployment to enhance farm productivity, sustainability, and livelihoods.

The synergistic approach involving innovative strategies for utilizing plant genetic resources (PGR), farmers' innovations, expert knowledge, applied mathematics, and omics is intended to address the complexity inherent in genetic resources with the aim to contribute to the needed increase in agricultural production in a time of rapidly changing conditions. Such synergistic approaches between innovations and theories, between practice and models, have helped enormously in accelerating progress made in a number of other disciplines such as physics, medicine, and information technology. This is even more important in agriculture, in order to offset the effects of climate change while achieving the dual goals of increased production coupled with sustainable intensification to meet the global increase in food demands (FAO 2011).

Genetic resources and genetic improvement have provided 50% of the increase in yields achieved over the past few decades in major global crops such as wheat, rice, and maize, with the other 50% coming from improved management and use of inputs (Byerlee et al. 1999, FAO 2011, Smith et al. 2014). They also helped in closing the yield gaps by generating cultivars adapted to local conditions and by making them more resilient to biotic (e.g., insects, diseases, and viruses) and abiotic stresses such as droughts and floods (FAO 2009).

The challenge of climate change emphasizes the value of these resources for farming in the future, as stated at the 2014 Lillehammer International Conference on *Genetic Resources for Food and Agriculture in a Changing Climate*. This conference asserted that genetic resources are more important for the future of farming than any other factor, because they contain the genes that will facilitate adaptation (Præbel and Groeneveld 2014).

To effectively contribute to the sustainable crop production and intensification process (FAO 2011, CGIAR 2015), farmers will also need a new and genetically diverse portfolio of improved cultivars with adaptive traits that allow the crops to provide higher yields under changing conditions of drought, heat, and increasingly virulent pests and diseases (FAO 2011). In addition, there is an urgency to accelerate the delivery of new cultivars by strengthening the connections among PGR, plant breeding, and seed delivery (FAO 2011). The timing issue addressed in early findings suggested that the speed by which novel trait variation is found is as important as the process of incorporating such novel variation into an improved genetic background. The need to shorten the time to deliver these improved cultivars, while also involving farming communities, is as crucial as the development of improved cultivars. Recent studies have shown that farmers are active in responding to changing climate conditions by applying a systems approach combining agronomic practices with the use of alternative species or cultivars (GFA 2014, Dawson et al. 2015).

The application of mathematics to capture genetic patterns, such as those present in genetic resources, was recognized in 1948 by Gustave Malécot (in English: Malécot 1969) under his probabilistic theory underlying genetic differentiation and spatial genetic structuring as a result of stochastic processes, following Sewall Wright's statistical theory (Heywood 1991, Ishida 2009). Recent studies on marine species have also revealed the presence of such patterns because of ecological and oceanographic factors. These patterns were previously dismissed as *chaotic* (Selkoe et al. 2010).

The synergistic approach aims thus to explore genetic resources and exploit the patterns of adaptation displayed by these resources, including patterns induced by climate change, while involving the perspective of farmers. Such approaches and patterns have helped in tracing the origin and diversity of crops and in locating new and agronomically important trait variation (Bari et al. 2012). The presence of such patterns and *a priori* information implies the possibility of prediction to locate and identify adaptive and rare traits, with the aim to achieve the dual goals of crop production and intensification.

This chapter introduces and presents current progress made by combining practical innovations, including farmers' innovations, with applied mathematics to accelerate the process in identifying traits related to climate change. The chapter presents a synergistic approach to explore patterns and *a priori* information, combining mathematics with omics to value genetic resources. It also discusses future prospects in addressing uncertainties in the shifts in phenological as well as physiological traits and changes in plant responses induced by climate change.

Evaluating genetic resources with mathematical models, while involving farmers, could be of tremendous relevance to decision making (Kotschi 2007). A tight and remarkable interdependency between advances made in medicine and those made in mathematics has been reported and demonstrated to be beneficial to both of these disciplines (Glamore et al. 2013). In terms of farmers' perspectives, Soleri and Cleveland (2001) included social science, as well as a new method using hypothetical scenarios based on the biological model and the farmers' own experiences to explore their perceptions of genetics, with particular regard to their knowledge of genetic variation and its relation to environmental variation and heritability. Their hypotheses in terms of heritability and genetic variation of two traits of maize provided insights into the nature of farmer knowledge that is of particular relevance to the theoretical basis for exploring genetic variation (Soleri and Cleveland 2001).

WHY A SYNERGISTIC APPROACH?

The objective of this chapter is to demonstrate the importance of a synergistic approach in utilizing PGR through combining innovative ideas of farmers with mathematical model-based procedures and omics. Using a synergistic approach between theory and know-how, between the rational and the experimental (Bachelard 1968, Sensevy et al. 2008), has helped to accelerate progress in chemistry, physics, ecology, medicine, and information technology, and has led to rapid reshaping of their structure (Cramer 2004, Cagnaccil et al. 2010). According to Cramer (2004), it is difficult to imagine making progress without having this combination or link between the practical aspects and mathematical theoretical frameworks. The synergistic approach has been reported as also helping to ensure more reliable guidance for future research.

This anticipated link between the experimental (data), on the one hand, and the theoretical or mathematical frameworks, on the other hand, was first established by Gustave Bachelard (1968) under his epistemology framework, where he considered scientific knowledge of itself as an object open to further logical reasoning and thinking to reach an ultimate and practical solution. He highlighted the dynamics in the interchange between empirical data and theoretical frameworks. In essence, empirical data are not pre-set, but are also the results of this interchange between alternating practical provisions and theoretical presuppositions, between realism and rationalism, leading to remarkably rapid advances in many disciplines (Sismondo 1999, Ebke 2012). The development of theoretical aspects can be in a reverse order from that of the empirical aspects or vice versa, for example, the support vector machine (SVM) models used in this book evolved from theory to practice, while neural network (NN) models originated were first deployed before the development of their theoretical frameworks (Kecman 2005).

This interchange between practice and theories may lead to new paradigms and new discoveries. The development of Thomas Kuhn's theory of paradigm shift was based on Bachelard's rationalism, which tends to reflect the empirical practices through a rationalist reconstruction of the theoretical practices in the natural sciences (Kuhn 1962, Vandenberghe 1999).

The chapters in this book intend to refer to and highlight the changes in paradigms with regard to the utilization of genetic resources, spanning from the search within genetic resources collections for desirable traits, capture of complex traits, use of omics technologies, identification of genes of adaptation and mitigation, role of domestication in adaptation, designing of genetic resources trials, and the innovative transfer from and to farming communities. This follows the work of Weitzman (1993) that provided a paradigmatic case in relation to *diversity theory*, developed in support of rational preservation policies. The book focuses on the new paradigm shifts in support of rational utilization of genetic resources rather than simply their conservation, combining theoretical and practical aspects with knowledge to speed up the search for adaptive traits and to rationalize the utilization of genetic resources.

To describe and classify scale-related phenomena in life sciences, from molecular to ecosystem levels of organization, mathematical fractal theories have helped in capturing and dealing with the complexity (Maurer 1994, Kenkeland Walker 1996, Ritchie and Olff 1999, Klar 2002). The presence of patterns from gene expression, or trait level, to population level may involve complex networks of genes, including genotype-by-environment interactions and epistatic interactions between genes regulating trait variation, requiring elaborated mathematical solutions (Cooper et al. 1999, Dennis 2002).

In the context of genetic resources, the elaboration of mathematical frameworks needs to consider the time issue also in the context of climate change. As genebanks will be under pressure to provide germplasm with needed traits in a timely fashion because of changing climatic conditions, the identification of the components of the traits and the germplasm carrying them will need to be quick and efficient. In their simulation model, Koo and Wright (2000) found that early identification of valuable crop traits in genetic resources, such as resistance to diseases, is of equal importance to the process of transferring these traits into an improved background. They also found that *a priori* information, which they called a *specialized knowledge case*, can contribute positively to expected net benefits, due to the increased probability of finding the desirable material and the associated cost savings. By analogy to physics, the *a priori* information or prior knowledge could be considered in this context as Euclidean geometry was in the context of general relativity (Reichenbach 1965). The *a priori* concept has evolved to an optional concept as per Poincaré's notion of convention, which is subject to choice and relative to its specific context of research or study (Nickles 2003).

In terms of *a priori* information, recent evidence suggests that climate change is proceeding more quickly than adaptation can occur (Hurme et al. 2000). Studies on climate change show that adaptation can proceed rapidly, over as few as 20 generations in the case of *Anthoxanthum odoratum* and *Agrostis tenuis* (Shaw and Ettersen 2012). The changes included those in physiological traits that exhibited plasticity in response to drought, and the potential of increased fitness in response to elevated CO_2 was also reported. The patterns of response were found to be similar to the patterns of evolution (Berger et al. 2013). Nevertheless, as indicated by Shaw and Ettersen (2012), the rates at which adaptive evolution occur on such a geographic range, relative to the rate of climate change, are crucial as many plant species lack the capacity to keep pace with the changes in conditions.

In addition to mathematical theories dealing with prediction and inference, the scope of applied mathematics could expand to deal with the uncertainty about the details of climate change, the measurement of complex traits, and assigning fuzzy values to trait states. Thus, a genebank accession with the desirable trait could belong to more than one trait state. Such trait variation can be considered as a response to stochastic ecological and co-evolutionary processes, where a stochastic process is a collection of random variables such as climate variables (Dieckmann and Law 1996, Insua et al. 2012). The presence of such patterns of trait variation driven by ecological and co-evolutionary processes implies the possibility of quantification and prediction (MacArthur 1972, Bari et al. 2006, Anderssen and Edwards 2012, Chave 2013).

For example, the stochastic ecological and co-evolutionary processes can be considered as a directed random walk in a trait's space which can, in turn, be described by a canonical diffusion mathematical equation to capture the rate of change over time of the expected trait value (Dieckmann and Law 1996, Champagnat et al. 2001, Champagnat and Lambert 2007). The trait can also be understood as representing the additive influence of many genes with small effects (Brown et al. 1996).

The synergistic approach coupled with the availability of massive environmental data presents an opportunity to delve into new frontiers of applied mathematics in agricultural research to further increase the utilization of PGR. It allows *in silico* evaluation that will reduce costs, with the potential to integrate new information with a mathematized language into modules that attempt to mimic biological processes such as complex networks of genes (Vignais and Vignais 2010). The *in silico* evaluation approach can include predicting key physiological traits that are difficult to measure (such as photosynthetic capacity) using qualitative combinations of plant trait characteristics that are easily measured (*soft*) and climate variables (Reich et al. 2007).

DATA—LARGE DATASETS

The datasets available have expanded tremendously and include large datasets in different formats, ranging from textual and qualitative data to numerical and digitized raster data. PGR data consist of both environmental data, denoted as X, and the trait (or evaluation) data, denoted as Y. These datasets are linked through spatial geographical coordinates, which provide common information on the site where the accessions were originally sampled (or evolved). Trait data (Y) consist of passport data and evaluation data. The passport data are composed of a standard set of descriptive information for each recorded accession held in genebanks, including the geographical location where the accession was sampled. Further Y data include observations taken on a number of traits such as reaction to a given disease or abiotic stress. The environmental data (X) include the long-term climate data of the collection sites and derived data such as an evapotranspiration and moisture index or an aridity index. X data are the set of variables that contain explanatory variables or predictors and Y data are either categorical (label) or numerical responses (trait descriptor states).

Data preparation may require transformation, of which several are possible, including the Tukey and Box–Cox transformations (Box and Cox 1964, Osborne 2010). The aim of Tukey's transformation is to avoid the discrepancy in variance between large and small values. The transformation also eliminates the multiplicative effect using the (generalized) power relationship between the spread and the mean (or the median) of the data. The power-law transformation is the Box–Cox transformation

$$y = \frac{\left(x^{\alpha} - 1\right)}{\alpha}$$

In Figure 10.1, aridity data have been transformed with $\alpha = 0.2$.

Prior to the dimension reduction, variables can also be standardized to a mean of 0 and a standard deviation of 1. In contrast to transformation, standardization is a process by which each observation is *transformed* with respect to all the

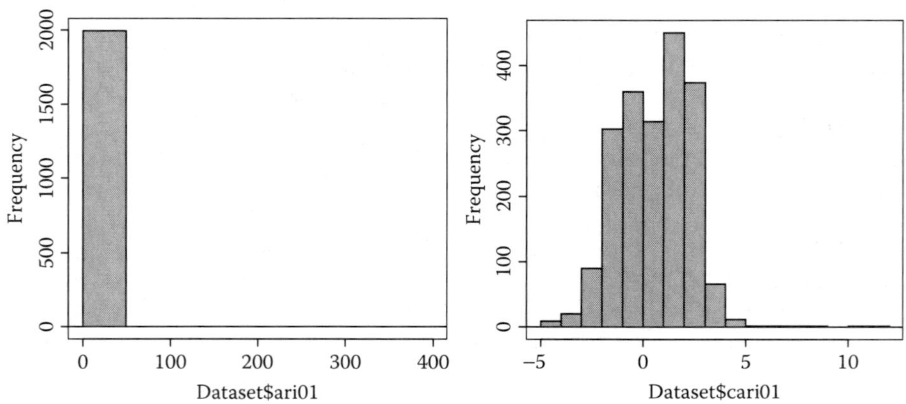

FIGURE 10.1 Histograms of variables before transformation (*left*) and after transformation (*right*).

observations within the variable. In terms of data transformation, there is also the possibility to eliminate unwanted variation because of the differences due to phenology (Ramsay et al. 2010). The characteristic of the site-climate variable is periodic, with an approximately sinusoidal pattern that can be represented as follows:

$$\mathrm{Clim}_i(t) \approx c_{i1} + c_{i2}\sin\left(\frac{\pi t}{6}\right) + c_{i3}\cos\left(\frac{\pi t}{6}\right)$$

where:

Clim_i is the climate function for the ith site
(c_{i1}, c_{i2}, c_{i3}) is a vector associated with the ith site (Ramsay et al. 2010)

CONCEPTS AND ANALYTICAL CONCEPTUAL FRAMEWORKS

The concepts and conceptual frameworks envisaged in the context of a synergistic approach are intended here as mathematical backing with regard to uncertainty and prediction. These concepts and conceptual frameworks are intended to develop models combining predictive accuracy with ease in interpretability.

BAYES–LAPLACE

In terms of uncertainty, a reference will be made to the Bayes–Laplace theorem that considers the probability of causes in relation to their effects, in contrast to the probability of effects in relation to their causes (Fisher 1930, 1936):

$$P(\mathrm{C} \mid \mathrm{E}) = \frac{P(\mathrm{E} \mid \mathrm{C})}{\sum_k P(\mathrm{E} \mid \mathrm{C} = k)}$$

where:

$P(\mathrm{C} \mid \mathrm{E})$ is the probability of cause (given the event or datum)
$P(\mathrm{E} \mid \mathrm{C})$ is the probability of an event or datum (given the cause)

To simplify, we will consider the case of a binary trait (Y) with two states, either 0 (absent) or 1 (present, such as heat tolerance). This can be formulated as a logistic regression model (Pohlman and Leitner 2003) for observation Y of accession i as

$$\mathrm{logit}(p_i) = \beta_0 + \sum_{j=1}^{n} \beta_j x_{ij}$$

where:

β_0 is an intercept
β_j are regression coefficients

Here data consist of $\{Y_i, X_i = (x_{i1}, \ldots, x_{in})\}$, with the first N_1 accessions $i = 1, \ldots, N_1$ as a training set, and the models are tested on an unknown set of accessions of which the predictive probabilities $P(Y_i = 1|X_i)$, $i = N_1 + 1, \ldots, N$ are compared to the evaluation data (Y_i), $i = N_1 + 1, \ldots, N$ (Borsuk 2008).

Under the logistic equation, the response variable Y is modeled by the logit(p) link function so that $p = P(Y = 1|X)$. The logit stands for the logarithm of odds (Pohlman and Leitner 2003):

$$\text{logit}(p) = \ln\left(\frac{p}{1-p}\right) = \beta_0 + \sum_{j=1}^{n} \beta_j X_j$$

which in turn leads to the mathematical expression

$$p = P(Y = 1 | X) = \frac{\exp\left(\beta_0 + \sum_{j=1}^{n} \beta_j X_j\right)}{1 + \exp\left(\beta_0 + \sum_{j=1}^{n} \beta_j X_j\right)}$$

This transformation assumes a linear relationship between the logit of the probability of $Y = 1$ and the climate variables X. The above equation describes Y as a Bernoulli-distributed variable (Gollin et al. 2000) that can be alternatively modeled by the probit link function (Feelders 1999).

In the Bayes–Laplace context, *a priori* information about the parameters of the mathematical model in the form of a probability distribution (prior probability) is superposed with the likelihood of the data to generate *a posteriori* information (posterior probability distribution). The mathematical probability model is described by θ, giving rise to a set of observations X obtained by random sampling. The posterior (predicted) probability is thus equal to the prior probability times the likelihood (Insua et al. 2012):

$$f(\theta | X) = \frac{f(\theta) f(X | \theta)}{\int f(\theta) f(X | \theta) d\theta}$$

where:

$f(X | \theta)$, the likelihood function, is the probability of the data X for a given parameter θ

$f(\theta)$ represents the prior distribution

To predict the values for the unevaluated accessions Y, the posterior predictive distribution based on the known value of θ is as follows:

$$P(Y | X) = \int f(Y | \theta, X) f(\theta | X) d\theta$$

Neural Networks

Here, we will refer to the modeling based on the NN methodology as a more general framework to formulate the model for the logistic regression, and the Bayes–Laplace inverse problem as a special case of logistic regression that uses Bayes's rule and conditional probabilities to determine the weights of the regression.

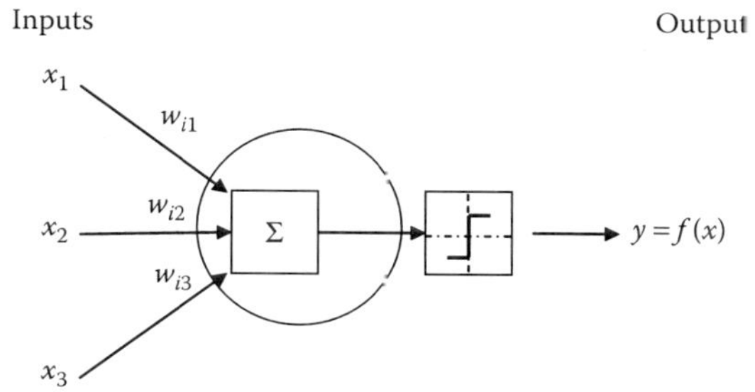

FIGURE 10.2 McCulloch–Pitts model neuron (i) with three inputs (dendrites) and one output (axon).

NNs are mathematical tools made of many processing element (PE) units (Figure 10.2) linked to communicate and carry out computations in parallel, independently from each other. This parallelism and high connectivity help in defying the assumptions that are usually required in the case of simple models, such as simple linear regression.

When all the neurons are interconnected, the system becomes an artificial neural network (ANN). As a technique inspired by the biological structure of a neuron, it has *dendrites* that receive impulses (input) and an *axon* that transmits impulses (output) (Golden 1996) (Figure 10.2). A neuron is described by its weight w_k and transfer function $f(x)$. The neuron receives a set of numbers (a vector) x_k as input and generates a number y as output. In Figure 10.2, each synapse (junction) has a weight: w_{i1}, w_{i2}, and w_{i3}. The neuron receives information as the weighted sum ($x = w_{i1}x_1 + w_{i2}x_2 + w_{i3}x_3$) of the three inputs ($x_1$, x_2, and x_3), which in turn is passed through a discontinuous threshold sigmoid nonlinearly to obtain a new activation (new input) value, denoted by $y = f(x) = f(w_{i1}x_1 + w_{i2}x_2 + w_{i3}x_3)$ (Golden 1996).

The ANN can also *learn*. The training of a network consists of multiple examples that are fed into it. There are different ways of feeding a network and there are different types of NNs to accommodate these different feeding methods. In practice, different types of networks are used for different purposes.

Among the ANN often used for PGR data are the radial basis function (RBF) networks (Figure 10.3), which have a strong mathematical base like SVMs, and are nonlinear hybrid networks typically containing a single hidden layer of PE. This hidden layer uses Gaussian transfer functions rather than the sigmoidal functions employed by maximum likelihood programs. The centers and widths of the Gaussian curves are set by unsupervised learning rules, and supervised learning is applied to the output layer.

RBF first carries out unsupervised clustering based on the k-nearest neighbor rule, and then applies a supervised classification using the information of cluster number and width (radius, hence the name *radial*). The estimation of the centers and widths of the RBFs is carried out according to the unsupervised k-nearest neighbor rule. The input space is first split into k clusters and the size of each cluster is obtained from the structure of the input data. The centers of the clusters give the centers of the RBFs, while the distance between the clusters provides their widths (Silipo 1999).

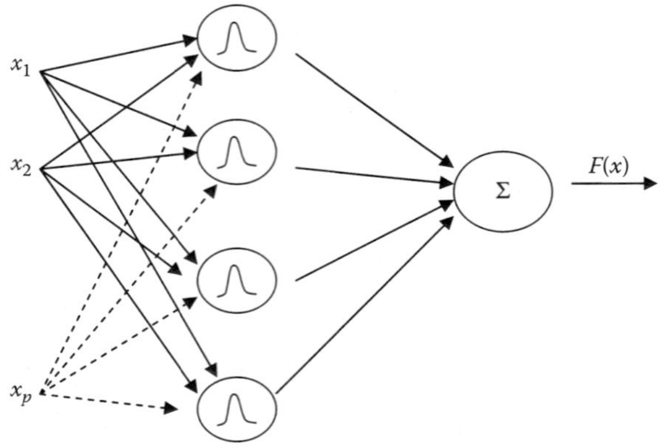

FIGURE 10.3 A radial basis function (RBF) network. The first layer contains simple neurons that transmit input. The second *hidden* layer contains the RBF neurons (Gaussian). The third layer (the output) contains neurons as simple linear units.

A typical RBF consists of three layers, where each layer is connected in a many-to-many structure to the next one with simple first-order connections (Figure 10.3). Every input component (x_p) is connected to the layer of hidden nodes. Each node in the hidden layer is a p multivariate Gaussian function (RBF), where:

$$G(x;x_i) = exp\left[\frac{-1}{2\sigma_i^2}\sum_{k=1}^{p}(x_k - x_{ik})^2\right]$$

Each of the second-layer nodes is connected to the third layer, where the RBF linearly weights the output of the hidden nodes to obtain the function:

$$F(x) = \sum_{i}^{N} w_i\left(G(x;x_i)\right)$$

Support Vector Machines

The SVM is also a learning-based technique that maps input data to a high-dimensional space, and then optimally separates it into respective classes by isolating those inputs that fall close to the data boundaries (Cortes and Vapnik 1995, Principe et al. 2000). The mapping of input data to high-dimensional space is carried out through functions known as *kernel functions* that allow the separation of data. The mathematical conceptual framework based on the learning approach considers as a learning set $L = \{(X_1, Y_1), \ldots, (X_n, Y_n)\}$ of a random vector (X_i, Y_i). X_i is the set of climate variables and Y_i is the presence or absence of the traits (trait descriptor states) at a given site *i*.

SVM optimally separates mapped input into respective classes:

$$x \in R^l \rightarrow \Phi(x) \in R^K$$

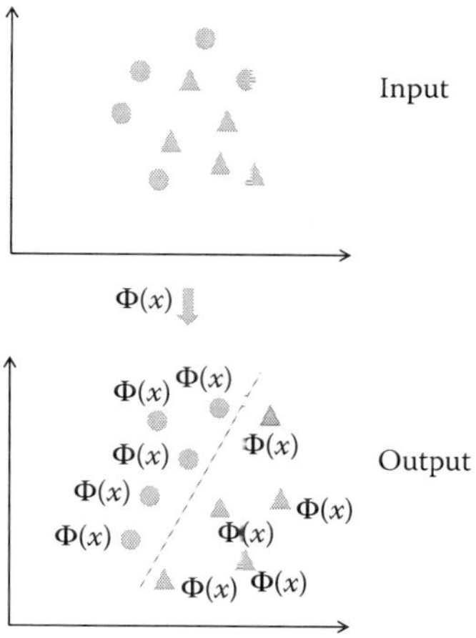

FIGURE 10.4 Schematic of the action of support vector machines as input/output machines that separate original input data (nonlinear space) into categories (output).

from l-dimensional space (input variable space) into k-dimensional space, where k is greater than l (Figure 10.4).

Random Forest

The random forest (RF) is a type of recursive partitioning and clustering algorithm that does not require normality assumptions and deals well with a large number of variables. RF can address complex interactions and can cope with highly correlated predictor variables (Strobl et al. 2008). RF differs from a standard tree classifier in that it *grows* many classification trees in the process. An object from an input vector is classified by all trees in the forest. Each tree gives a classification, and we say the tree *votes* for that class. The forest chooses the classification of a given object having the most votes over all the trees in the forest, correctly classifying a site as likely to experience heat, leading to accessions tolerant to heat.

The mathematical conceptual framework based on the learning approach considers as a learning set $L = \{(X_1, Y_1), \ldots, (X_n, Y_n)\}$ of a random vector (X_i, Y_i), namely (environment, trait) at a site i.

RF acts like an ensemble of classifiers in the form of

$$g = \left(g\left(x_1\right), \ldots, g\left(\mathrm{x}_k\right)\right)$$

The data in the RF module are split intrinsically, one part becoming a *training set* as a result of bootstrap sampling with replacement. The data that are not sampled to be part of the training set are called the *out-of-bag* (OOB) *set*, and it is used to test the predictive power of the RF module. A number, *mtrv*, is specified, which is less than the number of input variables (in this case the climatic parameters), such that at each node of the tree an *mtrv* number of variables is selected at random from the original variable set and the best split on these randomly selected variables is used to split the node. Each tree in the *forest* is grown to the largest extent possible without pruning until there are, *n tree*, number of trees. The model is optimized by varying the *mtrv* and *n tree* values. The optimization of the *mtrv* and *n tree* parameters is driven by monitoring the size of the mean square prediction error (rate of classification error) observed in the OOB set, that is, the ability of each iteration to correctly classify a site as subject to or free of the stress in question.

The climate variables that have a strong relationship to the trait would be those that split the site data correctly based on the following formula:

$$1-\frac{SS_r+SS_l}{SS_p}$$

where:

SS is the sum-of-squares that is equal to $\sum_j (X_j-\bar{X})^2$
subscript r indicates right of the split
subscript l indicates left of the split
p indicates the parent node that is split

At the split, the variable that produces less *entropy* measured using either information theory (Shannon index) or the Gini index (known as impurity measure) is ranked first (Battail 1997, Raileanu and Stoffel 2004). A reduction in the impurity is a prerequisite for the variable ranking/importance.

Accuracy Metrics

The accuracy of the different mathematical models can be tested using parameters derived from a 2 × 2 confusion matrix (Table 10.1). Among these parameters are the sensitivity, which is defined by $T_{00}/(T_{00}+T_{01})$, and the specificity defined by $T_{11}/(T_{10}+T_{11})$. These two parameters are indicators of the model's ability to correctly classify observations as either lacking or having the desired trait. The higher the values of sensitivity and specificity, the lower the error and thus the better the discriminating power of the model. An error occurs when the trait is present ($T = 1$) but classified as absent ($T^* = 0$) and vice versa. The former is a conditional probability notated by $P(T^* = 0|\, T = 1)$ while the latter is denoted by $P(T^* = 1 \mid T = 0)$. Thus, sensitivity can be defined by $P(T^* = 1 \mid T = 1)$ and specificity by $P(T^* = 0|T = 0)$.

TABLE 10.1
Confusion Matrix Presented as a 2 × 2 Contingency Table

		Observed Value/Score	
		Absent (0)	**Present (1)**
Predicted Value/Score	Absent (0)	T_{00}	T_{10}
	Present (1)	T_{01}	T_{11}

Further metrics commonly used as *accuracy metrics* of the models are kappa and *area under the curve* (AUC) values, the latter being derived from the *receiver operating characteristics* (ROC) curve (Swets et al. 2000, Fawcett 2006). The kappa value is a measure of agreement between the prediction and actual observations from the confusion matrix table. A value of kappa below 0.4 is an indication of poor agreement and a value of 0.4 and above is an indication of good agreement (Landis and Koch 1977). Thus, a high value is an indication that the performance of the model is adequate for prediction purposes.

Since kappa can be a threshold-dependent *metric* parameter, the AUC of the ROC plots is also used as a measure of the model's accuracy (Freeman and Moisen 2008). This is also known as the plot of true positive rate versus false positive rate, where the true positive rate is sensitivity and the false positive rate is (1 – specificity), to assess improvement over randomness. An AUC value of 0.5 represents randomness, while a value of 1 represents peak model performance (Fawcett 2006).

CONCLUSION

The availability of massive data in conjunction with the development of mathematical concepts and innovative approaches including farmers' innovations can help crop production to adapt to climate change. In terms of innovation transfer from and to the farming community, Soleri and Cleveland (2001) argued that a profound understanding of farmers' genetic perceptions could contribute to better transfer of technologies and provide farmers with conceptual tools they can use to adapt or develop their own innovations to best meet their needs. The combination of local knowledge with theoretical aspects has led to the development of a directed sampling strategy used successfully in sampling and conserving biodiversity (Haenn et al. 2014).

Evidence of the presence of patterns with respect to adaptive evolution in response to spatial gradients in climate across contemporary ranges is in support of the concept of tolerance to climate change (Rehfeldt et al. 1999, Davis et al. 2005, Shaw and Etterson 2012). This is also supported by the theoretical investigations of the evolutionary dynamics that, however, depend largely on genetic variation. The understanding of genetic patterns began with Malecot under his probabilistic theory underlying genetic differentiation and spatial genetic structuring as a result of stochastic processes (Heywood 1991, Ishida 2009). Since then, additional mathematical frameworks followed (May 1981), and Benoit Mandelbrot's (1977) mathematical fractal theories in dealing with complexity have helped in the description and classification of scale-related phenomena in life sciences, from molecular to ecosystem levels of organization (Maurer 1994, Kenkeland Walker 1996, Ritchie and Olff 1999).

REFERENCES

Anderssen, R.S. and M.P. Edwards. 2012. Mathematical modelling in the science and technology of plant breeding. *Int J Numer Anal Model B* 3(3):242–258.

Bachelard, S. 1968. Épistémologie et histoire des sciences. (XIIe Congrès international d'histoire des sciences, Paris, 1968), Revue de synthèse, IIIe série, 49–52.

Bari, A., A. Ayad, S. Padulosi et al. 2006. Analysis of geographical distribution patterns in plants using fractals. In *Complexus mundi, emergent patterns in nature*, ed. M. Novak, 287–296. Singapore: World Scientific.

Bari, A., K. Street, M. Mackay, D.T.F. Endresen, E. De Pauw, and A. Amri. 2012. Focused identification of germplasm strategy (FIGS) detects wheat stem rust resistance linked to environmental variables. *Genet Resour Crop Ev* 59:1465–1481.

Battail, G. 1997. *Théorie de l'Information: Application aux techniques de communication*. Paris, France: Masson.

Berger, D., E. Postma, W.U. Blanckenhorn, and R.J. Walters. 2013. Quantitative genetic divergence and standing genetic (co)variance in thermal reaction norms along latitude. *Evolution* 67:2385–2399. doi:10.1111/evo.12138.

Borsuk, M.E. 2008. Statistical prediction. In *Ecological models, encyclopedia of ecology 4*, eds. S.E. Jørgensen and B.D. Fath, 3362–3373. Oxford: Elsevier.

Box, G.E.P. and D.R. Cox. 1964. An analysis of transformations. *J Roy Stat Soc B* 26:211–252.

Brown, J.H., G.C. Stevens, and D.M. Kaufman. 1996. The geographic range: Size, shape, boundaries, and internal structure. *Annu Rev Ecol Systemat* 27:597–623.

Byerlee, D., P. Heisey, and P. Pingali. 1999. *Realizing yield gains for food staples in developing countries in the early 21st century: Prospects and challenges*. Paper presented to the Study Week on Food Needs of the Developing World in the Early 21st Century, The Vatican, January 27–30.

Cagnacci, F., L. Boitani, R.A. Powell, and M.S. Boyce. 2010. Animal ecology meets GPS-based radiotelemetry: A perfect storm of opportunities and challenges. *Philos T Roy Soc B* 365:2157–2162. doi:10.1098/rstb.2010.0107.

CGIAR. 2015. Research program on dryland systems. http://drylandsystems.cgiar.org/.

Champagnat, N., R. Ferriere, and G. Ben Arous. 2001. The canonical equation of adaptive dynamics: A mathematical view. *Selection* 2:73–84.

Champagnat, N. and A. Lambert. 2007. Evolution of discrete populations and the canonical diffusion of adaptive dynamics. *Ann Appl Probab* 17:102–155.

Chave, J. 2013. The problem of pattern and scale in ecology: What have we learned in 20 years? *Ecol Lett* 16:4–16.

Cooper, M., D.W. Podlich, and S.C. Chapman. 1999. Computer simulation linked to gene information databases as a strategic research tool to evaluate molecular approaches for genetic improvement of crops. In *Molecular approaches for the genetic improvement of cereals for stable production in water-limited environments*, eds. J.-M. Ribaut and D. Poland, 162–166. Mexico City, Mexico: CIMMYT.

Cortes, C. and V. Vapnik, 1995. Support-vector networks. *Mach Learn* 20(3):273–297. doi:10.1007/BF00994018.

Cramer, C.J. 2004. *Essentials of computational chemistry: Theories and models*. New York: John Wiley & Sons.

Darwin, C. 1859. *On the origin of species*. London: John Murray.

Davis, M.B., R.G. Shaw, and J.R. Etterson. 2005. Evolutionary responses to changing climate. *Ecology* 86:1704–1714. doi:10.1890/03-0788.

Dawson, I.K., J. Russell, W. Powell, B. Steffenson, W.T.B. Thomas, and R. Waugh. 2015. Barley: A translational model for adaptation to climate change. *New Phytol* 206(3):913–931. doi:10.1111/nph.13266.

Dennis, C. 2002. The brave new world of RNA: News features. *Nature* 418:122–124.

Dieckmann, U. and R. Law. 1996. The dynamical theory of coevolution: A derivation from stochastic ecological processes. *J Math Biol* 34:579–612.

Ebke, A.T. 2012. Monika Wulz: Erkenntnisagenten. Gaston Bachelard und die Reorganization des Wissens. *Stud E Eur Thought* 64(1–2):143–148. doi:10.1007/s11212-012-9164-4.

FAO. 2009. *How to feed the world in 2050: Expert Paper*. High-Level Expert Forum, October 12–13, 2009, Rome, Italy: Food and Agriculture Organization of the United Nations. http://www.fao.org/fileadmin/templates/wsfs/docs/expert_paper/How_to_Feed_the_World_in_2050.pdf.

FAO. 2010. *The second report on the state of the world's plant genetic resources for food and agriculture*. Rome, Italy: Food and Agriculture Organization of the United Nations.

FAO. 2011. *Save and grow: A policymaker's guide to the sustainable intensification of smallholder crop production*. Rome, Italy: Food and Agriculture Organization of the United Nations. http://www.fao.org/docrep/014/i2215e/i2215e.pdf.

Fawcett, T. 2006. An introduction to ROC analysis. *Pattern Recogn Lett* 27:861–874. doi:10.1016/j.patrec.2005.10.010.

Feelders, A.J. 1999. Statistical concepts. In *Intelligent data analysis: An introduction*, eds. M. Berthold and D.J. Hand, 15–66. Berlin, Germany: Springer-Verlag.

Fisher, R.A. 1930. Inverse probability. *Proc Cambridge Philos Soc* 26:528–535.

Fisher, R.A. 1936. The use of multiple measurements in taxonomic problems. *Ann Eugenics* 7:179–188. doi:10.1111/j.1469-1809.1936.tb02137.x.

Freeman, E.A and G.G. Moisen. 2008. A comparison of the performance of threshold criteria for binary classification in terms of predicted prevalence and kappa. *Ecol Model* 217:48–58.

GFA. 2014. *Innovation transfer into agriculture-adaptation to climate change (ITAACC). Bridging the gap between agricultural research and farmers' practice*. Final International Workshop—Findings and lessons learnt. May 7–8, 2014, Nairobi, Kenya/Hamburg, Germany: GFA Consulting Group GmbH.

Glamore, M.J., J.L. West, and J.P. O'Leary. 2013. Some observations on the interdigitation of advances in medical science and mathematics. *Am Surgeon* 79(12):1231–1234.

Golden, R.M. 1996. *Mathematical methods for neural network analysis and design*. Cambridge, MA: Massachusetts Institute of Technology.

Gollin, D., M. Smale, and B. Skovmand. 2000. Searching an ex situ collection of wheat genetic resources. *Am J Agric Econ* 82:812–827.

Haenn, N., B. Schmook, Y. Reyes, and S. Calmé. 2014. Improving conservation outcomes with insights from local experts and bureaucracies. *Conserv Biol* 28:951–958. doi:10.1111/cobi.12265.

Harlan, J.R. 1992. *Crops and man*, 2nd ed. Madison, WI: American Society of Agronomy-Crop Science Society.

Heywood, V.H. 1991. Developing a strategy for germplasm conservation in botanic gardens. In *Tropical botanic gardens: Their role in conservation and development*, eds. V.H. Heywood and P.S. Wyse-Jackson, 11–23. London: Academic Press.

Hurme, P., M.J. Sillanpää, E. Arjas, T. Repo, and O. Savolainen. 2000. Genetic basis of climatic adaptation in Scots pine by Bayesian quantitative trait locus analysis. *Genetics* 156:1309–1322.

Insua, D.R., F. Ruggeri, and M.P. Wiper. 2012. *Bayesian analysis of stochastic process models*. Chichester: John Wiley & Sons.

Ishida. Y. 2009. Sewall Wright and Gustave Malécot on isolation by distance. *Philos Sci* 76(5):784–796.

Kecman, V. 2005. Support vector machine—An introduction. In *Support vector machines: Theory and applications*, ed. L. Wang, 1–47. Berlin, Germany: Springer-Verlag.

Kenkel, N.C. and D.J. Walker. 1996. Fractals in the biological sciences. *Coenoses* 11:77–100.

Klar, A.S. 2002. Plant mathematics: Fibonacci's flowers. *Nature* 417:595.

Koo, B. and B.D. Wright. 2000. The optimal timing of evaluation of genebank accessions and the effects of biotechnology. *Am J Agric Econ* 82:797–811.

Kotschi, J. 2007. Agricultural biodiversity is essential for adapting to climate change. *GAIA—Ecol Perspect Sci Soc* 16(2):98–101.

Kuhn, T.S. 1962. *The structure of scientific revolution*. Chicago, IL: University of Chicago Press.

Landis, J.R. and G.G. Koch. 1977. The measurement of observer agreement for categorical data. *Biometrics* 33:159–174.

MacArthur, R.H. 1972. *Geographical ecology: Patterns in the distribution of species*. New York: Harper & Row.

Malécot, G. 1969. *The mathematics of heredity*. San Francisco, CA: W.H. Freeman.

Mandelbrot, B.B. 1977. *Fractals, form, chance, and dimension*. San Francisco, CA: W.H. Freeman.

Maurer, B.A., 1994. *Geographical population analysis: Tools for the analysis of biodiversity*. Oxford: Blackwell Science.

May, R.M. (ed.). 1981. *Theoretical ecology, principles and applications*, 2nd Ed. Oxford: Blackwell Science.

Nickles, T. 2003. *Thomas Kuhn*. Cambridge: Cambridge University Press.

Osborne, J.W. 2010. Improving your data transformations: Applying Box-Cox transformations as a best practice. *Pract Assess Res Eval* 15:1–9.

Pohlman, J.T. and D.W. Leitner. 2003. A comparison of ordinary least squares and logistic regression. *Ohio J Sci* 103:118–125.

Præbel, A. and L.F. Groeneveld (eds.). 2014. *Book of abstracts: Genetic resources for food and agriculture in a changing climate*, January 27–29, 2014, Lillehammer, Norway. Ås, Norway: The Nordic Genetic Resource Center.

Principe, J.C., N.R. Euliano, and W.C. Lefebvre. 2000. *Neural and adaptive systems: Fundamentals through simulations*. New York: John Wiley & Sons.

Raileanu, L.E. and K. Stoffel. 2004. Theoretical comparison between the Gini index and information gain criteria. *Ann Math Artif Intel* 41:77–93.

Ramsay, J.O., G. Hooker, and S. Graves. 2010. *Functional data analysis with R and Matlab*. New York: Springer.

Rehfeldt, G.E., C.C. Ying, D.L. Spittlehouse, and D.A. Hamilton, Jr. 1999. Genetic responses to climate for *Pinuscontorta* in British Columbia: Niche breadth, climate change, and reforestation. *Ecol Monogr* 69:375–407.

Reich, P.B., I.J. Wright, and C.H. Lusk. 2007. Predicting leaf physiology from simple plant and climate attributes: A global GLOPNET analysis. *Ecol Appl* 17:1982–1988.

Reichenbach, H. 1965. *The theory of relativity and* a priori *knowledge* (English translation of the 1920 *Relativitätstheorie und Erkenntnisapriori*). Berkeley, CA: University of California Press.

Ritchie, M.E. and H. Olff. 1999. Spatial scaling laws yield a synthetic theory of biodiversity. *Nature* 400:557–560.

Selkoe, K.A., J.R. Watson, C. White et al. 2010. Taking the chaos out of genetic patchiness: Seascape genetics reveals ecological and oceanographic drivers of genetic patterns in three temperate reef species. *Mol Ecol* 19:3708–3726.

Sensevy, G., A. Tiberghien, J. Santini, S. Laube, and P. Griggs. 2008. An epistemological approach to modeling: Cases studies and implications for science teaching. *Sci Educ* 92(3):424–446.

Shaw, R.G. and J.R. Etterson. 2012. Rapid climate change and the rate of adaptation: Insight from experimental quantitative genetics. *New Phytol* 195:752–765.

Silipo, R. 1999. Neural networks. In *Intelligent data analysis: An introduction*, eds. M. Berthold and D.J. Hand, 217–268. Berlin, Germany: Springer-Verlag.

Sismondo, S. 1999. Models, simulations, and their objects. *Sci Context* 12:247–260. doi:10.1017/S0269889700003409.

Smith, S., M. Cooper, J. Gogerty, C. Löffler, D. Borcherding, and K. Wright. 2014. Maize. In *Yield gains in major U.S. field crops*, eds. S. Smith, J. Specht, B. Diers, and B. Carver, 125–171. CSSA Special Publication 33. Madison, WI: ASA, CSSA, and SSSA.

Soleri, D. and D.A. Cleveland. 2001. Farmers' genetic perceptions regarding their crop populations: An example with maize in the central valleys of Oaxaca, Mexico. *Econ Bot* 55(1):106–128.

Strobl, C., A.-L. Boulesteix, T. Kneib, T. Augustin, and A. Zeileis. 2008. Conditional variable importance for random forests. *BMC Bioinform* 9:307.

Swets, J.A., R.M. Dawes, and J. Monahan. 2000. Better decisions through science. *Sci Am* 283:82–87.

Vandenberghe, F. 1999. "The real is relational": An epistemological analysis of Pierre Bourdieu's generative structuralism. *Sociol Theor* 17:32–67. doi:10.1111/0735-2751.00064.

Vignais, P.V. and P.M. Vignais. 2010. *Discovering life, manufacturing life: How the experimental method shaped life sciences.* Dordrecht, The Netherlands: Springer Science+Business. doi:10.1007/978-90-481-3767-1_5.

Weitzman, M. 1993. What to preserve: An application of diversity theory to crane conservation. *Q J Econ* 108(1):157–183.

11 Power Transformations

Application for Symmetrizing the Distribution of Sample Coefficient of Variation from Inverse Gaussian Populations

Y.P. Chaubey, A. Sarker, and M. Singh

CONTENTS

The coefficient of variation (CV) of a random variable (or that of the corresponding population), defined to be the ratio of the standard deviation to the mean of the corresponding population, has been used in wide-ranging applications in many areas of applied research including agro-biological, industrial, social, and economic research (Johnson et al. 1994, Chapter 15). In these applications, the random variable of interest is assumed to follow a Gaussian distribution that is symmetric and has support on the whole real number line (see Laubscher 1960, Singh 1993, Johnson et al. 1994, Chaubey et al. 2013). However, in many of these applications, the random variable may be more appropriately modeled by a distribution, which is positively skewed and is supported on the positive half. To model such situations, use of an inverse Gaussian (*IG*) distribution is often more justified compared to lognormal, gamma, and Weibull distributions (see Chhikara and Folks 1977, 1989, Kumagai et al. 1996, Takagi et al. 1997). More recently, Mudholkar and Natarajan (2002) discussed comparisons of shape of the *IG* distribution with the Gaussian distribution.

Since the distribution of the sample CV, in general, is not easy to handle, various approximations, mostly centered around the Gaussian distribution, have been discussed in the literature; see Banik and Kibria (2011) for a comprehensive review and comparison of various approximations. Recently, Chaubey et al. (2013) have investigated an approximately normalizing transformation of CV associated with a Gaussian population and contrasted its performance with the variance-stabilizing transformation (VST) that is often employed in this context. The likelihood ratio test for CV of an *IG* population has been investigated by Hsieh (1990), and more recently Chaubey et al. (2014a) have demonstrated that this test is *best invariant* under the group of scale transformations.

The purpose of this chapter is to review the properties of variance stabilizing and skewness-reducing transformations for CV in the context of the *IG* population as investigated recently by Chaubey et al. (2014b). The variables observed for evaluation of genetic resources and modeling climate data often need transformation so that the associated assumptions in applying the statistical methods are tenable. For analysis of variance of a single variable, Bartlett (1947) introduced VSTs on response variables from Poisson and binomial distributions. Further improvement was provided by Anscomb (1948). In the context of association between two variables measured by correlation coefficients, the $tanh^{-1}$ transformation due to Fisher (1915, 1921) is normally used. A number of other transformations for correlation coefficients between pairs of bivariate *t*, chi-square, and some discrete distributions, were studied by Kocherlakota and Singh (1982a,b) and for outlier-contaminated bivariate normal distribution by Gupta and Singh (1992). A number of transformations of genotypic correlation coefficients estimated from variety trials have been studied by Singh and Hinkelmann (1992). The general approach on transformation for symmetrizing the distribution of an intended statistic has been developed in Chaubey and Mudholkar (1983), while the distributional behavior of variance-stabilizing and symmetrizing transformations have been studied using a simulation approach for the CV for normal populations in Chaubey et al. (2013).

The *IG* distribution is found to have an application in the context of the genotypic variability changes with the environment. Variability in a genetic pool determines the extent to which genetic improvement can be made in a trait of interest. The environment differences may arise due to spatial variability or temporal variability and may reflect climate changes. Lentil (*Lens culinaris* Medikus ssp. *culinaris*), a rich source of protein, is generally grown in Mediterranean, sub-tropical, temperate, and nontropical dry environments globally, predominantly grown and consumed in South and West Asia, and East and North Africa, and contributes to nutritional security (Bhatty 1988). Based on randomly selected genotypes from the Lentil International Elite Nursery (LIEN), the statistical distribution of seed yield has been examined at two locations, New Delhi, India, and Larissa, Greece, during 2000. Details of the trials from where the datasets have been extracted are available in Sarker et al. (2010).

The organization of this chapter is as follows. In the next section, we list some basic properties of the *IG* distribution along with that of the corresponding sample CV. In the section after that, the transformations recently assessed in Chaubey et al. (2014b) have been cataloged and their comparison has been reproduced here, which demonstrates that the power transformation provides an excellent approximation.

The final section applies the symmetrizing power transformation in order to compare CV parameters in a hypothesis-testing framework.

IG DISTRIBUTION AND ESTIMATION OF CV

The probability density function *(pdf)* of the *IG* random variable X is given by

$$f(x|\mu,\lambda)=\left\{\frac{\lambda}{2\pi x^3}\right\}^{1/2}\exp\left\{-\frac{\lambda}{2\mu^2 x}(x-\mu)^2\right\} \quad x>0,\mu>0,\lambda>0 \tag{11.1}$$

to be denoted by $IG(\mu,\lambda)$, where μ is the mean of the distribution and $1/\lambda$ is known as the dispersion parameter, with mean and variance given, respectively, by μ and μ^3/λ. For a broad review and applications of the *IG* family and other related results, the reader may refer to Chhikara and Folks (1989) and Seshadri (1993, 1998). The CV for a population with the pdf of Equation 11.1 is given by $\gamma=\sqrt{\mu/\lambda}$. For our purpose we will consider the parameter $\phi=\gamma^2=\mu/\lambda$.

For a random sample X_1, X_2, ..., X_n from $IG(\mu,\lambda)$, we have two standard results concerning the distributions of the sample mean $\overline{X}=n^{-1}\sum_{i=1}^{n}X_i$ and $U=(n-1)^{-1}\sum_{i=1}^{n}\left(\frac{1}{X_i}-\frac{1}{\overline{X}}\right)$, namely

$$\text{(i) } \overline{X}\sim IG(\mu,n\lambda) \text{ and (ii) } (n-1)\lambda U\sim x_{n-1}^2 \tag{11.2}$$

Further, the random variables $\overline{X}$ and U are independent; thus, an unbiased estimator of ϕ is given by

$$\hat{\phi}=\hat{\mu}\widehat{\left(\frac{1}{\lambda}\right)}=\overline{X}U \tag{11.3}$$

The moments of $\hat{\phi}$ may now be obtained using the distributional properties of $\overline{X}$ and U; the following three central moments, which may be required for further computations, are given by

$$\mu_2(\phi)=E\left(\hat{\phi}-\phi\right)^2=\phi^2\left[\frac{2}{\nu}+\left(1+\frac{2}{\nu}\right)\frac{\phi}{n}\right] \tag{11.4}$$

$$\mu_3(\phi)=E\left(\hat{\phi}-\phi\right)^3=\phi^3\left[\frac{8}{\nu^2}+\frac{12}{\nu}\left(1+\frac{2}{\nu}\right)\frac{\phi}{n}+3\left(1+\frac{6}{\nu}+\frac{8}{\nu^2}\right)\left(\frac{\phi}{n}\right)^2\right] \tag{11.5}$$

$$\begin{aligned}\mu_4(\phi)=E\left(\hat{\phi}-\phi\right)^4=\phi^4&\left[\frac{12}{\nu^2}\left(1+\frac{4}{\nu}\right)+\frac{12}{\nu}\left(1+\frac{14}{\nu}+\frac{24}{\nu^2}\right)\left(\frac{\phi}{n}\right)\right.\\&\left.+3\left(1+\frac{36}{\nu}+\frac{188}{\nu^2}+\frac{240}{\nu^3}\right)\left(\frac{\phi}{n}\right)^2+15\left(1+\frac{12}{\nu}+\frac{44}{\nu^2}+\frac{48}{\nu^3}\right)\left(\frac{\phi}{n}\right)^3\right]\end{aligned} \tag{11.6}$$

VARIANCE-STABILIZING AND SYMMETRIZING TRANSFORMATIONS FOR $\hat{\phi}$

Chaubey et al. (2014b) derived the *VST*. Denoting the *VST* by $g_v(\hat{\phi})$ and considering the moments in approximations up to order $O(1/n^2)$, it is given by

$$g_v(\phi) = \sinh^{-1}\left(\frac{B}{\sqrt{\phi}}\right) = \ln\left[\frac{B}{\sqrt{\phi}} + \sqrt{1+\frac{B^2}{\phi}}\right], \quad \text{where } B = \sqrt{\frac{2n}{n+1}} \tag{11.7}$$

They also obtained the symmetrizing transformation (*ST*) of $\hat{\phi}$, denoted by $g_s(\hat{\phi})$, using the expressions for moments up to the same order as above, that is given by the integral

$$g_s(\phi) = \int e^{-a(\phi)} d\phi \tag{11.8}$$

where

$$a(\phi) = \frac{2}{3}\int \left\{\frac{f_1(\phi)}{f_2(\phi)}\right\} d\phi \tag{11.9}$$

with $f_1(\phi)$ and $f_2(\phi)$ defined as

$$f_1(\phi) = \mu_3(\phi) \tag{11.10}$$

and

$$f_2(\phi) = \mu_4(\phi) - \mu_2^2(\phi) \tag{11.11}$$

The integrals required in finding the *ST* are not available in explicit forms; a numerical integration procedure is recommended in Chaubey et al. (2014b). However, a simpler procedure using the power transformation as proposed in Jensen and Solomon (1972) for approximating the distribution of a quadratic form (see also Mudholkar and Trivedi 1981) has been adapted to the distribution of CV from an *IG* population in Chaubey et al. (2014b). This procedure considers $\left(\hat{\phi}/\phi\right)^{h_0}$ to be normally distributed where

$$h_0 = 1 - \frac{\kappa_1\kappa_3}{3\kappa_2^2} \tag{11.12}$$

where

$$\kappa_1 = \nu \tag{11.13}$$

$$\kappa_2 = \nu\left[2 + (\nu + 2)\frac{\phi}{n}\right] \tag{11.14}$$

$$\kappa_3 = \nu\left[8 + 12(\nu + 2)\frac{\phi}{n} + 3(\nu^2 + 6\nu + 8)\left(\frac{\phi}{n}\right)^2\right] \tag{11.15}$$

and $\nu = n - 1$. The mean $\mu(h_0)$ and variance $\sigma^2(h_0)$ of the approximating normal distribution are, respectively, given by

$$\mu(h_0) = 1 + \frac{h(h-1)\kappa_2}{2\kappa_1^2} + \frac{h(h-1)(h-2)}{24\kappa_1^3}\left[4\kappa_3 + 3(h-3)\frac{\kappa_2^2}{\kappa_1}\right] \tag{11.16}$$

and

$$\sigma^2(h_0) = \frac{h^2\kappa_2}{\kappa_1^2} + \frac{h^2(h-1)}{2\kappa_1^3}\left[2\kappa_3 + (3h-5)\frac{\kappa_2^2}{\kappa_1}\right] \tag{11.17}$$

The approximation afforded by the power transformation family has been demonstrated by Chaubey et al. (2014b) to provide an excellent approximation for computing the distribution of the CV. They question *the numerical effort* in using the general transformation g_s *over the simplicity of the explicit formulae using the power family of transformations.* We reproduce their figures comparing these two approximations along with the normal approximations using the *VST* and the untransformed CV.

Figures 11.1 and 11.2 plot the distribution functions of $\hat{\phi}$ along with various approximations, where the exact values are computed using an integral formula outlined in Chaubey et al. (2014b). The box plots of the errors, as displayed on the right in each figure, clearly demonstrate that the normal approximations rendered by the untransformed statistic and the *VST* show the same performance whereas the symmetrizing transformation gives a significant improvement. Moreover, the power transformation even gives a better performance than the general symmetrizing transformation. Due to the simple nature of the power transformation and its accuracy in approximating the probabilities, we recommend its use in inference about the CV, which is further illustrated in the following section.

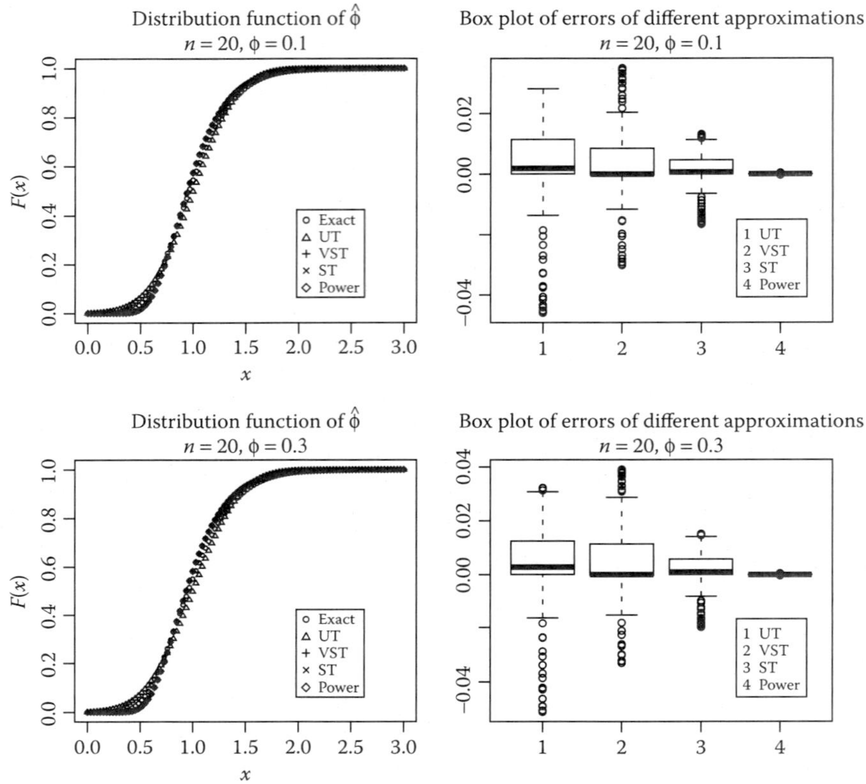

FIGURE 11.1 Normal approximation for various transformations, $n = 20$; UT: Untransformed, VST: Variance stabilizing transformation, ST: Symmetrizing transformation, Power: Power transformation.

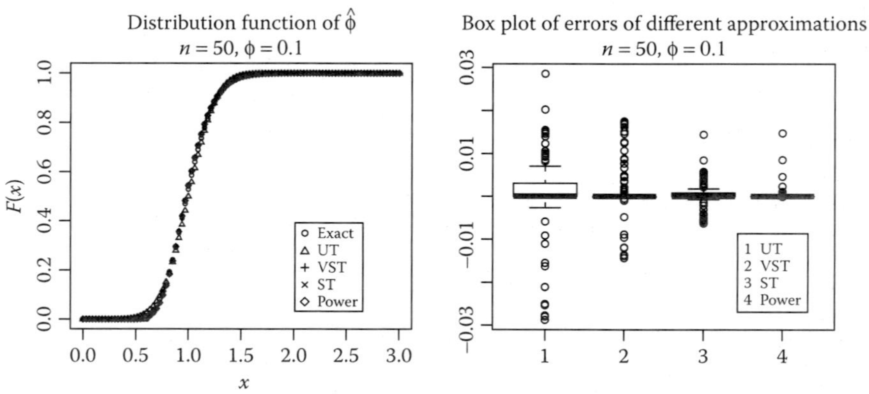

FIGURE 11.2 Normal approximation for various transformations, $n = 50$; UT: Untransformed, VST: Variance stabilizing transformation, ST: Symmetrizing transformation, Power: Power transformation.

(Continued)

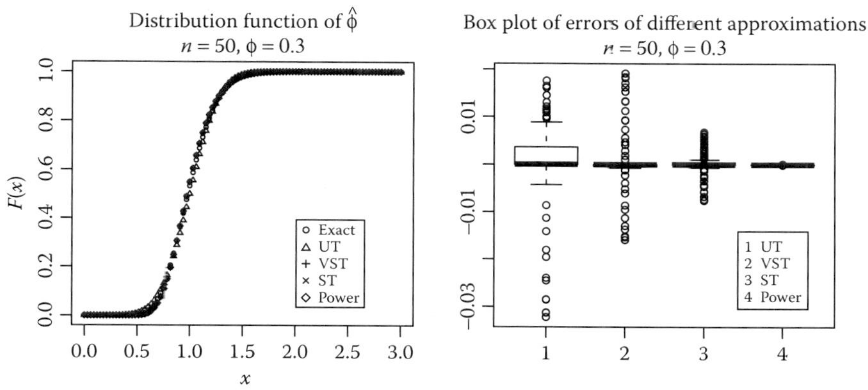

FIGURE 11.2 (Continued) Normal approximation for various transformations, $n = 50$; UT: Untransformed, VST: Variance stabilizing transformation, ST: Symmetrizing transformation, Power: Power transformation.

AN APPLICATION: COMPARING UNIFORMITY OF A GENOTYPE AT TWO OR MORE LOCATIONS

With the goal of estimating genotypic variability, 81 randomly selected genotypes from the LIEN were evaluated in square lattices with two replications, with a local check at several locations. However, the data from the predicted responses at two locations, New Delhi, India, and Larissa, Greece, during 2000, are considered here for the purpose of evaluation. The datasets on seed-yield response (kg/ha) evaluated as best linear-unbiased predicted values from the analysis of lattice experiment in each of the two environments are used here (Sarker et al. 2010).

The graphs in Figures 11.3 and 11.4 show the distribution fitted for 79 observations of a particular genetic material at two locations in Greece and India based on experiments conducted by International Center for Agricultural Research in the Dry Areas

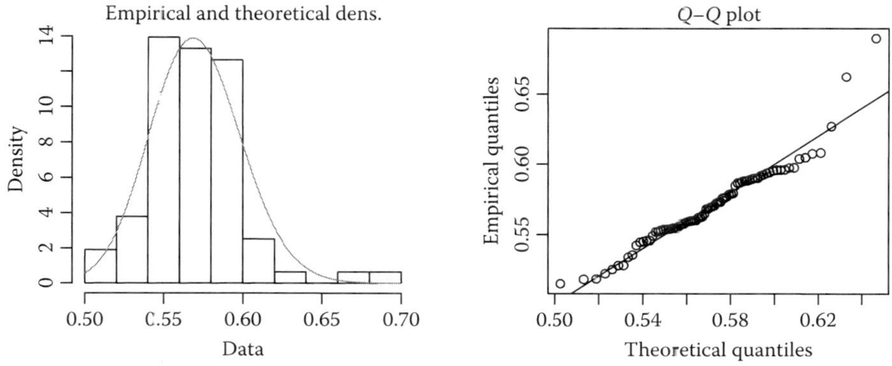

FIGURE 11.3 Distributional fit for the square root transformation of observations at Larissa, Greece. *(Continued)*

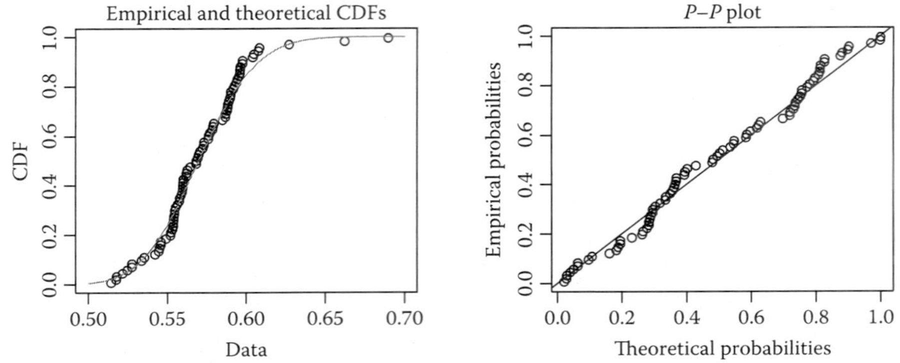

FIGURE 11.3 (Continued) Distributional fit for the square root transformation of observations at Larissa, Greece.

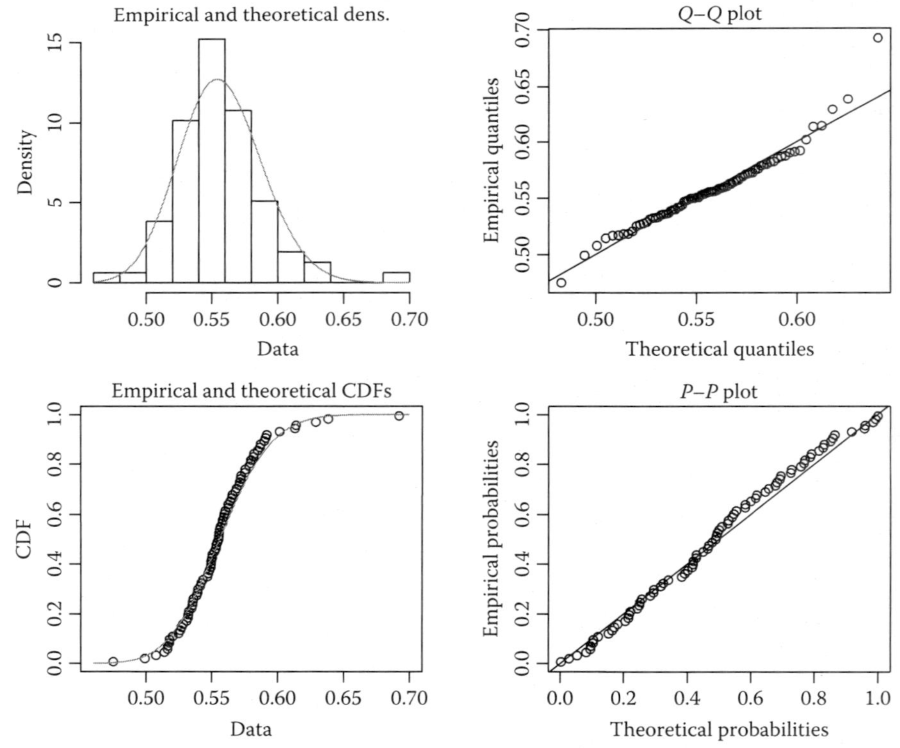

FIGURE 11.4 Distributional fit for the square root transformation of observations at New Delhi, India.

TABLE 11.1
Summary Statistics for the Two Samples

Location	Sample Size	Mean	Dispersion	$\hat{\phi}$
Greece	79	17.1386	0.003876	0.0664
India	79	19.5429	0.004316	0.0843

(ICARDA), where a square root transformation of the original observations is found to provide a good fit to the *IG* distribution.

Table 11.1 gives the maximum likelihood estimators of the mean and dispersion ($1/\lambda$) for the two samples.

The pooled value of $\hat{\phi}$ based on the rescaled samples by their means, thus based on 158 observations, is $\hat{\phi}=\left(\nu_1\hat{\phi}_1+\nu_2\hat{\phi}_2\right)/\left(\nu_1+\nu_2\right)=0.0753$. This gives the optimum power transformation for $h_0 = 0.3094$. We form the following test statistic to test the null hypothesis H_0: $\phi_1 = \phi_2$ that would be considered to signify the same diversity for the two accessions:

$$Z=\frac{\hat{\phi}_1^{h_0}-\hat{\phi}_2^{h_0}-\left[\mu_1(h_0)-\mu_2(h_0)\right]}{\hat{\phi}^{h_0}\sqrt{\sigma_1^2(h_0)+\sigma_2^2(h_0)}}$$

where $\mu_i(h_0)$ and $\sigma_i^2(h_0)$, i = 1,2, are obtained from Equations 11.16 and 11.17 with n replaced by n_i and ϕ replaced by the pooled value $\hat{\phi}$. Since the two sample sizes are equal, we have under the null hypothesis, $\mu_1(h_0) = \mu_2(h_0)$, and

$$\sigma_1^2(h_0)=\sigma_2^2(h_0)=0.002546864$$

Therefore, the observed value of the test statistic is given by

$$Z_{\text{obs}}=\frac{0.0664^{0.3094}-0.0843^{0.3094}}{0.0753^{0.3094}\sqrt{2\times 0.00254686}}=-1.0329$$

The corresponding two-sided p-value equals P = 0.3016, which is too high at the commonly used 5% level of significance, hence we accept the hypothesis H_0: $\phi_1 = \phi_2$ and conclude that the two populations are equally diverse and the diversity as indicated by the CV is not significantly affected by the environmental changes comprising the two geographical locations.

In terms of more than two locations involving large numbers of accessions as would be encountered when screening collections of genetic resources, we may study the performance of the power transformation used in the Wald test for homogeneity of the transformed parameters.

ACKNOWLEDGMENTS

We acknowledge the partial support of this research through a Discovery Grant from NSERC, Canada, and thank Dr. Abdallah Bari for useful suggestions on an earlier draft of the manuscript.

REFERENCES

Anscomb, F.J. 1948. The transformation of Poisson, binomial and negative-binomial data. *Biometrika* 35:246–254.

Banik, S. and B.M.G. Kibria. 2011. Estimating the population coefficient of variation by confidence intervals. *Commun Stat-Simul Comp* 40:1236–1261.

Bartlett, M.S. 1947. The use of transformations. *Biometrics* 3(1):39–52.

Bhatty, R.S. 1988. Composition and quality of lentil (*Lens culinaris* Medik.): A review. *Can I Food Sc Tech J* 21(2):144–160.

Chaubey, Y.P. and G.S. Mudholkar. 1983. *On the symmetrizing transformations of random variables* (Preprint). Montreal, Canada: Concordia University. http://spectrum.library.concordia.ca/973582/

Chaubey, Y.P., D. Sen, and K.K. Saha. 2014a. On testing the coefficient of variation in an inverse Gaussian population. *Stat Probab Lett* 90:121–128.

Chaubey, Y.P., M. Singh, and D. Sen. 2013. On symmetrizing transformation of the sample coefficient of variation from a normal population. *Commun Stat-Simul Comp* 42:2118–2134.

Chaubey, Y.P., M. Singh, and D. Sen. 2014b. *Symmetrizing and variance stabilizing transformations of sample coefficient of variation from inverse Gaussian distribution.* Technical Report. New Delhi, India: Indian Statistical Institute, Delhi Centre.

Chhikara, R.S. and J.L. Folks. 1977. The inverse Gaussian distribution as a lifetime model. *Technometrics* 19:461–468.

Chhikara, R.S. and J.L. Folks. 1989. *The inverse Gaussian distribution*. New York: Marcel Dekker.

Fisher, R.A. 1915. Frequency distribution of the values of the correlation coefficient in samples of an indefinitely large population. *Biometrika* 10(4):507–521.

Fisher, R.A. 1921. On the probable error of a coefficient of correlation deduced from a small sample. *Metron* 1:3–32.

Gupta, S.C. and M. Singh. 1992. Behaviour of correlation coefficient and its transforms from bivariate normal sample contaminated with bivariate normal outliers. *Sankhya Ser B* 54(2):184–199.

Hsieh, H.K. 1990. Inferences on the coefficient of variation of an inverse Gaussian distribution. *Commun Stat – Theor Meth* 19:1589–1605.

Jensen, D.R. and H. Solomon. 1972. A Gaussian approximation to the distribution of a definite quadratic form. *J Am Stat Assoc* 67:898–902.

Johnson, N.L., S. Kotz, and N. Balakrishnan. 1994. *Distributions in statistics: Continuous univariate distributions*, Vol. 1, 2nd Ed. New York: John Wiley & Sons.

Kocherlakota, S. and M. Singh. 1982a. On the behaviour of some transforms of the sample correlation coefficient: Discrete bivariate populations. *Commun Stat – Theor Meth* 11(18):2017–2043.

Kocherlakota, S. and M. Singh. 1982b. On the behaviour of some transforms of the sample correlation coefficient in samples from the bivariate t and the bivariate chi-square distributions. *Commun Stat – Theor Meth* 11(18):2045–2060.

Kumagai, S., I. Matsunaga, Y. Kusaka, and K. Takagi. 1996. Fitness of occupational exposure data to inverse Gaussian distribution. *Environ Model Assess* 1:277–280.

Laubscher, N.F. 1960. Normalizing the noncentral t and F-distributions. *Ann Math Stat* 31:1105–1112.

Mudholkar, G.S. and R. Natarajan. 2002. The inverse Gaussian models: Analogues of symmetry, skewness and kurtosis. *Ann I Stat Math* 54:138–154.

Mudholkar, G.S. and M.C. Trivedi. 1981. A Gaussian approximation to the distribution of the sample variance for nonnormal populations. *J Am Stat Assoc* 76:479–485.

Sarker, A., M. Singh, S. Rajaram, and W. Erskine. 2010. Adaptation of small-seeded red lentil (*Lens culinaris* Medikus subsp. *culinaris*) to diverse environments. *Crop Sci* 50:1250–1259.

Seshadri, V. 1993. *The inverse Gaussian distribution: A case study in exponential families.* Oxford: Clarenden Press.

Seshadri, V. 1998. *The inverse Gaussian distribution: Statistical theory and applications.* New York: Springer-Verlag.

Singh, M. 1993. Behavior of sample coefficient of variation drawn from several distributions. *Sankhya Ser B* 55:65–76.

Singh, M. and K. Hinkelmann. 1992. Distribution of genotypic correlation coefficient and its transforms for non-normal populations. *Sankhya Ser B* 54(1):42–66.

Takagi, K., S. Kumagai, I. Matsunaga, and Y. Kusaka. 1997. Application of inverse Gaussian distribution to occupational exposure data. *Ann Occup Hyg* 41:505–514.

12 Toward More Effective Discovery and Deployment of Novel Plant Genetic Variation

Reflection and Future Directions

M.C. Mackay, K.A. Street, and L.T. Hickey

CONTENTS

Humans began to domesticate plants and animals, and thus began to live with agriculture, around 10,000 years ago (Lee and DeVore 1968). It is only within the last 160 years, following the publication of naturalist Charles Darwin's *On the Origin of Species by Means of Natural Selection, or the Preservation of Favoured Races in the Struggle for Life* in 1859 and the subsequent rediscovery of Mendel's laws of inheritance around 1900, following his earlier reports of experiments with hybridizing peas (published 1866), that humankind has been able to move into its present manner of existence with modern agriculture and industrialized societies. The dawn of plant breeding as a science is recorded by Biffen (1905), who describes the late nineteenth-century and early twentieth-century beginnings of wheat breeding, whereby various traits were introgressed into new varieties by means of hybridization and selection. N.I. Vavilov, in his studies of the origin and geography of cultivated plants (Vavilov 1926) and agroecographical surveys of the main field crops (Vavilov 1957), identified patterns of linkages between adaptive traits in plants and the environment where they were endemic along with the need to start with the *right material* in the hybridization and selection breeding processes. Despite these advances and related knowledge being readily available, the investment in developing effective and efficient ways to identify parental material (possessing the novel alleles so necessary to breeding programs) appears miniscule in contrast to that invested in the collection, acquisition, and subsequent conservation of plant genetic resources (PGR) during the second half of the twentieth century. While calamities such as the Irish potato famine of the eighteenth century and the 1969–1970 epidemic of Southern corn leaf blight in North America vindicated the need for investing in PGR acquisition and conservation, there was still the need to ensure effective methods of discovery to be developed.

APPLYING AVAILABLE KNOWLEDGE TO PLANT IMPROVEMENT

By the mid-twentieth century, substantial collections of PGR had been established in recognition of their necessity as the source of novel genetic variation for breeding programs. This was particularly the case following a technical conference on the "Exploitation, Utilization and Conservation of Plant Genetic Resources" conducted by the Food and Agriculture Organization of the United Nations (FAO) (Bennett 1968). Acquisition through collection and the subsequent conservation of PGR was a growth industry for a period of some 25 years prior to the 1990s.

Access to data detailing the characteristics of PGR stored in *ex situ* genebanks has been (Frankel 1977), and still is (Khoury et al. 2010), accepted as a prerequisite to facilitating their use in plant breeding. Nevertheless, despite many years spent trying to enhance the use of global data standards (e.g., IPGRI/Bioversity Descriptors[*]) to facilitate access to and availability of such data, more deployment is significantly required. In recent times, the efforts of the Global Crop Diversity Trust[†] (GCDT), the Secretariat of the International Treaty on Plant Genetic Resources for Food and Agriculture[‡] (ITPGRFA), and most of the Consultative Group on International Agricultural Research[§] (CGIAR) centers joined forces to establish a single global portal (Genesys: https://www.genesys-pgr.org/welcome). At this site, genebanks can share relevant information about the accessions they conserve, and germplasm users are able to search for material for their research and breeding programs. Therefore, while having access to some data is far better than having no data, significant effort and investment is still required to encourage, enable, and/or convince genebanks, plant improvement scientists, and breeders to make all the data at their disposal available to the Genesys portal. As indicated above, data about accessions is a prerequisite to the effective identification and deployment of new genetic variation.

A literature search (Mackay 2011) revealed that, apart from systematically evaluating whole collections or chancing upon accessions with desirable alleles, no attempts to develop innovative and more effective methods of discovering new genetic variation among the millions of accessions that had been accumulated in genebanks could be found prior to the core collection concept proposed by Frankel (1984). The way in

[*] IPGRI/Bioversity International Descriptors: http://www.bioversityinternational.org/research-portfolio/information-systems-for-plant-diversity/descriptors/.

[†] GCDT: http://www.croptrust.org/.

[‡] ITPGRFA: http://www.planttreaty.org/.

[§] CGIAR: http://www.cgiar.org/.

which this approach addressed the discovery objective was to concentrate as much as possible on the genetic variation in the whole collection into a smaller (5%–10% of the whole) subset, thereby reducing the number of accessions to evaluate. There have been, however, numerous diversity studies later that mention the potential to improve utilization (e.g., Diwan et al. 1994, Bisht et al. 1998) rather than actually demonstrating the potential or describing effective new methods of identifying novel genetic variation.

EFFECTIVE DISCOVERY AND DEPLOYMENT OF PGR

Earlier, in describing how wheat breeders and others used the Australian Winter Cereals Collection (AWCC), Mackay (1986) introduced the concept of the *predictive* value of some descriptors. A combination of geographic origin, growth habit, grain color, ear emergence, and plant height are few examples of descriptors that could be used to infer the type of environment in which an accession might have evolved or undergone passive and/or active selection. At the time, the AWCC was considering two options to enhance utilization:

1. A *predictive* approach using some descriptors to establish a small *core* of accessions
2. A co-ancestry sieve to eliminate closely related genotypes from evaluation studies

This concept was further progressed by Mackay (1990), who forecast that the potential of the predictive approach "will not be fully realized until more suitable tools for linking environments to accessions are developed, such as a database which cross-references soil types and climate with geographic regions" (p. 22). A simple example was given to demonstrate how a small, trait-specific, *targeted* subset of 42 accessions was developed to successfully identify Russian wheat aphid (*Diuraphis noxia*) resistance in a hexaploid bread wheat (*Triticum aestivum* var. *vavilovii* Jakubz and Puchalski).

It was from these unpretentious beginnings that the focused identification of germplasm strategy (FIGS) emerged as an approach to target the specific and novel genetic variation required by breeders to overcome contemporary challenges to plant productivity. The principles behind developing a FIGS subset involved

- Collating the available information to understand better the specific trait and/or trait variant sought, including if and where previously effective instances of expression were discovered.
- Drawing feasible inferences from other descriptors where possible.
- Using any other available tools such as GIS and statistics to eliminate accessions of no interest and/or identify accessions more likely to possess the targeted expression.

Crucial to this approach being effective, or actually identifying a subset of accessions that actually contains the required genetic variation, is a partnership between the genebank and the plant breeder to establish the fundamental inferences and assumptions. That is, to determine and use any relationship(s) between the desired trait and other available relevant information on which the search for the *best bet* accessions is based. The more data that are available about the conserved accessions, the more successful one could expect the outcome to be. Clearly, the more inferences required to construct the FIGS subset, the larger number of accessions that would be required. Conversely, the more relevant information available and the greater the knowledge of the trait concerned, the richer one could expect the expression of the desired trait within the FIGS subset, the latter being the more efficient outcome in terms of numbers of accessions, cost of verification, and time involved to identify and deploy the newly identified allele(s).

FIGS IN ACTION

Thus, with the shift of thought from the general utilization of PGR toward how to actually identify the parental material required by contemporary breeding programs, came the need to research and develop the processes involved in selecting germplasm for evaluation. This is expected to lead toward practicable solutions to identify more effectively novel variation in less adapted material such as landraces and crop wild relatives (CWR).

The first serious attempt to apply such a strategy was a project with the objective of developing a subset of bread wheat landrace accessions that concentrated traits associated with drought tolerance. Supported by the Grains Research and Development Corporation (GRDC, Canberra, Australia), the project entitled "Technologies for the targeted exploitation of the N I Vavilov Institute of Plant Industry (VIR), ICARDA and Australian bread wheat landrace germplasm for the benefit of the wheat breeding programs of the partners" commenced in 2002. The partners undertaking this research were the International Center for Agricultural Research in the Dry Areas (ICARDA), VIR, and the AWCC. Determining the geo-references of some 16,500 bread wheat landraces held by the three partners prior to using several statistical methods, together with GIS technologies to link PGR and agroclimatic data with the geo-references, a subset of some 700 accessions was selected as being more likely to possess drought-tolerant traits. As a control, another subset of a similar number of accessions was selected using published core collection construction methodologies. It was at this time, in 2003, that the approach was coined "the focused identification of germplasm strategy." A comparison of the FIGS drought set and the core collection was made and the following conclusions were drawn:

1. Landrace accessions in the FIGS drought set originated from within a distinct range of latitudes ranging from North Africa through the Middle East to Central Asia. By contrast, the core collection accessions were from locations representing the full range of the source material.
2. Principal component analysis was conducted to gain a comprehensive picture of climatic influences in the environments from which the accessions in this study were collected. In a plot of the first two principal components, the first component explained 47% of the variation between sites, while the

> second component explained a further 21% of the total variation. This plot clearly illustrated that the core collection is composed of accessions from sites that are influenced by a wide range of climates, from lower temperatures and higher precipitation to more arid environments. By contrast, the FIGS drought set is from environments predominantly experiencing lower precipitation, higher temperatures, higher potential evapotranspiration, and that are generally more arid.

A general conclusion drawn was that the chances of finding novel genetic variation for traits conferring adaptation to hot, dry conditions would be higher in the FIGS drought set rather than in a core collection; further details of this early application of the FIGS approach is available in Mackay (2011).

A subsequent application of FIGS to the identification of new variation for the disease powdery mildew in wheat (*Blumeria graminis* f. sp. *tritici*) was undertaken in collaboration with the University of Zurich and the USDA-ARS National Small Grain Collection (Aberdeen, Idaho), the latter supplying evaluation data relating to the reaction of numerous bread wheat landraces to a suite of powdery mildew pathogen races. Through a process of profiling the environmental characteristics of sites where tolerant or resistance had previously been identified to the range of pathogen races and identifying similar environmental profiles among the 16,500 source landrace accessions, some 1300 were sampled for further evaluation. These were subsequently evaluated for their reaction to a number of known powdery mildew isolates based on their pathogenicity to known alleles from the *Pm3* locus, and 211 accessions were found to be either resistant or have intermediate resistance to at least one of the isolates used. Molecular characterization allowed the identification of 111 candidate accessions for which isolation of *Pm3* alleles was carried out only on the resistant landrace accessions, and coding sequences were amplified from 45 of these before being cloned and sequenced. Further molecular analyses identified seven novel functional *Pm3* alleles (Bhullar et al. 2010). In this case, molecular techniques were added to those previously used to facilitate the effective identification of parental material for breeding purposes.

Bari et al. (2012, 2014) used a modeling framework to define, within a nonlinear framework, the potential trait–environment relationship, where recursive learning techniques are used to overcome the problem of restrictive parametric paradigms, on the one hand, and the prerequisite distribution assumptions on the other (Drake et al. 2006). This study agrees with the work done by Endresen et al. (2011, 2012), who used other nonlinear models to show the relationship between disease resistance and long-term climatic conditions at collection sites of genebank accessions. The contribution of modeling approaches is clearly established where evaluation data exist for adaptive traits.

Further examples of the assembly of FIGS sets targeting biotic and abiotic traits are readily available in the literature. These include Sunn pest (El-Bouhssini et al. 2009, 2012) and Russian wheat aphid (El-Bouhssini et al. 2011, 2012) in wheat, drought resistance traits in faba bean (Khazaei et al. 2013), and net-blotch in barley

(Endresen et al. 2011). Another example, dubbed as *predictive characterization*, is described by Thormann et al. (2014), whereby the FIGS approach has been applied to minor crop landraces and CWR, in contrast to those of major crops.

CURRENT AND FUTURE DEVELOPMENTS

A number of areas whereby the FIGS approach, which in many respects can still be considered as in its formative stage, could be enhanced so as to yield targeted subsets of accessions with even greater effectiveness, and with greater efficiency, are currently being addressed or under consideration. These include sourcing and facilitating the availability of further data categories (including genetic data), adding to the classes of environmental data and improving their resolution, capturing the georeferences of collected material in existing collections, aligning climatic data with crop-growing seasons, and developing online tools, which guide breeders through the procedures involved in developing customized FIGS subsets.

Classes of environmental data that could be added to the FIGS arsenal include anecdotal data such as breeder or farmer information about crop attributes, scalar data such as aspect or slope, binary data such as irrigated or nonirrigated, and percentage data such as cloud cover. To date, the environmental data used to assist in choosing accessions have been limited to the WorldClim (http://www.worldclim.org/) and ICARDA databases along with a few purpose-developed continuous surface layers of data generated by ICARDA scientists. These data are available at different spatial resolutions, ranging from just under 1 km^2 to over 340 km^2 at the equator. Obviously, the greater the resolution the more effective one could expect the resulting FIGS subset to be. The use of daily data averages, rather than monthly averages, would also improve the effectiveness as would the alignment of the data harvested with the approximate growing season of the species involved to permit the use of within-season parameters so as to choose accessions for specific attributes at specific developmental stages. For example, one could be targeting frost tolerance during the grain-fill period or a specific temperature-humidity regime at a time critical to the spread of a fungal pathogen. While ICARDA is currently addressing some of these challenges, the resources to address others need to be obtained.

Initially, FIGS sets were largely developed as a result of the research interests of PGR scientists and the relationship some of these had with the user community. However, the interests of the user community have become increasingly more influential with regard to the specific traits being targeted and the availability of resources to undertake the development of FIGS subsets. For various reasons, since PGR scientists are usually so focused on in-house responsibilities, such as managing conservation and distribution activities, they rarely possess the time or resources to participate in utilization. Consequently, the development of online tools and decision support software to enable pre-breeders and/or breeders to develop custom FIGS subsets addressing the specific traits their particular breeding situation necessitates, would increase the effectiveness of mining PGR at a lower cost. By linking such an online resource with the source(s) of information required to construct the FIGS subsets, even greater efficiencies are possible. Such sources could be Genesys, the global portal to PGR information in support of the Global Information System (http://www.planttreaty.org/content/gis) of the ITPGRFA, as well as DivSeek, the Diversity Seek Initiative (http://www.divseek.org/).

FIGS IN THE PLANT IMPROVEMENT DOMAIN

There are at least three distinct but interrelated phases to the plant improvement domain:

1. *Conservation*: The acquisition, conservation, and management of PGR.
2. *Discovery*: The identification and selection of suitable parental material.
3. *Breeding*: The incorporation of desirable genetic traits into new cultivars while embodying the necessary agronomic background for the intended production environment.

While advances can be made in each phase, we have demonstrated here that investment in the neglected discovery phase offers significant untapped opportunities for streamlining and effectiveness. This is despite Vavilov's posthumous publication (1957) clearly demonstrating the linkages between adaptive traits and environmental conditions, along with the need to start with the right material in the hybridization and selection process. There is evidence that the investment in the conservation phase is being addressed, at least partly, through the development and deployment of facilitating information systems, such as Genesys and DivSeek, mentioned above.

We see significant potential for more strategic investment in the breeding phase and outline a practical pathway here. Plant breeding can be considered a victim of its own success; the past 100 years of successful breeding has resulted in significant yield improvements, but has resulted in a bottleneck in terms of the genetic diversity available in elite germplasm. Thus, opportunities for further genetic improvement within modern germplasm pools are reaching a limit (Haudry et al. 2007, White et al. 2008). To overcome this plateau, breeders need to explore alternative sources of genetic diversity. As outlined above, FIGS provides a powerful strategy to accelerate the identification of useful genetic diversity. However, a degree of pre-breeding is often required before this diversity will be accepted by a breeding program. Accessions such as landraces or progenitor species often carry highly desirable traits of interest, but lack quality characteristics preferred by industry and are unadapted to the target environment. Pre-breeding is performed by public organizations such as universities or national or state government research bodies, or sometimes conducted by breeders separate to their core breeding activities. Pre-breeding efforts generally focus on three main objectives to enable breeders to introduce and manipulate new genes more effectively in their breeding material:

1. Understanding the genetic control of the trait and developing associated DNA markers
2. Improving the adaptation package of the trait of interest
3. Validating the trait of interest in backgrounds relevant to breeding

Traditionally, pre-breeders employed bi-parental quantitative trait locus (QTL) mapping to dissect the genetic control of a trait and to identify DNA markers linked to the gene(s) of interest. This involves constructing a mapping population derived from two parents: one that carries the trait of interest and another that lacks the trait. The population is then phenotyped and genotyped and a map is constructed. A statistical analysis can be performed to position the potential QTLs for the trait of interest. This process is time consuming. By the time populations are developed and adequate phenotype data are collected across 2 or 3 years, this initial phase can take 5 or more years. However, following QTL discovery, a validation step is often required before breeders are willing to use the newly identified markers. A validation approach often involves backcrossing the QTL of interest into several cultivars, to evaluate trait expression in different genetic backgrounds that are of relevance to the breeders. This can take an additional 3–5 years, making the entire pre-breeding process a lengthy one, often exceeding 10 years.

The study of isolated bi-parental populations is considerably expensive, as it requires the development and evaluation of a purpose-built population that has very little commercial value. Further, the discoveries made in a pre-breeding bi-parental population are often limited to a restricted germplasm pool and lack relevance to actual breeding populations. It is important to keep in mind that there is no *absolute* relative merit value for a gene or trait or genotype. The value of a gene always depends on the context. Such context dependencies occur at all scales and generate major challenges for crop improvement programs (Cooper et al. 2009).

To enhance the efficiency and effectiveness of pre-breeding efforts, mapping populations can be genetically connected to one another or to breeding populations. A framework, such as nested-association mapping (NAM), could provide a powerful platform for dissecting the genetics of target traits. NAM is a recently developed strategy for plant genetic analysis that combines advantages of defined population structure, as used for traditional linkage analysis approaches, with the power of association mapping to encapsulate high levels of genetic diversity and exploit historic recombination. To develop a platform for NAM, a panel of *donor* accessions is crossed to a *reference* accession. Small sub-populations of recombinant inbred lines (RILs) are then developed from each donor–reference combination. Detailed understanding of the genetic relationship of RILs within a NAM platform, combined with extensive allelic diversity contributed by donor lines, provides great power to dissect the genetic architecture of complex traits. NAM was originally conceived by Buckler et al. (2009), and it has been most extensively applied in maize. This strategy has proven highly effective in identifying even small-effect QTLs, for flowering time (Buckler et al. 2009), kernel composition (Cook et al. 2012), and disease resistance (Kump et al. 2011). NAM has been rapidly recognized as a broadly relevant approach, and platforms are in development for other major crop species, including barley and soybean (Guo et al. 2013).

Typically, the reference accession used in a NAM population is a cultivar or elite breeding line widely adapted to the target population environment or preferred by industry. During RIL development, some selection can be applied for plant height and maturity. This can improve the quality of lines and provide an *appropriate agronomic window* to examine the traits of interest. In this way, genetic diversity can be evaluated within a context more relevant to breeding. The elimination of tall and late maturity types also improves the precision of phenotyping for some traits.

A NAM population is considered an *immortal resource* that can be repeatedly evaluated for limitless traits and can be expanded for genetic diversity over time. For instance, as new sources of genetic diversity are identified, they can be crossed into the NAM resource for genetic analysis. This flexible framework is a real advantage over other strategies, such as multiparent advanced generation inter-cross (MAGIC; Huang et al. 2012), where founders are fixed and genetic diversity cannot be expanded once population development is finalized. Adoption of multiple adapted reference accessions in a NAM population can also permit validation of QTL effects in multiple genetic backgrounds. This can significantly reduce the timeframe for pre-breeding, where both the discovery and validation phases are performed simultaneously.

In addition to linking genetic mapping populations, integration across traits would also speed up deployment in new cultivars. If a coordinated effort across traits were employed, it would be possible to combine trait packages, making the material more valuable and desirable from a breeding perspective. If pre-breeders can deliver trait packages (e.g., tolerance to frost and drought), it would reduce the number of breeding cycles required to combine these traits in elite material. Again, a NAM pre-breeding framework would help facilitate delivery of germplasm incorporating *trait packages*, rather than a single trait. Negative or positive trait interactions could also be detected and elite genotypes that carry multiple traits could be more efficiently identified.

For inbred cereal crops such as wheat, barley, and oats, breeding programs typically perform one generation of crossing and four-to-six generations of self-pollination before seed increase and yield testing. With off-season nurseries, this process takes a minimum of 3 years, limiting rates of genetic gain for yield. Doubled haploids can shorten the time required to begin yield testing; however, doubled haploid technology is costly, requires specialized facilities, only works for selected germplasm, and precludes early generation selection for high-heritability traits. A new method for rapid generation advance, called *speed breeding*, has considerable advantages for spring wheat, because it provides increased recombination during line development and enables selection in early generations for some traits. The technique utilizes controlled temperature regimes and 24-h light to accelerate plant growth and development (Hickey et al. 2012). The low-cost management system enables up to six plant generations of wheat annually, just like *Arabidopsis*. The speed-breeding technology has been successfully applied to other crops, including canola, barley (Hickey et al. 2011), and groundnuts (O'Connor et al. 2013). This provides a useful tool to accelerate backcrossing of traits into adapted germplasm or development of NAM populations for pre-breeding studies. Moreover, speed breeding has been integrated with phenotypic screening for some target traits in wheat, such as grain dormancy and rust resistance (Hickey et al. 2009, 2010, 2012). Rapid generation advance coupled with phenotypic screening provides an opportunity to rapidly backcross desirable traits without the need for population or marker development.

CONCLUSION

Substantive investments have been made in many aspects of the plant improvement domain; however, it is advocated here that the discovery phase, where new genetic variation is identified as parental material, has largely been neglected and opportunities to redress this are outlined. Furthermore, within the *big picture* of plant improvement, a far more integrated strategy is required to realize fully the effectiveness and efficiency of plant improvement from the *conservation*, through the *discovery*, to the *breeding* phases. The primary basis for conserving plant genetic resources is to facilitate their access, use, and deployment as improved cultivars. This can only be effectively and efficiently achieved by means of making information about the characteristics of PGR available for the discovery phase and the development of methods to rapidly identify sought-after genetic variation, such as the FIGS approach. Additional gains in the successful, economical, and timely introgression of such genetic variation are also available via the use of innovative pre-breeding strategies. A scheme of integrated investment across the plant improvement domain would facilitate the involvement of all stakeholders, thereby returning greater efficacy in outcome and cost to all stakeholders and investors.

REFERENCES

Bari, A., A. Amri, K. Street et al. 2014. Predicting resistance to stripe (yellow) rust (*Puccinia striiformis*) in wheat genetic resources using focused identification of germplasm strategy. *J Agr Sci Cambridge* 152(6):906–916. doi:10.1017/S0021859613000543.

Bari A., K. Street, M. Mackay, D.T.F. Endresen, E. De Pauw, and A. Amri. 2012. Focused identification of germplasm strategy (FIGS) detects wheat stem rust resistance linked to environmental variables. *Genet Resour Crop Ev* 59:1465–1481. doi:10.1007/s10722-011-9775-5.

Bennett, E.E. 1968. *Record of the FAO/IBP Technical Conference on the Exploration, Utilization and Conservation of Plant Genetic Resources.* Rome, Italy, September 18–26, 1967. Food and Agriculture Organization of the United Nations. Rome, Italy: FAO.

Bhullar, N.K., Z. Zhang, T. Wicker, and B. Keller. 2010. Wheat gene bank accessions as a source of new alleles of the powdery mildew resistance gene *Pm3*: A large scale allele mining project. *BMC Plant Biol* 10:88. doi:10.1186/1471-2229-10-88.

Biffen, R.H. 1905. Mendel's laws of inheritance and wheat breeding. *J Agr Sci Cambridge* 1:4–48. doi:10.1017/S0021859600000137.

Bisht, I.S., R.K. Mahajan, and D.P. Patel. 1998. The use of characterisation data to establish the Indian mungbean core collection and assessment of genetic diversity. *Genet Resour Crop Ev* 45:127–133.

Buckler, E.S., J.B. Holland, P.J. Bradbury et al. 2009. The genetic architecture of maize flowering time. *Science* 325(5941):714–718.

Cook, J.P., M.D. McMullen, J.B. Holland et al. 2012. Genetic architecture of maize kernel composition in the nested association mapping and inbred association panels. *Plant Physiol* 158:824–834.

Cooper, M., F.A. van Eeuwijk, G.L. Hammer, D.W. Podlich, and C. Messina. 2009. Modeling QTL for complex traits: Detection and context for plant breeding. *Curr Opin Plant Biol* 12:231–240.

Diwan, N., G.R. Bauchan, and M.S. McIntosh. 1994. A core collection for the United-States annual medicago germplasm collection. *Crop Sci* 34:279–285.

Drake, J.M., C. Randin, and A. Guisan. 2006. Modelling ecological niches with support vector machines. *J Appl Ecol* 43:424–432.

El-Bouhssini, M., F. Ogbonnaya, M. Chen, S. Lhaloui, F. Rihawi, and A. Dabbous. 2012. Sources of resistance in primary synthetic hexaploid wheat (*Triticum aestivum* L.) to insect pests: Hessian fly, Russian wheat aphid and Sunn pest in the fertile crescent. *Genet Resour Crop Ev* 60:621–627.

El-Bouhssini, M., K. Street, A. Amri et al. 2011. Sources of resistance in bread wheat to Russian wheat aphid (*Diuraphis noxia*) in Syria identified using the focused identification of germplasm strategy (FIGS). *Plant Breed* 130:96–97.

El-Bouhssini, M., K. Street, A. Joubi, Z. Ibrahim, and F. Rihawi. 2009. Sources of wheat resistance to Sunn pest, *Eurygaster integriceps* Puton, in Syria. *Genet Resour Crop Ev* 56(8):1065–1069.

Endresen, D.T.F., K. Street, M. Mackay, A. Bari, and E. De Pauw. 2011 Predictive association between biotic stress traits and ecogeographic data for wheat and barley landraces. *Crop Sci* 51:2036–2055.

Endresen, D.T.F., K. Street, M. Mackay et al. 2012. Sources of resistance to stem rust (Ug99) in bread wheat and durum wheat identified using focused identification of germplasm strategy. *Crop Sci* 52:764–773.

Frankel, O.H. 1977. Genetic resources. *Ann NY Acad Sci* 287:332–344.

Frankel, O.H. 1984. Genetic perspectives of germplasm conservation. In *Genetic manipulation: Impact on man and society*, eds. W. Arber, K. Illmensee, W.J. Peacock, and P. Starlinger, 161–170. Cambridge: Cambridge University Press.

Guo, B., D. Wang. Z. Guo, and W. Beavis. 2013. Family-based association mapping in crop species. *Theor Appl Genet* 126(6):1419–1430.

Haudry, A., A. Cenci, C. Ravel et al. 2007. Grinding up wheat: A massive loss of nucleotide diversity since domestication. *Mol Biol Evol* 24:1506–1517.

Hickey, L.T., M.J. Dieters, I.H. DeLacy, M.J. Christopher, O.Y. Kravchuk, and P.M. Banks. 2010. Screening for grain dormancy in segregating generations of dormant × non-dormant crosses in white-grained wheat (*Triticum aestivum* L.). *Euphytica* 172:183–195.

Hickey, L.T., M.J. Dieters, I.H. DeLacy, O.Y. Kravchuk, D.J. Mares, and P.M. Banks. 2009. Grain dormancy in fixed lines of white-grained wheat (*Triticum aestivum* L.) grown under controlled environmental conditions. *Euphytica* 168:303–310.

Hickey, L.T., W. Lawson, G.J. Platz et al. 2011. Mapping *Rph20*: A gene conferring adult plant resistance to *Puccinia hordei* in barley. *Theor Appl Genet* 123:55–68.

Hickey, L.T., P.M. Wilkinson, C.R. Knight et al. 2012. Rapid phenotyping for adult plant resistance to stripe rust in wheat. *Plant Breed* 131:54–61.

Huang, B.E., A.W. George, K.L. Forrest et al. 2012. A multiparent advanced generation intercross population for genetic analysis in wheat. *Plant Biotech J* 10(7):826–839.

Khazaei, H., K. Street, A. Bari, M. Mackay, and F.L. Stoddard. 2013. The FIGS (focused identification of germplasm strategy) approach identifies traits related to drought adaptation in *Vicia faba* genetic resources. *PLoS ONE* 8(5):e63107. doi:10.1371/journal.pone.0063107.

Khoury, C., B. Laliberté, and L. Guarino. 2010. Trends in *ex situ* conservation of plant genetic resources: A review of global crop and regional conservation strategies. *Genet Resour Crop Ev* 57:625–639.

Kump, K.L., P.J. Bradbury, R.J. Wisser et al. 2011. Genome-wide association study of quantitative resistance to southern leaf blight in the maize nested association mapping population. *Nat Genet* 43:163–168.

Lee, R. and I.E. DeVore. 1968. *Man the hunter*. Chicago, IL: Aldine Publishing.

Mackay, M.C. 1986. Utilizing wheat genetic resources in Australia. In *Proceedings of the 5th assembly, Wheat Breeding Society of Australia*, ed. R. McLean, 56–61. Perth, Australia: Western Australian Department of Agriculture.

Mackay, M.C. 1990. Strategic planning for effective evaluation of plant germplasm. In *Wheat genetic resources: Meeting diverse needs*, eds. J.P. Srivastava and A.B. Damania, 21–25. Chichester, UK: John Wiley & Sons.

Mackay, M.C. 2011. Surfing the genepool: The effective and efficient use of plant genetic *resources*. Acta Universitatis agriculturae Sueciae Doctoral Thesis No. 2011:90.

O'Connor, D.J., G.C. Wright, M.J. Dieters et al. 2013. Development and application of speed breeding technologies in a commercial peanut breeding program. *Peanut Sci* 40:107–114.

Thormann, I., M. Parra-Quijano, D.T.F. Endresen, M.L. Rubio-Teso, M.J. Iriondo, and N. Maxted. 2014. *Predictive characterization of crop wild relatives and land races. Technical guidelines version 1.* Rome, Italy: Bioversity International.

Vavilov, N. 1957. *Agroecological survey of the main field crops.* Moscow, Russia: The Academy of Sciences of the USSR.

Vavilov, N.I. 1926. Studies on the origin of cultivated plants (Russian with English summary). *Bull Appl Bot Plant Breed (Leningrad)* 16(2):139–248.

White, J., J.R. Law, I. MacKay et al. 2008. The genetic diversity of UK, US and Australian cultivars of *Triticum aestivum* measured by DArT markers and considered by genome. *Theor Appl Genet* 116:439–453.

13 Identifying Climate Patterns during the Crop-Growing Cycle from 30 Years of CIMMYT Elite Spring Wheat International Yield Trials

Z. Kehel, J. Crossa, and M. Reynolds

CONTENTS

Earth has experienced a series of warm and cold periods throughout its history, with climate and temperature changes showing temporal and spatial distribution. Over the past 100 years, Earth's average surface temperature increased by about 0.8°C (1.4°F) and the rate of temperature increase accelerated toward the end of that time frame. Climate model projections by the Intergovernmental Panel on Climate Change

(IPCC) indicate that during the twenty-first century, global surface temperature is likely to rise a further 1.1°C–2.9°C (2°F–5.2°F) for their lowest emissions scenario, and 2.4°C–6.4°C (4.3°F–11.5°F) for their highest. The range of these estimates arises from the use of models with differing sensitivity to greenhouse gas concentrations.

Rising temperatures due to increasing greenhouse gas concentrations have produced distinct warming patterns on the Earth's surface, with greater warming over most land areas and significant seasonal differences in warming in the Arctic. For example, during the second half of the twentieth century, there was intense winter warming across parts of Canada, Alaska, northern Europe, and Asia, while summer warming was particularly strong across the Mediterranean region, the Middle East, and various other places, including parts of western United States. The incidence of heat waves and record high temperatures has increased across most regions of the world, while the incidence of cold snaps and record cold temperatures has decreased.

Crop production is naturally sensitive to climate variability as plants tend to grow more quickly at higher temperatures, leading to shorter growing periods with less time to produce biomass and grains. The changing climate will also bring other hazards, including greater water stress and the risk of higher temperature peaks that can quickly damage crops (National Academy of Sciences 2012: www.nas.edu/climatechange). Some of the early studies on the impact of climate change on crop production highlighted that changes in crop development at warmer temperatures are important in determining the impact of climate change on crop yield. For example, wheat yield declined by approximately 5%–8% (Wheeler et al. 1996) or 10% (Mitchell et al. 1993) per 1°C rise in mean seasonal temperature.

Wheat is the most widely grown crop in terms of area harvested and provides nearly 20% of the calories consumed by humans, making it imperative to study the impact of climate on wheat production. A recent study by Gourdji et al. (2013) indicated that wheat yield exhibited the most sensitivity to warming during grain-filling and that 95% of International Center for Maize and Wheat Improvement (CIMMYT)'s international wheat nursery locations would have a lower mean yield due to an increase in temperature of 2°C; however, actual climate projections with regional and seasonal variations were not used in that study. There is also strong evidence that the wheat growth cycle is shortening and that, as a result, grain yield is reduced (Mitchell et al. 1993). The impact of climate change on crop productivity will be greatly influenced by how climate affects the rate of crop development and, hence, the duration of crop growth.

Flowering is a critical stage of wheat development because seed number is determined during that stage. Wheat is also particularly sensitive during this period to temperature and water stress, and to biotic constraints (Curtis 1968). For example, in many annual crops, brief periods of high temperatures can greatly reduce seed set and, hence, crop yield, if they coincide with a brief critical period of only 1–3 days around flowering time (Matsui et al. 1997, Vara Prasad et al. 2000, Wheeler et al. 2000, Jagadish et al. 2008). Controlling crop development stages will be critical to reduce the impact of climate change on Yield. This can be done through the determination of season length and hence the availability of solar radiation, water, and nutrients for growth and by changing the exposure of the crop to climate extremes.

Adaptation to moderate changes in climate that influence temperature, season length, and planting dates, as well as the occurrence of abiotic stress, can be achieved by selecting varieties with appropriate flowering time and crop duration (Ludlow and Muchow 1990, Richards 2006). The life cycle and phenology of landraces and new cultivars have been very successfully selected/manipulated to maximize the range of environments in which crops as well as their yields grow (Evans 1993, Roberts et al. 1996), at least in current climates. A major challenge for plant breeders is how to plan for future climate change.

Vegetation phenological seasons have been shown to change spatially and temporally in response to trends in climate change across the Northern Hemisphere (Myneni et al. 1997, Menzel et al. 2001, Zhang et al. 2004). Nevertheless, some growth stages of annual crops, such as heading, flowering, and maturity date, were found to be strongly affected by weather and climate (Chmielewski et al. 2004, Hu et al. 2005, Tao et al. 2006, Peltonen-Sainio and Rajala 2007, Estrella et al. 2009, Sacks and Kucharik 2011, Siebert and Ewert 2012). Length of growing period showed a decreasing trend in most of the studied regions of China. They also found that mean temperature during the growing period was increasing significantly. Mean day length during the vegetative growing period showed a decreasing trend in most of the studied stations, resulting from the delay of sowing date and/or advancement of heading date. Furthermore, thermal requirements during grain-filling increased significantly (Tao et al. 2012).

Weather and climatic data can exhibit important spatial and temporal correlations that can be assessed using different statistical methodologies. For example, principal component analysis (PCA), which has been utilized extensively in meteorology, has revealed important features such as climate patterns and variability. However, problems encountered when applying PCA to climate disciplines are whether or not the spatial patterns extracted by PCA are in real physical mode, and whether the extracted principal components (PCs) can be interpreted as biophysical entities. In meteorology, data used for analysis are often described as spatial data and/or a combination of many variables; therefore, multivariate techniques such as PCA have been widely used.

Since weather data are generally extensive, they can be mined for occurrence of particular patterns that distinguish specific weather phenomena. It is therefore possible to extract spatio-temporal signals using data mining techniques such as the self-organized map (SOM), which is a particular application of artificial neural networks (Kohonen 1995). Currently, the SOM is often used as a statistical tool for multivariate analysis, because it is both a projection method that maps high-dimensional data to low-dimensional spaces and a clustering and classification method that orders similar data patterns into neighboring SOM units. SOM has gained popularity in the meteorology and oceanography communities as a powerful pattern-recognition and feature-extraction method and has been applied to a variety of datasets in those fields (Liu and Weisberg 2011).

The capability of SOM for detecting new patterns while keeping pre-existing patterns suggests that it would be useful for investigating climate change. This capability allows SOM to compare past patterns to new patterns and to analyze the changes that have occurred from the past to the present. If the patterns associated with climate change are orthogonal to the dominant pattern in the past, PCA would be able to extract new climate patterns as PCs; however, patterns associated with climate change are not likely to be orthogonal to past climate patterns, and it will take a long time for the new patterns to occupy a significant amount of the total variance. The use of SOM for extracting past and new patterns in climate data and for comparing the extracted patterns is likely to provide useful information.

In this chapter, we present an investigation of spatio-temporal changes in the climate profiles of CIMMYT Elite Spring Wheat International Yield Trials (ESWYT) locations since 1986 during three wheat physiological stages: vegetative, reproductive, and grain-filling. The main output of this research was an assessment of the shift in global climatic variables over the last three decades during wheat growth stages using 954 site × year combinations from 1986 to 2009 and collecting 22 climatic variables during the vegetative, reproductive, and grain-filling periods. This provided two main outcomes: first, an improved understanding of when and where most climate change occurred in major wheat-growing regions, and second, identification of climate analog sites that will help wheat scientists plan and develop specific breeding programs for different regions of the world. In this study, we applied three different statistical methods to identify climate patterns at ESWYT locations: regression and spatial statistics for assessing possible spatial and temporal trends, PCA for reducing highly dimensional data, and a nonlinear neural network or SOM that represents highly dimensional patterns in a reduced two-dimensional map.

MATERIALS AND METHODS

DATASETS

In this study, 954 site × year combinations from 1986 to 2009 were used. The number of locations per year fluctuated from 22 locations in 1992 to 62 in 1998, with an average of 42 locations per year. Locations in India, Mexico, Pakistan, Iran, Spain, Egypt, Afghanistan, Chile, China, and Brazil were present in all years. Three phases (vegetative [VEG], reproductive [REP], and grain-filling [GF]) of wheat's developmental cycle were used, as explained in Gourdji et al. (2013).

Climatic variables used in this study were minimum (*tmin*), maximum (*tmax*), and average temperature (*tavg*), vapor pressure deficit (*vpd*), relative humidity (*rhum*), growing degree days (*gdd*), solar radiation (*srad*), precipitation (*prec*), day length (*dl*), diurnal temperature range (*dtr*), number of days with *tmin* < 0°C (frost days *FD*), number of days with *tmax* < 0°C (icing days *ID*), number of days with *tmax* > 30°C (summer days *SU*), number of days with *tmin* > 25°C (tropical nights *TR*), maximum length of dry spell which is the longest period of consecutive days with no or less than 1 mm precipitation (*MLDS*), maximum length of wet spell, which is the longest period of consecutive days with at least 1 mm of precipitation (*MLWS*), number of days when precipitation >5 mm (*R5mm*), number of days when precipitation >10 mm (*R10mm*), simple precipitation intensity index describing the mean daily precipitation amount during a time period (*SDII*), and maximum 1-day (*Rx1day*) and 5-day (*Rx5days*) precipitation (http://www.icdc.zmaw.de/415.html?&L=1#c2509). Means of all the predefined climatic variables were calculated for the three stages of wheat growth (VEG, REP, and GF) and for the whole length of the wheat cycle for the period 1986–2009. Seasonality (i.e., standard deviation × 100) of the first 10 variables was also computed (*seas*). Daily data were used to calculate all climatic indices (Gourdji et al. 2013). Moreover, length in days of vegetative (*veg-days*), reproductive (*rep-days*), and grain-filling (*gf-days*) periods were also used as environmental variables (Table 13.1).

Statistical Methods

Linear Regression

Data were preprocessed by first fitting a fixed-effect model in order to remove the location effects from climatic variables. This was done to account for the heterogeneity bias in the estimate of regression parameters caused by having more locations from hot areas in some years. Trends since 1986 for all predicted climatic variables were then investigated using linear regression; statistical significance was tested using the two-tailed *t*-test.

Spatial Statistics

Spatial autocorrelation (SAU) was used to assess whether a climatic variable within each year is clustered, dispersed, or random over space. We used the Moran *I* (Moran 1950) as a measure of SAU. If one computes the SAU and finds that the data are clustered, one question comes to mind: How are the data clustered? Methods developed by Getis and Ord (1992, 1996) not only allow hypothesis testing to determine whether clustering has occurred within the data but also provide information on the extent to which above- and below-average values cluster more strongly and identify local concentrations of clustering (Laffan 2006, Mueller-Warrant et al. 2008). The *G* statistic is computed as follows:

$$G = \frac{\sum_{i=1}^{n}\sum_{j=1}^{n} w_{i,j}x_i x_j}{\sum_{i=1}^{n}\sum_{j=1}^{n} x_i x_j}, \forall j \neq i$$

where:

x_i represents the value of location i
x_j represents the value of location j
w_{ij} is the weight assigned to each pair of features x_i, x_j

The weight w_{ij} was computed using a distance between locations (Legendre and Legendre 1998).

The *z*-score statistics is computed as follows:

$$z_G = \frac{G - E[G]}{\sqrt{V[G]}}$$

where:

$$E[G] = \frac{\sum_{i=1}^{n}\sum_{j=1}^{n} w_{i,j}}{n(n-1)}, \forall j \neq i$$

$$V[G] = E\left[G^2\right] - E[G]^2$$

The z-score is highly positive (negative) and significant at 1%, 5%, or 10%, which means that high (low) values of the data are clustered spatially together and that there is a less than 1%, 5%, or 10% likelihood that this high-clustered (low-clustered) pattern could be the result of a random process.

The Getis-Ord G_i^* test statistic is a local adaptation of global Getis-Ord General G and seeks to identify areas of hot and cold clustering based on local neighboring values (Getis and Ord 1996, Laffan 2006). The G_i^* statistic is calculated as the sum of the differences between local sample values and the mean, and is observed as standard normal distribution of z-score values:

$$G_i^* = \frac{\sum_{j=1}^{n} w_{i,j}x_j - \overline{X}\sum_{j=1}^{n} w_{i,j}}{S\sqrt{\frac{\left[n\sum_{j=1}^{n} w_{i,j}^2 - \left(\sum_{j=1}^{n} w_{i,j}\right)^2\right]}{n-1}}}$$

where:

x_j is the attribute value for location j

w_{ij} is the spatial weight between locations i and j

n is the total number of features

$$\overline{X} = \frac{\sum_{j=1}^{n} x_j}{n}$$

$$S = \sqrt{\frac{\sum_{j-1}^{n} x_j^2}{n} - \left(\overline{X}\right)^2}$$

The results of the Getis-Ord G_i^* test may best be visualized in a cartographic output format to easily identify local variation within the data. The resulting z-scores (G_i^*) and p-values reveal where locations with either high or low values cluster spatially. These scores work by looking at each location within the context of neighboring locations. A location with a high (low) value is interesting but may not be a statistically significant hot (low) spot. To be a statistically significant high HH (low LL) spot, a location must have a high (low) value and be surrounded by locations with high (low) values as well. This analysis can also identify outliers: high–low (HL) clusters are locations of high values surrounded by locations of low values or low–high (LH) clusters for the opposite pattern. A geographic region with concentration of outliers is a region with unpredictable climate. For all spatial statistical analyses, we used only maximum and minimum temperatures during the three physiological stages, and the first five PCs resulting from PCA. Spatial analysis and mapping were done using the software ArcMap 10.1 (ESRI 2011).

Principal Component Analysis

PCA has been widely used as a multivariate technique that reduces multivariate data into a set of orthogonal axes along which data variation is maximized. In this study, we ran PCA for each year of our dataset. First, the mean of each climatic variable was subtracted so that the average became 0. We then stored the first five PCs and the ratio of the variance explained by each PC to total variance as a proportion of total variance. We also ran PCA using the complete dataset (location × year combinations). Weights of the climatic variables on the PCs were computed, and Pearson correlations between PCs and climatic variables were calculated. We used the dudi.pca function of the ade4 package (Dray et al. 2007) developed under R (R Development Core Team 2009).

Self-Organized Maps

SOMs were introduced by Kohonen (1989). A SOM is an unsupervised nonlinear neural network used to find representative patterns in high-dimensional data, which projects the extracted patterns on regularly arranged two-dimensional grids called SOM-maps. Each grid of a SOM-map is a node or neuron. Each node has one reference vector, which shows the extracted pattern and has the same dimension as the input vector. Vectors that are closely located on a SOM-map show similar patterns, while reference vectors that are located separately show different patterns. Each input vector will be compared to all reference vectors and assigned to the node where the most similar reference vector belongs. In our study, the data to be analyzed are matrices of 959 vectors (location × year combinations). Each vector is characterized by 70 climatic variables (variable × stage combinations).

The first step of the SOM process is to define map size; we selected a medium SOM-map size of 160 grids (20 × 8) that represents 17% of the initial data sample. We then initiated the reference vectors linearly along the two greatest eigenvectors on the SOM-map, that is, the first and second PCs obtained from the input data can be used as the initial values of the reference vectors. One of the input vectors is then selected and the winner node is identified by computing the Euclidean distance between the selected input vector and all reference vectors. We proceeded to do the scaling using the linear scale, so their variances were equal and thus did not influence the computation of Euclidean distance between vectors. The winner node is the one that minimizes the distance between the initial and reference vectors.

The reference vector of the winning node is then updated by adjusting it slightly to reduce the difference between the node reference vector and the input data vector. This adjustment is scaled according to the topological distance from the winning node. In this process, the surrounding nodes are also adjusted. The procedure continues with each successive data record and is then repeated for a number of iterations through the dataset until there is no further change in the SOM reference vectors. Once the training is complete, each of the SOM nodes has a reference vector that describes a vector in the original 70-dimensional data space. Moreover, samples are placed in the most similar neuron or node of the SOM-map; in this way the data structure can be visualized, and the role of climatic variables in defining the data structure can be revealed visually. Hits markers and PCA projections comparing the initial data and the SOM-map grid were used to verify that the SOM-map represents all the patterns in our data.

We then computed the U-matrix, which was introduced by Ultsch and Roske (2002) and is one of the most popular and useful ways of visualizing clusters with a SOM. A U-matrix is obtained by computing the distance in the input space between units that are neighbors in the output space. The U-matrix has, in our case, a dimension of (39×15); $u(i,j)$ is the distance between map units $m(i)$ and $m(j)$, and $u(k)$ is the mean $u(k) = \{u(k-1,k) + u(k,k+1)\}/2$. The U-matrix visualizes distances between neighboring map units, and thus shows the cluster structure of the map: high values of the U-matrix indicate a cluster border, and uniform areas of low values indicate the clusters themselves. If these differences are small, it means that the units are close together and that there is a cluster of similar data in that region of the input space.

Best matching units (BMUs) for the 959 initial locations from the resulting SOM-map were computed. To compute the BMU for a vector, the Euclidean distance between each node's weight vector and the current input vector is computed. The node with a weight vector closest to the input vector is tagged as the BMU. We then computed *k*-mean clustering with cluster numbers ranging from 2 to 30, and selected the optimum cluster number as the one minimizing the Davies–Bouldin index (Davies and Bouldin 1979). Each node was then assigned to a cluster; using the BMU for the initial vectors, we also allocated them to a cluster (the same cluster as the best matching node). All SOM analyses were done using the SOM toolbox developed by Vesanto et al. (1999) (http://www.cis.hut.fi/projects/somtoolbox) for MATLAB® 7 (http://www.mathworks.com).

RESULTS

In the following three subsections, we describe the historical changes in climatic variables affecting the three main wheat development stages (VEG, REP, and GF) by means of regression and spatial analyses as well as PCA and SOM-maps.

Linear Regression and Spatial Analyses

Both location and year effects were highly significant for all climatic variables. Regression of adjusted climatic variables on year showed several significant slopes. For VEG and REP, only *tmax* had a significant trend across years. All temperatures (*tmax*, *tmin*, and *tavg*) had a significant trend for the GF period. The trend was always positive and higher for GF than that computed for REP and VEG. Relative humidity presented a negative temporal trend but the trend only in VEG stage was significant. More decrease was noticed for VEG, GF, and then for REP. Days to heading as well as the crop cycle are generally getting shorter; VEG and REP lengths are reduced but the cycle is gaining days in the GF stage. The trends for *dl* were significantly negative and small for the three stages (Table 13.1).

Most climatic indices showed a significant spatial pattern across years, especially temperatures (*tmax*, *tmin*, and *tavg*) during the three plant development stages. When we used Getis and Ord's general score for minimum and maximum temperatures

TABLE 13.1
Regression with Climatic Data Corrected for Location Effects

Climatic Variable	Intercept	Slope	*P*(Intercept)	*P*(Slope)	R2
tmin_veg	13.61	−0.003	0.52	0.78	0.32
tmin_rep	−13.40	0.012	0.57	0.31	4.23
tmin_gf	−33.92	0.024	0.12	0.03	18.04
tmin_seas	−35.12	0.023	0.05	0.01	23.76
tmax_veg	−39.25	0.030	0.09	0.01	22.55
tmax_rep	−41.28	0.032	0.19	0.04	15.96
tmax_gf	−64.12	0.046	0.04	0.00	29.31
tmax_seas	−80.13	0.052	0.00	0.00	56.44
dtr_veg	−52.87	0.033	0.02	0.00	29.12
dtr_rep	−27.88	0.020	0.25	0.09	11.18
dtr_gf	−30.19	0.022	0.04	0.00	28.88
dtr_seas	−45.01	0.029	0.00	0.00	43.37
tavg_veg	−12.82	0.013	0.50	0.17	7.73
tavg_rep	−27.34	0.022	0.27	0.08	12.18
tavg_gf	−49.02	0.035	0.06	0.01	25.58
tavg_seas	−57.62	0.037	0.00	0.00	46.55
srad_veg	0.76	0.007	0.98	0.59	1.22
srad_rep	68.20	−0.025	0.15	0.28	4.80
srad_gf	−13.45	0.018	0.62	0.20	6.61
srad_seas	−26.33	0.022	0.22	0.05	15.63
vpd_veg	−7254.53	4.018	0.03	0.02	20.31
vpd_rep	−6303.50	3.687	0.26	0.20	6.89
vpd_gf	−15138.76	8.380	0.04	0.02	19.47
vpd_seas	−11316.95	6.195	0.01	0.00	29.15
dl_veg	24.28	−0.007	0.00	0.00	30.62
dl_rep	39.87	−0.014	0.00	0.00	65.58
dl_gf	24.37	−0.006	0.00	0.00	54.54
dl_seas	15.95	−0.002	0.00	0.20	6.88
rhum_veg	499.79	−0.221	0.01	0.01	22.73
rhum_rep	90.47	−0.020	0.66	0.85	0.15
rhum_gf	143.06	−0.049	0.35	0.52	1.71
rhum_seas	344.06	−0.146	0.04	0.07	13.02
gdd_veg	2619.92	−0.859	0.14	0.33	3.98
gdd_rep	644.12	−0.124	0.00	0.08	12.32
gdd_gf	−12196.80	6.717	0.01	0.00	32.38
gdd_seas	−8765.28	5.667	0.07	0.02	19.98
dth_mean	531.85	−0.221	0.00	0.01	24.27
veg_days	485.83	−0.208	0.01	0.02	20.19
rep_days	82.27	−0.028	0.13	0.29	4.66
gf_days	−576.07	0.1201	0.02	0.01	24.61

during VEG, REP, and GF and across 15 years (1995–2009), *tmin* exhibited the least significant spatial clusters across years compared to *tmax* at all stages. Nevertheless, GF showed significantly high clusters in more years than REP, and REP showed high clusters in more years than VEG (Table 13.2).

Getis and Ord local scores for *tmax* at VEG, REP, and GF identified significant clusters and significant outliers for this variable. Almost no outliers were found for the vegetative stage (Figure 13.1). Significantly high clusters (black dots in Figure 13.1) were found to be located in some regions of Mexico, Brazil, southern and eastern Africa, and India. Significantly low clusters (dark gray dots) were found mainly in Europe, the Mediterranean region, Argentina, Iran, Afghanistan, and eastern China. For the REP stage (Figure 13.2), fewer significantly high clusters were found compared to the VEG stage and were found in the same regions as for vegetative stage with the exception of southern and eastern Africa. More outliers were identified and located in regions with low clusters for *tmax* during VEG mainly to the east of the Mediterranean Sea and along the Nile River. During the GF stage (Figure 13.3), the region with significantly high clusters was extended to include some locations in Afghanistan, Iran, the Middle East, and Myanmar (Burma). No other regions represented a significantly high cluster, except one in Sudan and Ciudad Obregón in northern Mexico. Moreover, almost all locations in Latin America, southern and eastern Africa, Europe, and the Mediterranean region showed significantly low clusters. Most outliers, meaning unpredictable *tmax*, were found for *tmax* during REP stage. These sites were mainly in the Middle East, Central Europe, and China and along the Nile River.

Principal Component Analysis

PCA performed on the matrix of climatic data for each year showed similar patterns for the first five eigenvalues. They explained around 90% of the total yearly climate variability, with the first eigenvalue explaining around 70%, the second 10%, the third 5%, the fourth 3%, and the fifth 2.5%. The minimum amount of climatic variability explained by the first five eigenvalues was observed for years 2002 and 2004 (86%) and the maximum found for year 1989 (96%). The average of variance explained using classes of years was 93.2% for 1986–1990, 91.7% for 1991–2000, and 88.5% for 2001–2009. Running the PCA on all the datasets produced the same patterns of eigen structure; the first five components explained 90% of total climatic variability (Figure 13.4).

Climatic variables showed similar weight patterns as the first PCs from year to year. The Spearman correlation was used to assess the relationships between PCs (1–5) and climatic variables (Figure 13.5). The first principal component (PC1) showed a positive correlation with *rhum* during VEG, REP, and GF, and also with *rep-days*. PC1 was negatively associated with water availability (*R5mm*, *R10mm*, and *Rx5days*) during VEG stage and also during the whole cycle. PC2 was positively linked to all temperatures (*tmax*, *tmin*, and *tavg*) and *vpd* during all three stages plus seasonality. Strong correlation was found between *SU* (summer days) during VEG, REP, and cycle length (CL) and tropical nights (*TR*) during the GF stage and CL. The association was negative for PC2 with water availability (*R5mm*,

TABLE 13.2
Spatial Clusters (High and Low) for *tmin* and *tmax* during VEG, REP, and GF

	tmin_veg		*tmax_veg*		*tmin_rep*		*tmax_rep*		*tmin_gf*		*tmax_gf*	
Year	***p*-Value**	**Cluster**	***p*-Value**	**Cluster**	***p*-Value**	**Cluster**	***p*-Value**	**Cluster**	***p*-Value**	**Cluster**	***p*-Value**	**Cluster**
1995	NaN	NaN	.3392	rand.	.3658	rand.	.1679	rand.	.0000	high	.0001	high
1996	.5036	rand.	.0501	high	.4336	rand.	.0815	high	.0003	high	.0000	high
1997	NaN	NaN	.1084	rand.	.3213	rand.	.8897	rand.	.0320	high	.0097	high
1998	NaN	NaN	.4388	rand.	NaN	NaN	.8988	rand.	.0003	high	.0000	high
1999	NaN	NaN	.0101	high	.0968	high	.0103	high	.0000	high	.0000	high
2000	.1365	rand.	.0028	high	.9595	rand.	.0859	high	.0021	high	.0000	high
2001	NaN	NaN	.1795	rand.	.0033	high	.0021	high	.0000	high	.0000	high
2002	.6714	rand.	.4215	rand.	.8590	rand.	.9247	rand.	.0010	high	.0000	high
2003	NaN	NaN	.2048	rand.	.1327	rand.	.0818	high	.1010	rand.	.1348	rand.
2004	NaN	NaN	.1121	rand.	.5892	rand.	.9132	rand.	.0000	high	.0000	high
2005	.4310	rand.	.0037	high	.0253	high	.0008	high	.0514	high	.0032	high
2006	NaN	NaN	.4108	rand.	.6581	rand.	.4251	rand.	.0001	high	.0000	high
2007	NaN	NaN	.5464	rand.	.2256	rand.	.1020	rand.	.0004	high	.0000	high
2008	NaN	NaN	.1207	rand.	.4245	rand	.0955	high	.0001	high	.0000	high
2009	.9666	rand.	.2163	rand.	.0148	high	.0011	high	.0000	high	.0000	high

rand. = random.

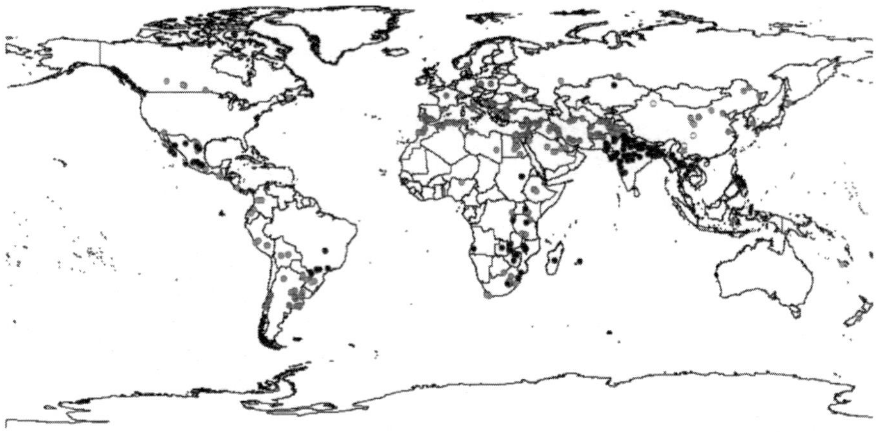

FIGURE 13.1 Clusters for maximum temperature during wheat's vegetative stage. Black dots are significantly high clusters, dark gray dots are significantly low clusters, light gray dots are high–low clusters, white dots (circles) are low–high clusters, and gray dots are nonsignificant clusters.

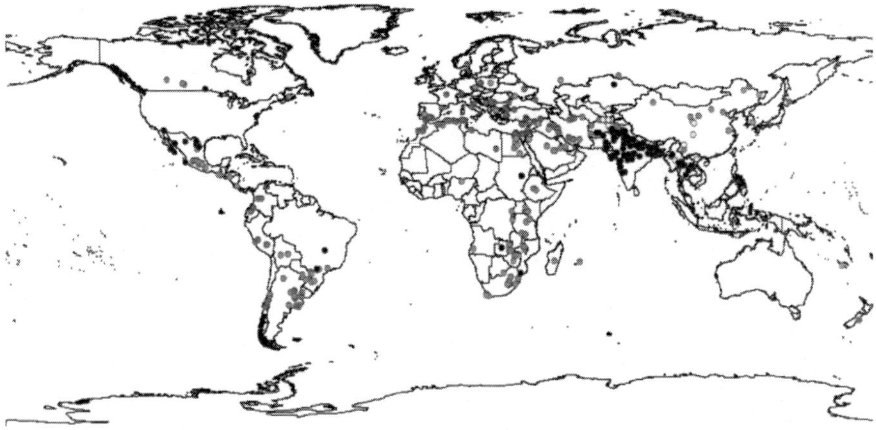

FIGURE 13.2 Clusters for maximum temperature during wheat's reproductive stage. Black dots are significantly high clusters, dark gray dots are significantly low clusters, light gray dots are high–low clusters, white dots (circles) are low–high clusters, and gray dots are nonsignificant clusters.

R10mm, and *Rx5days*) during VEG, GF, and CL. PC3 was positively correlated with temperature-related indices (*tmax*, *tmin*, *tavg*, *srad*, *vpd*, *dl*, and *SU*) during VEG. However, the correlation was negative with temperature-related indices (*tmax*, *tmin*, *tavg*, *srad*, *vpd*, *dl*, and *SU*) during GF and also with the number of cold days and nights (*ID* and *FD*) during VEG. As for PC4, a high positive correlation was found with temperatures (*tmax*, *tmin*, and *tavg*) during GF and *TR* during GF. PC4 was negatively correlated with *rhum* at all stages, *FD* during REP and

FIGURE 13.3 Clusters for maximum temperature during wheat's grain-filling stage. Black dots are significantly high clusters, dark gray dots are significantly low clusters, light gray dots are high–low clusters, white dots (circles) are low–high clusters, and gray dots are nonsignificant clusters.

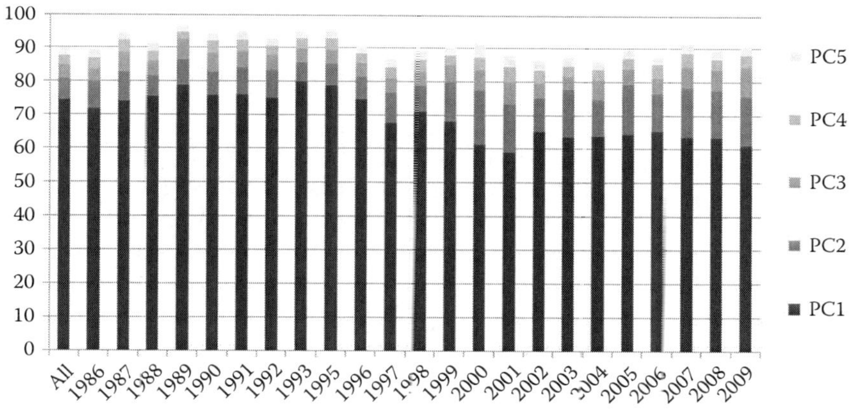

FIGURE 13.4 Contribution of the first five principal components to the total climate year variability. All indicate results of all years together.

the length of the reproductive stage. For PC5, positive association was found with the grain-filling climate profile especially temperatures, tropical nights (*TR*), and the length of GF. On the other hand, VEG was negatively linked with the fifth PC (temperatures, *dl*, *TR*, and *ID*).

Global spatial statistics (Moran's *I*) output showed that only PC1, PC3, and PC4 had significant spatial clusters. PC2 was significantly dispersed and PC5 had a random distribution across the world. Nevertheless, local spatial statistics (cluster analysis) revealed a large number of outliers for PC1, especially in South Asia (India, Pakistan, and Afghanistan) and northern Mexico. South Asia was also the region where most significantly low clusters were found. For PC2, no significantly low or

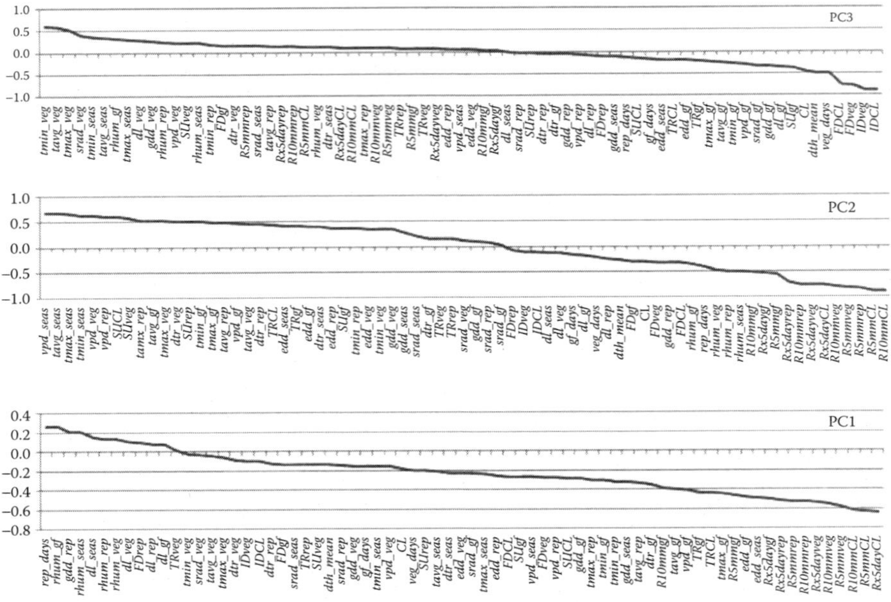

FIGURE 13.5 Plots of the correlations between the first three PCs and the climatic variables.

high clusters were found, but several outliers were present and dispersed around the world. The same conclusions were reached for PC3 and PC5. Only PC4 showed a concentration of significantly high clusters in some regions of India. This is in concordance with the high correlation between PC4 and temperatures at GF and also the high concentrations of high clusters for these temperatures at GF in the same region.

Self-Organized Maps

A SOM-map was constructed with 160 grids (20 rows × 8 columns). The U-matrix of the SOM-map grids showed patterns across the map, and some clusters could be identified visually (Figure 13.6). Running PCA and plotting PC1 against PC2 indicates that the SOM-map grids are well distributed along the initial data; they also represent the variability found in the data based on distance. The SOM-map nodes in the figure are distributed along the range of variables (from light [low values] to dark [high values]).

Also presented here are maps of different climatic variables for the grids. For minimum temperature (Figure 13.7) and during VEG, only a few grids showed high values but they were concentrated at the bottom of the map, and low values were distributed without any pattern across the map. More grids with high values for *tmin* during the REP stage than in VEG were found in the same region of the map. Low values were more concentrated on the left-hand side of the map. GF, however, presented a good pattern of minimum temperature, starting with low values at the top of the map and increasing until reaching high values at the bottom of the map. Also, we notice that more high values were present at GF compared to the other two stages.

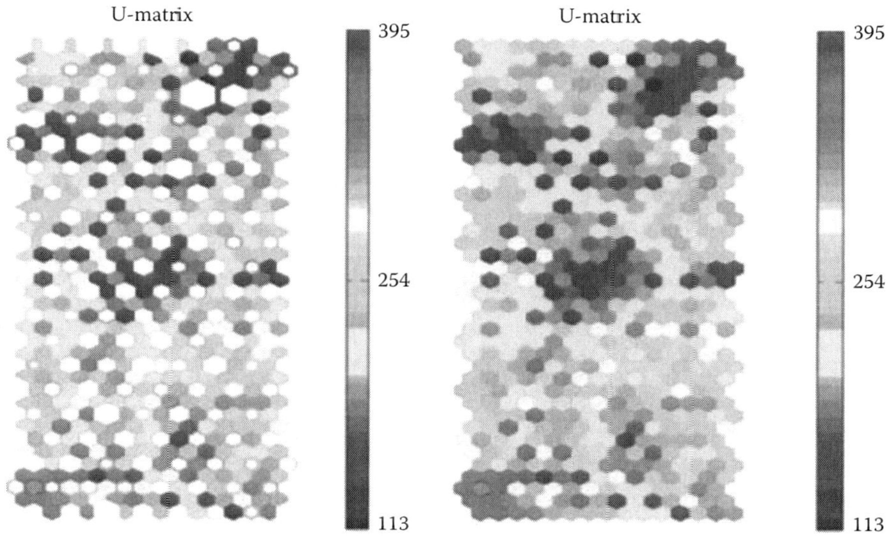

FIGURE 13.6 U-Matrix (*right*) and U-Matrix with hits SOM-map grids (*left*). Dark nodes show clusters and light nodes limits between clusters.

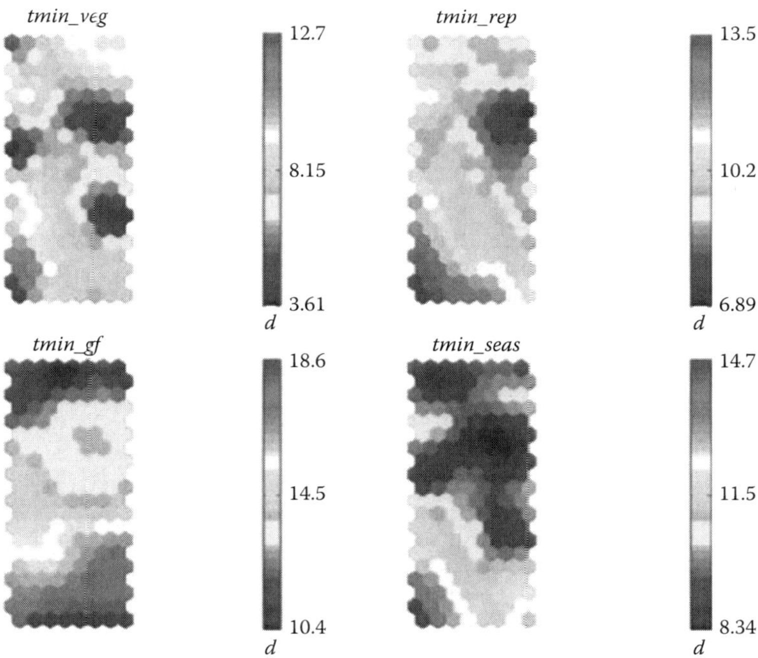

FIGURE 13.7 Minimum temperature patterns at different stages of the wheat cycle and seasonality using values of the SOM-map grid.

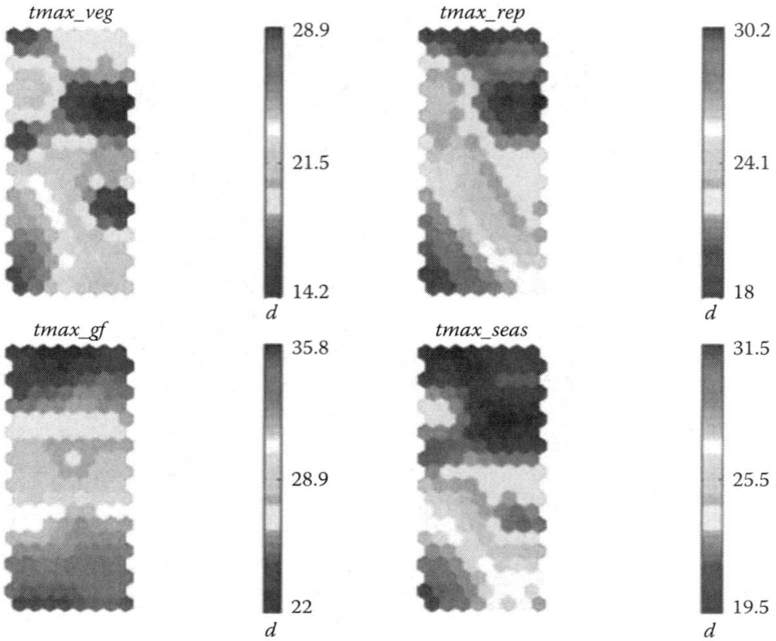

FIGURE 13.8 Maximum temperature patterns at different stages of the wheat cycle and seasonality.

The same patterns were present for maximum (Figure 13.8) temperature and average (Figure 13.9) temperature in the SOM-map, with the exception that more extreme values (high and low) were present for *tmax* and *tavg* than for minimal temperature. For *rhum* (Figure 13.10), patterns were the opposite of the temperature pattern. High *rhum* was found at the top of the SOM-map and decreased smoothly to reach lower values at the bottom of the map. More high values were identified for VEG, followed by GF, and then REP. For stage length, patterns in the SOM-map were different from stage to stage and also for CL. Most of the extremely high values (long periods) were found at VEG and then for CL. More than 50% of the map nodes had short grain-filling periods. For the four variables, patterns were left–right compared to the top–bottom patterns found for all other. No clear visual patterns were observed for *srad* across the wheat stages. Also, spatial variability within the SOM-map was different from one stage to another.

SOM identified 10 clusters that we named cluster1, cluster2, and so on (Figure 13.11a). Clusters are similar when they are mapped close to each other on the SOM-map. Figure 13.11b shows similarity between nodes using color-coded presentation. Labels of nodes are also presented for a visual assessment of the clustering patterns (Figure 13.11c).

The average percentage of grids assigned to clusters was 10%, with a minimum of 6.25% of grids in cluster3 and a maximum of 14.4% in cluster6. SOM computed BMU for each data unit. This assigns each data unit to a specific SOM-map grid. Because each map grid is assigned to a cluster, data units can be assigned to clusters as well. The maximum percentage of data units allocated to a cluster was 14.1%

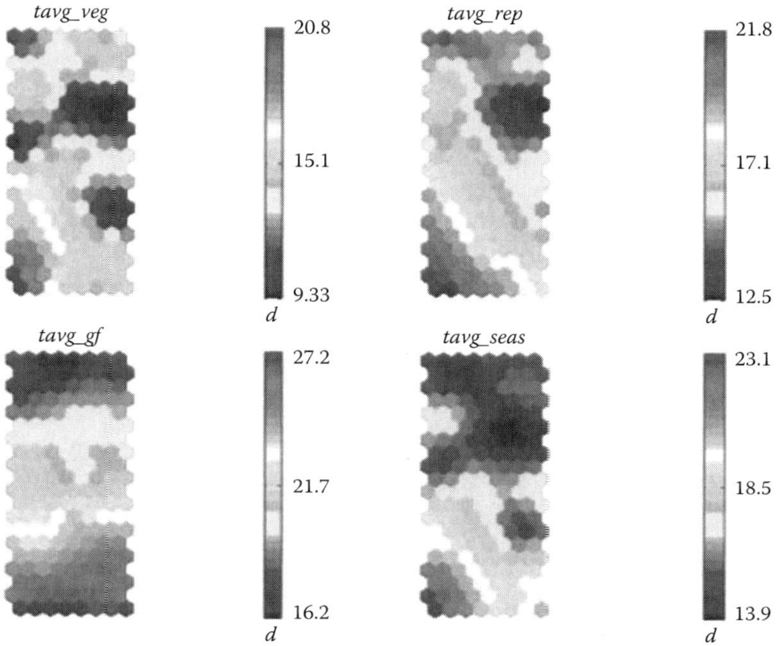

FIGURE 13.9 Average temperature patterns at different stages of the wheat cycle and seasonality.

for cluster10, and the minimum was allocated to cluster3, with 7.1% of data units. The average was 10% of data units per cluster. Clusters at the top of the SOM-map contain more locations from earlier years than clusters from the bottom of the map where locations from recent years were particularly more present (Table 13.3).

The BMU was computed for all 954 locations × year combinations in the original data. The locations were then assigned to the same cluster as the BMU. Plotting data units by cluster and by year showed (Figure 13.12) that, in general, first clusters had more locations from past years and that last clusters have more locations from recent years. In other words, we can see a temporal pattern in the SOM-map (Figure 13.11c) from the top (earlier years) to the bottom (recent years). Consequently, the SOM process could identify the temporal pattern existing in this dataset.

Linking this finding to the results found by plotting climatic variables in the SOM-map clearly showed, for example, that temperatures have been increasing recently. The *tmax* increased more than *tmin*, and this increase was more important during GF than at the other two stages (VEG and REP). Moreover, the *rhum* is decreasing but larger decreases were found during GF and REP, and then VEG. This is in agreement with the results found by linear regression. The *dtr* showed a greater increase during GF and VEG than during REP. The length of the REP and GF stages in general was greater in the first clusters than the last ones, demonstrating a decrease in the length of these two stages, as previously shown in the regression results. However, the increase in *veg days* resulting from regression was not clear in the SOM results.

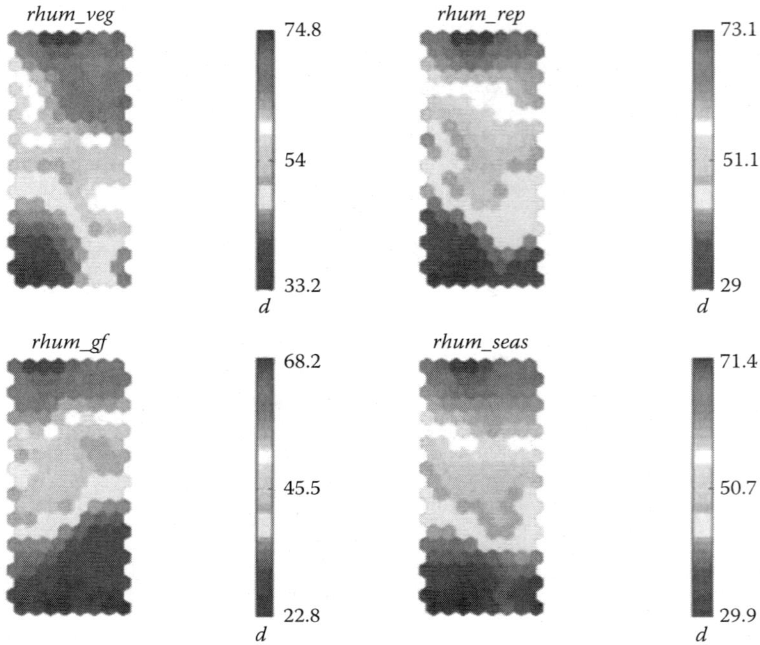

FIGURE 13.10 Relative humidity patterns at different stages of the wheat cycle and seasonality.

The Moran *I* index for cluster variables was 0.47 (*z*-score = 104.75, *p*-value = .00) demonstrating that cluster affectation for ESWYT locations is highly clustered across the world. The general *G z*-score was also very significantly positive (20.42) and showed that locations assigned to high clusters (→ 10) were grouped together. Cluster mapping showed that significantly high clusters were located mainly in India, Pakistan, Afghanistan, southern Iran, the Middle East, southern Egypt, and Sudan. Significantly low clusters were distributed across the globe and were not concentrated in a specific region of the world. Outliers were found in Mexico, China, and northern Iran. Neighboring locations were assigned to similar clusters by the SOM process, and high clusters (8, 9, and 10) were located in the same geographic regions and showed a very significant spatial pattern compared to low clusters (1, 2, 3, and 4).

In North and Central America, most locations had the same clustering patterns across years. Most of them were low (1, 2, and 5). The only exception was the region in northern Mexico (Ciudad Obregón). The same results were found for most locations in Latin America, with the exception of one location in Brazil and one in Uruguay. Southern Africa showed very different clusters from one region to another but locations maintained the same cluster pattern across years. The Nile Valley and the Middle East showed a significant number of locations with high clusters especially in locations along the Nile River, Saudi Arabia, and southern Iran. Locations in northern Iran, northern Syria, and southern Turkey were always assigned to low clusters. The Indo-Gangetic Plain (India, Pakistan, and Nepal) contained the

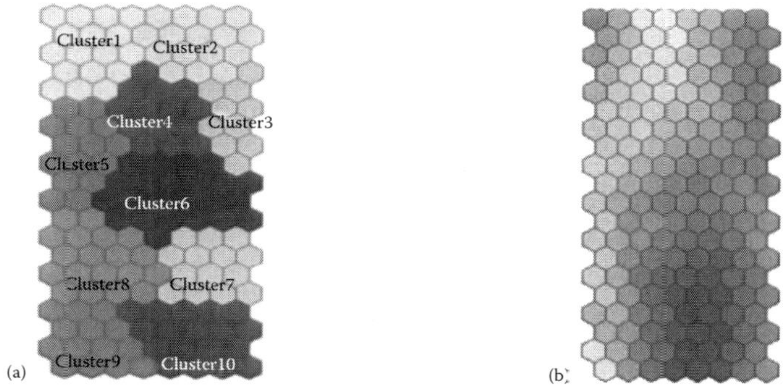

	86–90				86–90				
	24015chn89	19004dz86	12209eth87	24014chn90	12803tan89	10007za96	53201ecu87	19402tun92	
	42301mex96	25304sko98	10401mad90	65108ita92	12302ken86	19103egy0	19125egy96	19402tun88	91–00
86–90	24045chn98	211103syr93	10018za89	10002za87	19103egy1	10002za86	10505maw89		
	10002za89	42217mex98	14803con98	51004arg99	19303mor3		19103egy93	12804tan90	
	12809tan89	10002za98	10002za96	21201tur1	20309ira3	19005dz3	20309ira8	20203cyp88	
	12302ken91	50208bra97	22608pak92	19005dz95	19005dz91	21103syr86	20309ira91	12803tan92	91–00
	24001chn90	51003arg96	22210ind98	51004arg89	20009afg96	21103syr5	20003afg4	20012afg9	
91–00	27104tha87	20009afg2	51004arg3	22207ind98	21116syn8	51204chl93	51206chl98	19303mor2	
	20004afg2	19005dz99	21103syr89	19136egy3	19101egy2	19101egy95	51204chl98	51206chl89	
	42123mex98	11101zim97	10026za9	10502maw87	19101egy97	11103zim93	11101zim2	10901swa1	
	27104tha86	10504maw97	10026za3	11109zim86	11103zim97	19101egy89	11103zim1	11103zim87	01–09
	10002za93	22009ban97	11109zim88	11103zim91	19106egy0	20309ira95	22212ind4	19106egy88	
	10704moz6	11010zam1	11101zim92	22241ind3	22207ind2	20003afg3	20005afg9	20315ira3	
01–09	11101zim96	10504maw96	19110egy97	22210ind0	20004afg3	20004afg7	20006afg6	20011afg9	01–09
	19123egy99	22208ind97	11009zam4	19106egy6	22219ind99	20002afg5	20017afg5	20004afg5	
	22219indS	22001ban88	11009zam5	11006zam1	19105egy92	22213ind96	21123syr8	19106egy89	
	22206ind97	22209inds	19106egy9	19105egy91	21005sa86	22204ind0	19105egy98	19207lib89	
01–09	22222ind2	11011zam7	19105egy4	19105egy93	22220ind4	20301ira95	20303ira8	11009zam3	01–09
	12701sud6	22222ind96	21005sa87	22208ind7	19105egy6	20307ira8	19105egy95	19105egy99	
	12701sud91	19125egy99	21011sa4	21011sa3	222207ind1	20013afg98	20029afg4	20029afg9	
(c)	01–09				01–09				

FIGURE 13.11 Clusters (a), similarity (b), and labels (c) for the SOM-map. Labels are composed of (from left to right) a 5-digit unique location identifier + two or three letters identifying the country + 1 or 2 digits for the year.

TABLE 13.3
Percentage of Locations in a Given Class of Years for the 10 Clusters Found for SOM Grids

Classes	1986–1990	1991–2000	2001–2009
Cluster1	53	47	0
Cluster2	38	38	13
Cluster3	30	30	40
Cluster4	14	50	36
Cluster5	27	53	20
Cluster6	26	35	39
Cluster7	7	14	79
Cluster8	5	53	42
Cluster9	7	29	64
Cluster10	11	44	44

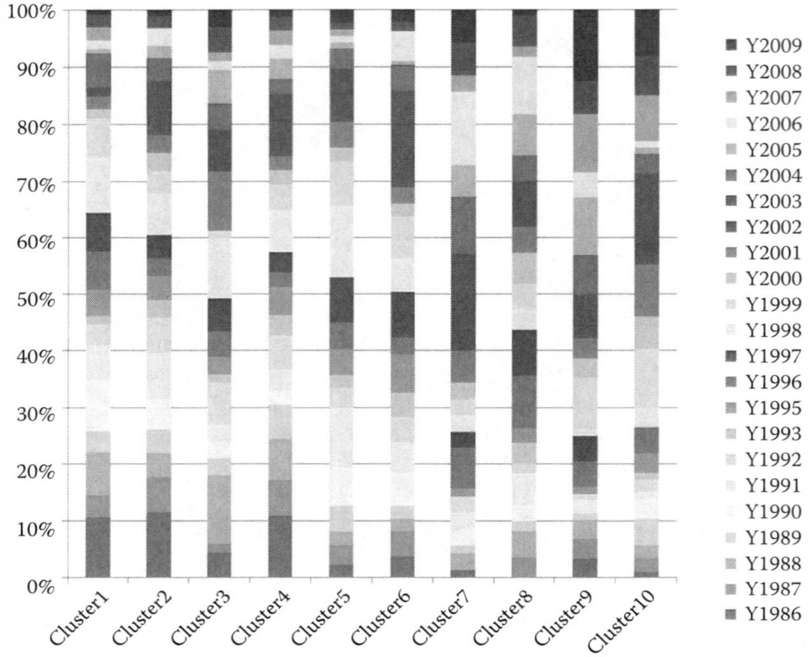

FIGURE 13.12 Percentage of locations in a given year for the 10 clusters found by SOM for all locations.

maximum number of locations assigned to high clusters. Allocation to clusters was independent of the year, though most locations were used in recent years. Locations in Europe and North Africa were mainly assigned to low clusters. Some locations were assigned to clusters in the middle of the SOM-map in recent years, proving that these locations were experiencing a rise in temperature, for example. Locations in China were characterized by cluster1 or cluster5, which were close together on the SOM-map and thus represented a similar climatic profile.

Plotting standardized climatic variables for each node of the SOM-map (Figure 13.13) could reveal the profile for each node and then for each cluster. Climatic variables showed different importance at different nodes. The difference in importance gradually changes from node to node across the SOM-map. For example, *tmin* during VEG was more important at nodes located at the top of the map and then in earlier years, while the importance of *tmin* during GF was minimal for the same nodes. At the bottom right-hand side of the map, GF was more important than the two other stages for minimum temperature. The same pattern was found for *tmax*. The importance of *rhum* was divided into three equal parts corresponding to the three stages at the top of the SOM-map. Recently, *rhum* was found to be more important during VEG or GF for some nodes on the left-hand side of the SOM-map (Figure 13.13). Consequently, we can conclude that all stages are important at ESWYT locations when it comes to temperatures (*tmin* and *tmax*), with more weight found for GF. As for *rhum*, more weight was recently found during VEG.

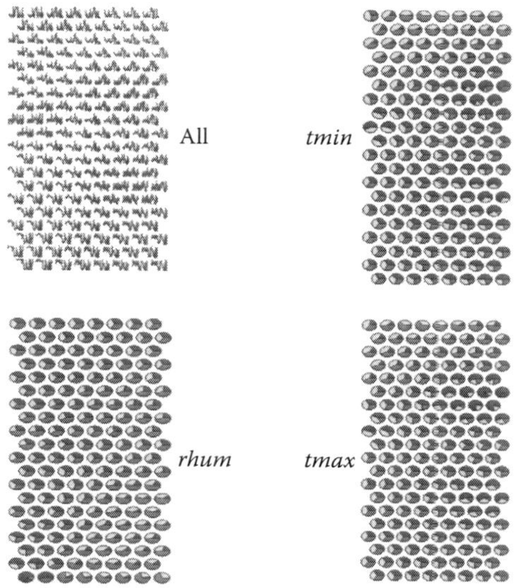

FIGURE 13.13 Relevance of all climatic variables for the SOM-map grids (all), *tmin*, *tmax*, and *rhum* at different wheat development stages (gray for VEG, white for REP, and black for GF).

DISCUSSION

In this study, we explored spatio-temporal patterns of climate profiles using almost 1000 location × year combinations from CIMMYT Elite Spring Wheat International Yield Trials since 1986. The patterns were studied using 22 climatic variables at 3 physiological stages of wheat development. Temporal trends were clear for most of the climatic variables, especially during GF, which implies a reduction in the GF period. The wheat cycle is facing a significant rise in temperature, especially *tmax*, more at the end of the cycle (GF) than at the beginning (VEG). Humidity and precipitation decreased over years at most ESWYT locations. Wheat is rarely affected by individual factors, and temperature stress is frequently associated with reduced water availability.

High temperature during GF is an important factor limiting wheat yield in many environments (Wardlaw and Wrigley 1994) and is expressed mainly as a reduction in grain size (Wardlaw et al. 1989). This could be caused by a number of physiological factors that are sensitive to high temperature, including the accelerated senescence of photosynthetic machinery and inhibition of enzymes in the starch-synthesis pathway (Cossani and Reynolds 2012). In the grain-filling stage, warming had a negative impact on yields across the full range of temperatures (Gourdji et al. 2013).

Neighboring locations had similar climatic variables. This is normal since climate presents a linear relationship with latitude and longitude. Moreover, daily data used in this study were generated using interpolation (Gourdji et al. 2013). The *tmin* is more randomly distributed in the world than *tmax*. Moreover, locations with high *tmax* are located in neighboring geographic regions more during GF than at the other two stages. Locations with high values for maximum temperature during grain-filling are more spatially concentrated than during the two other stages (REP and

VEG). Regions presenting this significant spatial pattern are the Middle East and South Asia (India, Pakistan, and Afghanistan). Climatic variables generally present strong spatial patterns, especially temperatures during GF. This result makes it easier to predict using spatially clustered climatic variables.

The PCA efficiently summarized the large number of climatic variables into five axes. The extracted PCs could be explained by climatic factors or/and by specific wheat development periods. The reduction of climatic multiscale data into a few axes can be used in other analyses, such as genotype × environment interaction analysis, or in genomic selection, particularly when PCs present a climatic or biological significance. Yet, this process was not able to identify any of the spatial or temporal patterns present in the data and hence cannot be used to investigate any climate change. It may be that patterns in climatic data are not orthogonal, and also it may be a long time until the new pattern occupies a significant amount of the total variance. Moreover, the five resulting axes showed a high number of outliers in the spatial clustering process, thereby making it difficult to predict based on PCA results.

CONCLUSION

We demonstrated the capability of SOM to detect weather patterns during wheat physiological stages around the globe since 1986. Our analytical goal here was to learn about spatial and temporal structure in the SOM created from the dataset. The approach was able, in a visual fashion, to identify regional and temporal change. Unfortunately, SOM is unable to predict future patterns or changes unless we assume that the patterns detected by SOM will continue to operate in the future (Sang et al. 2008). The SOM neural network is more appropriate for analyzing climate data because climate data are nonlinear in nature. We provided strong evidence of the spatial variation in climatic variables. Several climate models indicate that the spatial patterns will not change significantly even with the projected climate change (Chu et al. 2011). This would imply that the spatial patterns found in this study will still occur in future scenarios.

Crop-growth simulation models are useful tools to assess the impact of climate change and variability on growth and yield, especially for temperature. Good-quality phenotypic and physiological environmental data, along with detailed climatic profiles of trials, provide a valuable base to investigate sensitive stages and climatic variables driving yield variability in the context of a detailed G × E analysis (Reynolds et al. 2004). Results of a G × E analysis can identify the sensitivities of CIMMYT genotypes to different important climatic variables at wheat development stages specifically for each region. Patterns of change in climatic location profiles together with genotypic sensitivities to climate can help elaborate a specific breeding strategy for specific regions, and make use of thousands of genotypes held by the CIMMYT wheat genebank.

ACKNOWLEDGMENTS

This work was supported by the Climate Change, Agricultural and Food Security (CCAFS) program of the Consultative Group on International Agricultural Research (CGIAR).

REFERENCES

Chmielewski, F., A. Muller, and E. Bruns. 2004. Climate changes and trends in phenology of fruit trees and field crops in Germany, 1961–2000. *Agr Forest Meteorol* 121:69–78.

Chu, W., X. Gao, T.J. Phillips, and S. Sorooshian. 2011. Consistency of spatial patterns of the daily precipitation field in the western United States and its application to precipitation disaggregation. *Geophys Res Lett* 38:L04403.

Cossani, C.M. and M.P. Reynolds. 2012. Physiological traits for improving heat tolerance in wheat. *Plant Physiol* 160:1710–1718.

Curtis, D.L. 1968. The relation between the date of heading of Nigerian sorghums and the duration of the growing season. *J Appl Ecol* 5:215–226.

Davies, D.L. and D.W. Bouldin. 1979. A cluster separation measure. *IEEE T Pattern Anal* 2:224–227.

Dray, S., A.B. Dufour, and D. Chessel. 2007. The ade4 package—II: Two-table and K-table methods. *R News* 7(2):47–52.

ESRI. 2011. *ArcGIS Desktop: Release 10*. Redlands, CA: Environmental Systems Research Institute.

Estrella, N., T.H. Sparks, and A. Menzel. 2009. Trends and temperature response in the phenology of crops in Germany. *Global Change Biol* 13:1737–1747.

Evans, L.T. 1993. *Crop evolution, adaptation and yield*. Cambridge: Cambridge University Press.

Getis, A. and J.K. Ord. 1992. The analysis of spatial association by use of distance statistics. *Geogr Anal* 24:3.

Getis, A. and J.K. Ord. 1996. Local spatial statistics: An overview. In *Spatial analysis: Modelling in a GIS environment*, eds. P. Longley and M. Batty, 261–277. Cambridge: GeoInformation International.

Gourdji, S.M., K. Matthews, M.P. Reynolds, J. Crossa, and D.B. Lobell. 2013. An assessment of wheat breeding gains in hot environments. *Philos T Roy Soc B* 280:1752.

Hu, Q., A. Weiss, S. Feng, and P.S. Baenziger. 2005. Earlier winter wheat heading dates and warmer spring in the US Great Plains. *Agr Forest Meteorol* 135:284–290.

Jagadish, S.V.K., P.Q. Craufurd, and T.R. Wheeler. 2008. Phenotyping parents of mapping populations of rice for heat tolerance during anthesis. *Crop Sci* 48:1140–1146.

Kohonen, T. 1989. *Self-organization and associative memory*. 3rd ed. Berlin, Germany: Springer-Verlag.

Kohonen, T. 1995. *Self-organizing maps*. 1st ed. Berlin, Germany: Springer-Verlag.

Laffan, S.W. 2006. Assessing regional scale weed distributions, with an Australian example using *Nassella trichotoma*. *Weed Res* 46:194–206.

Legendre, P. and L. Legendre. 1998. *Numerical ecology*. Amsterdam, the Netherlands: Elsevier Science B.V.

Liu, Y. and R.H. Weisberg. 2011. A review of self-organizing map applications in meteorology and oceanography. In *Self-organizing maps—Applications and novel algorithm design*, ed. J.I. Mwasiagi, 253–272. Rijeka, Croatia: InTech.

Ludlow, M.M. and R.C. Muchow. 1990. A critical evaluation of traits for improving crop yields in water-limited environments. *Adv Agron* 43:107–149.

Matsui, T., K. Omasa, and T. Horie. 1997. High temperature-induced spikelet sterility of Japonica rice at flowering in relation to air temperature, humidity and wind velocity conditions. *Jpn J Crop Sci* 66:449–455.

Menzel, A., N. Estrella, and P. Fabian. 2001. Spatial and temporal variability of the phenological seasons in Germany from 1951–1996. *Global Change Biol* 7:657–666.

Mitchell, R.A.C., V. Mitchell, S.P. Driscoll, J. Franklin, and D.W. Lawlor. 1993. Effects of increased CO_2 concentration and temperature on growth and yield of winter wheat at two levels of nitrogen application. *Plant Cell Environ* 16:521–529.

Moran, P.A.P. 1950. Notes on continuous stochastic phenomena. *Biometrika* 37:17–23.

Mueller-Warrant, G.W., G.W. Whittaker, and W.C. Young. 2008. GIS analysis of spatial clustering and temporal change in weeds of grass seed crops. *Weed Sci* 56(5):647–669.

Myneni, R.B., C.D. Keeling, C.J. Tucker, G. Asrar, and R.R. Nemani. 1997. Increased plant growth in the northern high latitudes from 1981–1991. *Nature* 386:698–702.

Peltonen-Sainio, P. and A. Rajala. 2007. Duration of vegetative and generative development phases in oat cultivars released since 1921. *Field Crops Res* 101:72–79.

R Development Core Team. 2009. *R: A language and environment for statistical computing.* Vienna, Austria: R Foundation for Statistical Computing.

Reynolds, M.P., R. Trethowan, J. Crossa, M. Vargas, and K.D. Sayre. 2004. Physiological factors associated with genotype by environment interaction in wheat. *Field Crop Res* 85:253–274.

Richards, R.A. 2006. Physiological traits used in the breeding of new cultivars for water-scarce environments. *Agr Water Manage* 80:197–211.

Roberts, E.H., A. Qi, R. Ellis, R.J. Summerfield, R.J. Lawn, and S. Shanmugasundaram. 1996. Use of field observations to characterize genotypic flowering responses to photoperiod and temperature: A soyabean exemplar. *Theor Appl Genet* 93:519–533.

Sacks, W.J. and C.J. Kucharik. 2011. Crop management and phenology trends in the U.S. Corn Belt: Impacts on yields, evapotranspiration and energy balance. *Agr Forest Meteorol* 151:882–894.

Sang, H., A.E. Gelfand, C. Lennard, G. Hegerl, and B. Hewitson. 2008. Interpreting self-organizing maps through space–time data models. *Ann Appl Stat* 2(4):1194–1216.

Siebert, S. and F. Ewert. 2012. Spatio-temporal patterns of phenological development in Germany in relation to temperature and day length. *Agr Forest Meteorol* 152:44–57.

Tao, F., M. Yokozawa, Y. Xu, Y. Hayashi, and Z. Zhang. 2006. Climate changes and trends in phenology and yields of field crops in China, 1981–2000. *Agr Forest Meteorol* 138:82–92.

Tao, F., S. Zhang, and Z. Zhang. 2012. Spatiotemporal changes of wheat phenology in China under the effects of temperature, day length and cultivar thermal characteristics. *Eur J Agron* 43:201–212.

Ultsch, A. and F. Roske. 2002. Self-organizing feature maps predicting sea levels. *Inf Sci* 144(1–4):91–125.

Vara Prasad, P.V., P.Q. Craufurd, R.J. Summerfield, and T.R. Wheeler. 2000. Effects of short episodes of heat stress on flower production and fruit-set of groundnut (*Arachis hypogaea* L.). *J Exp Bot* 51:777–784.

Vesanto, J., J. Himberg, E. Alhoniemi, and J. Parhankangas. 1999. Self–organizing map in Matlab: The SOM toolbox. In *Proceedings of the Matlab DSP Conference 1999, Espoo, Finland*, 35–40.

Wardlaw, I.F., I.A. Dawson, P. Munibi, and R. Fewster. 1989. The tolerance of wheat to high temperatures during reproductive growth. I. Survey procedures and general response patterns. *Aust J Agric Res* 40:1–13.

Wardlaw, I.F. and C.W. Wrigley. 1994. Heat tolerance in temperate cereals: An overview. *Aust J Plant Physiol* 21:695–703.

Wheeler, T.R., G.R. Batts, R.H. Ellis, P. Hadley, and J.I.L. Morison. 1996. Growth and yield of winter wheat (*Triticum aestivum*) crops in response to CO_2 and temperature. *J Agr Sci Cambridge* 127:37–48.

Wheeler, T.R., P.Q. Craufurd, R.H. Ellis, J.R. Porter, and P.V. Vara Prasad. 2000. Temperature variability and the yield of annual crops. *Agr Ecosyst Environ* 82:159–167.

Zhang, X., M. Friedl, C.B. Schaaf, and A.H. Strahler. 2004. Climate controls on vegetation phenological patterns in northern mid- and high latitudes inferred from MODIS data. *Global Change Biol* 10:1133–1145.

14 Assessing Plant Genetic Resources for Climate-Change Adaptive Traits

Heat Traits

A. Bari, M. Inagaki, M. Nachit, A.B. Damania, H. Ouabbou, M. Karrou, C. Biradar, and B. Humeid

CONTENTS

The vulnerability of wheat is likely to increase due to an increase of heat stress according to recent wheat-production simulation models combined with local climate-change data. This is particularly an issue affecting dryland areas where global climate models all converge in forecasting temperature increase in these areas (Girvetz et al. 2009). The greenhouse gas emissions are causing the heating up of the atmosphere (Mendelsohn and Dinar 2009), and the combination of high temperatures and low humidity increases the risk of heat stress for crops (Gobin 2012). Recent model simulations with greenhouse gas emissions factored in have shown that the occurrence of recent extreme heat waves was likely due to climate change (Herring et al. 2014).

For a wheat crop, high temperatures prior to and at anthesis can affect spikelet formation and reduce grain number, which are important components of wheat yield (Farooq et al. 2011, Semenov and Shewry 2011, Curtis and Halford 2014). High temperature can also affect kernel set because temperatures above 35°C can be detrimental and lethal to pollen (Damania 2008). An increase in temperature is likely to also affect plant phenology by shortening crucial periods of crop growth such as the grain setting and filling periods, thus reducing grain yield proportionally to the shortened duration of this crucial period (Hatfield et al. 2008). Overall heat stress has been reported to be a problem of economic significance (Semenov and Halford 2009). For sustained crop production in the face of the effects of climate change, new germplasm with adaptive traits is necessary.

Plant genetic resources provide the opportunity to identify new germplasm with adaptive traits from landraces and wild relatives of wheat, such as species of the genus *Aegilops*, traits that may not be sufficiently available within the current breeding germplasm (Keilwagen et al. 2014). Semenov and Halford (2009) stated that crop-production simulation models combined with high-resolution climate-change scenarios may identify key high-level traits important under drought and high-temperature stress in wheat (see Table 14.1). Tolerance to heat was found to be associated with greater grain number per spike in wild wheat with a potential for improving wheat cultivars with higher grain number per spike at high temperature (Ehdaie and Waines 1992, Khanna-Chopra and Viswanathan 1999). The number of days to heading is also one of the major traits of adaptation to climate change. The genes regulating this trait were identified in wheat to be linked to intrinsic earliness, photoperiod sensitivity, and vernalization (Araus et al. 2008). Wild plant populations occur naturally over a gradient of environmental conditions, allowing natural selection to elicit adaptive traits to those different environments (Henry and Nevo 2014).

Renewal of wheat germplasm available to breeding programs by re-introducing novel genetic variation from *Aegilops tauschii* has been reported. To address the limitation of lack of information associated with the *A. tauschii* accessions, Mizuno et al. (2010) used Bayesian analysis combining genotypic data with geographic information selection for maximizing the variation (Jones et al. 2013).

Searching wheat genetic resources for adaptive traits can be, however, a daunting and costly process given the large size of wheat genetic resources collections and the large number of accessions that would have to be evaluated to pinpoint such traits. Searching for trait variation in plant genetic resources would also require more rapid approaches since early identification of important traits is of equal importance to their transfer into improved varieties (Gollin et al. 2000, Koo and Wright 2000). The approach presented

TABLE 14.1
Effect of Drought and Heat on Wheat

	Stress Factor (*SF*)	
Plant Organ/Trait	**Drought**	**Heat**[a]
Seed (kernels per spike, *kps* and thousand kernel weight, *tkw*)	Reduced size (it limits assimilate supply, which in turn affects seed filling/size) [*tkw*]	Reduced number (affects pollen or ovule function) [*kps*]
Grain filling period (*gfp*)	*gfp* decreases (with little effect on seed-filling rate after flowering)	Large decrease in *gfp* (accompanied with slight increase in seed-filling rate)

Source: Prasad, P.V.V. et al., *J Agron Crop Sci*, 197, 430–441, 2011.
[a] Heat directly affects seed-filling period when compared to drought.

in this chapter consists of simulating an evolutionary process that would exert heat stress and impose a selection pressure on genetic resources such as wild wheat genetic resources to yield the desired traits. This study was conducted first to detect the presence of patterns by linking the traits to climate variables and second to exploit such patterns by identifying climate-change effects related to traits of tolerance to heat.

METHODOLOGIES

Climate Data

The climate data were extracted from maps generated using terrain variables as auxiliary variables converted first into digital elevation models during the interpolation process. Parameter estimation was undertaken over a regular grid with the same dimensions and resolution as digital elevation models. The combination of point climatic data and terrain, in the form of a digital elevation models, allows generating spatially or temporally linked derived variables, such as potential evapotranspiration and aridity index. From the climate maps, 60 climatic variables were extracted for the sites from which the wild wheat accessions were originally sampled. These 60 variables represent monthly average minimum temperature (*tmin*), monthly average maximum temperature (*tmax*), monthly average precipitation (*prec*), monthly average evapotranspiration (*pet*), and monthly average aridity index (*ari*) (Table 14.2).

To detect the presence of relationships between traits of tolerance (phenotypic) and climate variables, while also selecting a set of climate variables that are more likely to predict the trait's expression, the modeling techniques used were based on a machine-learning approach involving a recursive partitioning model (Acharjee et al. 2013). Because collinearity among these variables was expected, linear models may not be able to detect patterns (Cutler et al. 2007). Random forests (RFs) are suitable for highly correlated data where interactions among variables are expected. In terms of climate data, the drought indices measured are based on precipitation and evapotranspiration (e.g., aridity index [*ari*]). For heat, the temperatures, in particular maximum temperature, have been used as predictors.

TABLE 14.2
Environment Variables Used in the Study

Name	Description	Unit	Count
pet	Monthly potential evapotranspiration	mm	12
ari	Monthly moisture index		12
prec	Monthly precipitation	mm	12
tmin	Monthly minimum temperature	°C	12
tmax	Monthly maximum temperature	°C	12

Source: De Pauw, E., Climatic and soil datasets for the ICARDA wheat genetic resource collections of the Eurasia Region Explanatory Notes, Technical Note, ICARDA GIS Unit, 2008.

Trait Data

Two hundred accessions have been evaluated for several traits, including days to heading (*dahe*) and yield components, in particular the number of spikelets per spike (*skps*) between 2008 and 2009. Tolerance to heat was found to be associated with greater spikelet formation and shorter grain filling period, which is associated with days-to-heading. Only 190 accessions were used in this study, as 10 accessions were lacking geographical coordinates.

To characterize the trait distribution pattern as either present or absent for an accession, the data related to heat traits were split into natural groups or clusters based on a clustering technique, whereby the maximum number of clusters was determined based on the *K*-means clustering approach. The approach splits data into *k* clusters by maximizing between-cluster variation relative to within-cluster variation. It splits one cluster into two, and then continues splitting until the within-group sum of squares can no longer be reduced. In this case, the trait dataset of observations ($y1, y2, \ldots, yn$), where the trait's measurements are a *d*-dimensional vector, the clustering process splits trait data into *k* clusters, $C = \{C1, C2, \ldots, Ck\}$, so as to minimize the within-cluster sum of square to find a solution of the equation:

$$\text{argmin}_C \sum_{i=1}^{k} \sum_{y \in C_i} \left\| y - \mu_i \right\|^2$$

where μ_i is the mean of cluster C_i.

Here, for illustration, we split the data into two groups. The membership of accessions in the two groups is based on a combination of the two major traits: number of spikelets and heading date.

MODELING FRAMEWORK

The modeling framework is based on the paradigm that the value of a trait state (Y) as a response variable, such as tolerance to heat, depends on the climatic variables (X), where $X = (x_1, \ldots, x_n)$. The aim of the RF model is to both predict the trait tolerance to heat and examine the variable importance. Climate variables that contribute more than others to the model performance could be more relevant to the heat-tolerant trait expression. RF is a type of recursive partitioning and clustering algorithm that does not require normality assumptions and deals well with a large number of variables. RF can address complex interactions and can cope with highly correlated predictor variables (Strobl et al. 2009).

RF differs from standard tree algorithms in that it acts like an ensemble classifier where the best splitters are randomly selected at each node among subset predictors (Liaw and Wiener 2002, Breiman 2004). RF *grows* many classification trees in the process where an object from an input vector is classified by all trees in the forest. Each tree gives a classification, and we say the tree *votes* for that class. The forest chooses the classification of a given object having the most votes over all the trees in the forest.

TABLE 14.3
Agreement between Predictions and Observations of the RF Model as Measured by the Kappa Statistic

		Observed Value/Score	
		Absent (0)	**Present (1)**
Predicted value/score	Absent (0)	T_{00}	T_{10}
	Present (1)	T_{01}	T_{11}

The mathematical conceptual framework based on the learning approach takes as a learning set $L = \{(X_1, Y_1), \ldots, (X_n, Y_n)\}$ of a random vector (X_i, Y_i), where X_i is the set of climate variables and Y_i is either presence or absence of the traits (trait descriptor states) at a given site *i*.

The accuracy of the RF model was measured based on values derived from a confusion-matrix table and the *area under the curve* (AUC) of the *receiver operating characteristics* (ROC) (Swets et al. 2000, Fawcett 2006). The kappa statistic was used also to measure the specific agreement between predictions and observations in the confusion-matrix table (Table 14.3). A value of kappa below 0.4 is an indication of poor agreement and a value of 0.4 and above is an indication of good agreement (Landis and Koch 1977). An AUC value of 0.5 is an indication of the presence of a pattern in the data with the possibility of a dependency between the trait and the environment. Thus, high values of both AUC and kappa are an indication that the model's performance is adequate for prediction purposes (Scott et al. 2002).

Both AUC and kappa values as well as other accuracy metrics such as sensitivity, defined by $T_{11}/(T_{11} + T_{01})$, and specificity, defined by $T_{00}/(T_{10} + T_{00})$, are indicators of the ability of model to correctly classify observations as either susceptible or tolerant to heat. The higher the values of sensitivity and specificity, the lower the error in differentiating between the accessions, that is, those that are likely to contain the traits from those that are less likely.

Variable Importance

In terms of variable importance, we used RF for conditional recursive partitioning (cforest). Conditional recursive partitioning was also explored to assist in both interpretability and reduction of bias vis-à-vis the selection of covariates/predictors that would be most relevant to the heat tolerance (Hothorn et al. 2006). The ease of interpretability under the tree-based approach is linked to the fact that the variables are identifiable because they are not reduced or projected such as in the case of PCA and PLS analyses.

At the split, the variable that produces less *entropy*, measured using either information theory (Shannon index) or the Gini index (known as impurity measure) is ranked first (Figure 14.1). A reduction in the impurity is a prerequisite for the variable ranking/importance, which can be best visualized in the graphs generated by algorithms.

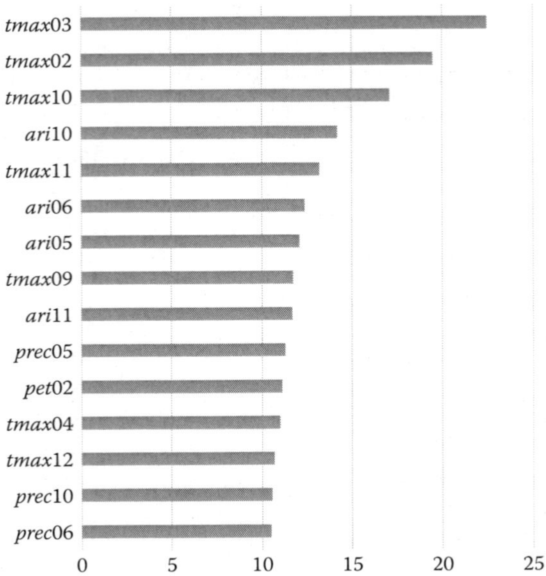

FIGURE 14.1 Gini index value of each of the climate variables with high values indicating variable importance vis-à-vis traits of number of spikelets and number of days to heading.

The importance of any variable X_j in a given tree t can be derived from the equation

$$VI^t(X_j) = \frac{\sum_{i \in \overline{\mathfrak{B}}^t} I\left(y_i = \hat{y}_i^t\right)}{\left|\overline{\mathfrak{B}}^t\right|} - \frac{\sum_{i \in \overline{\mathfrak{B}}^t} I\left(y_i = \hat{y}_{i,\pi_j}^t\right)}{\left|\overline{\mathfrak{B}}^t\right|}$$

where:

$\overline{\mathfrak{B}}^t$ is the out-of-bag sample (the data that are not selected to be part of the training set are referred to as the *out-of-bag* set) for a tree, with t values ranging from 1 to ntree (number of trees in the *forest*, which is left to grow to the largest extent possible without pruning until there are ntree)

$\hat{y}_i^t$ is the predicted class for observation i before permutation

$\hat{y}_{i,\pi_j}^t$ is the predicted class after permuting its value for variable X_j

The importance value or score for each variable is computed as the average importance over all the trees in the form of (Strobl et al. 2008, 2009):

$$VI(Xj) = \frac{\sum_i^{\text{ntree}} VI(X)}{\text{ntree}}$$

Each of the variables ranked high among the different groups (*tmax*, *tmin*, *prec*, *pet*, and *ari*) was also used separately to develop ROC values using a logarithmic equation where the response variable Y is adjusted to a response vector logit(p)

with $p = P(Y = 1)$. The logit stands for the logarithmic equation (Pohlmann and Leitner 2003):

$$\text{logit}(p) = \ln\left(\frac{p}{1-p}\right) = \beta_0 + \sum_{i=1}^{n} \beta_i X_i$$

which in turn leads to the mathematical expression of

$$p = \frac{\exp\left(\beta_0 + \sum_{i=1}^{n} \beta_i X_i\right)}{1 + \exp\left(\beta_0 + \sum_{i=1}^{n} \beta_i X_i\right)}$$

and this transformation assumes a linear relationship between the logit of the probability of $Y = 1$ and the climate variables. The above equation describes Y as a Bernoulli function (Gollin et al. 2000), of which the standard normal output is the Probit model and the logistic distribution is the Logit model (Feelders 1999).

RESULTS

In this study, genetic variability for grain number per spike (*kps*) was conducted and the results indicate a potential for locating wild wheat with higher grain numbers per spike at high temperature. The prediction was based on the quantification of trait-environment relationships as indicated by accuracy metrics in Figure 14.2 and Table 14.4. These accuracy metrics (AUC, sensitivity, specificity, correct classification, and

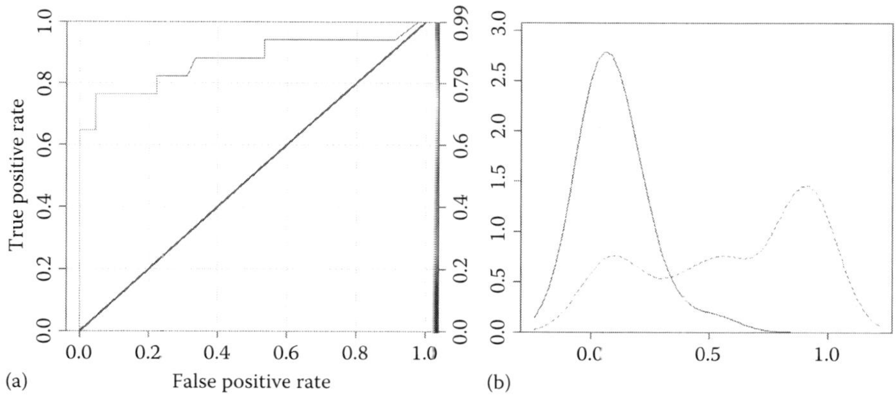

FIGURE 14.2 ROC plots (a) illustrates that the curve for the model was well above the diagonal line, as expected when the model is different from random and a pattern exists in the data. The vertical trend toward the left-hand side is also an indication that the models classified the resistant accessions more correctly with fewer false positive errors. The distribution density plot (b) illustrates the extent of separation between the two trait states (susceptible accessions: solid line; tolerant accessions: discontinuous line).

TABLE 14.4
Average Value of Accuracy Metrics over 10 Iterations on the Test Set (60 Climate Variables)

Iteration	AUC	Sensitivity	Specificity	Proportion Correct	Kappa
1	0.83	0.74	0.93	0.87	0.69
2	0.69	0.50	0.88	0.81	0.38
3	0.72	0.50	0.93	0.81	0.48
4	0.77	0.67	0.88	0.84	0.51
5	0.83	0.80	0.85	0.84	0.60
6	0.81	0.71	0.91	0.85	0.63
7	0.82	0.76	0.87	0.84	0.61
8	0.82	0.78	0.86	0.84	0.62
9	0.76	0.60	0.91	0.84	0.54
10	0.73	0.54	0.92	0.84	0.48
Mean	0.78	0.66	0.89	0.84	0.55

kappa) are all indicators that the RF model classified correctly the accessions as having traits either susceptible or tolerant to heat. In particular, AUC is 0.78 on average over 10 iterations, which is far over 0.5, which would be obtained if the accessions were random. The value of kappa is also high with an average of 0.55 over 10 iterations; a kappa value below 0.4 is an indication of poor agreement between the observations and the predictions.

The results in Figure 14.2 show the presence of a relationship between the current or past climate data and heat tolerance. Both the ROC values and the kappa values are all well above the acceptable values of 0.5 and 0.4, respectively, in Table 14.4. The ROC plots in Figure 14.2 illustrate also that the curve for the model was well above the diagonal line (Figure 14.2a), which is expected when the model is different from random and when a pattern exists in the data. The vertical trend toward the left-hand side in Figure 14.2a is also an indication that the models classified the resistant accessions more correctly with fewer false positive errors (Fawcett 2006). The graph in Figure 14.2b illustrates the distribution pattern of the accessions, with the accessions likely to have the heat traits marked with discontinuous line.

This distribution density plot (Figure 14.2b) illustrates further the extent of separation between the two trait states with those that are potentially tolerant with some overlapping of susceptible accessions. Overall, the model was also able to correctly classify the sites that yield either tolerant or susceptible genotypes with a high correct classification (Table 14.4).

Based on the Gini-index rank of each climate variable the average maximum temperature of March was the most important predictor variable, followed by average maximum temperatures of February, but also October monthly temperature (Figure 14.1). These are important periods as they coincide with the growing season of wild wheat and for the adequate sum and number of degree days to trigger its phenology stages.

DISCUSSION

Highly accurate values for metrics are an indication of the presence of a relationship between the heat-tolerance traits and climate data. The values also suggest that the pattern occurred more than would be expected at random or by chance alone. The values of AUC are above 0.7 (Table 14.4), indicating the presence of a pattern, whereas values of 0.5 and below suggest randomness.

Recent research on patterns reported that plant species display patterns in their phenotypic and phenological trait variation in response to climatic warming (Duckett et al. 2013, Zink 2014), including the distribution of yet-unknown traits of high value for developing or ongoing domestication processes (Davis et al. 2012, van Zonneveld et al. 2012, Schiffers et al. 2013, Henry and Nevo 2014, Read et al. 2014).

Adaptation occurred in the past and on a broader timescale, leading to speciation as a result of changes in climate (Mittelbach et al. 2007). Adaptation is also occurring to contemporary climate change; however, it is not so certain whether adaptation can proceed sufficiently rapid or fast enough to allow populations to adapt (Shaw and Etterson 2012).

For example, genetic variation in response to drought stress was found within both cultivated and wild relatives of lentil (*Lens culinaris* ssp. *orientalis*), a species that is confined to habitats with below-average rainfall (Erskine et al 2011). Quantitative relationships between traits and climate parameters have been established and suggested as a basis for selecting subsets as well as for modeling the impact of climate changes (Barboni et al. 2009).

This process has been reported to be effective in identifying accessions with functional diversity for a specific trait of interest at a time (Vasudevan et al. 2014) because it is based on the concept of co-evolution between the plant and its environment, identifying adaptive traits that can be linked to eco-climatic environmental parameters. Khazaei et al. (2013) demonstrated how prevailing climate environmental data from the collection sites of faba bean landraces was correlated to drought traits using nonlinear modeling techniques. This pattern has been found to occur among plants more often than would be expected by chance (Read et al. 2014). Plant species display changes in patterns of their phenotypic and their phenological trait variation in response to climatic warming (Thompson et al. 2013, Zink 2014), including the distribution of yet-unknown traits of high value for developing or ongoing domestication processes (van Zonneveld et al. 2012, Davis et al. 2012).

Other recent initiatives, where climate variables used in lieu of traits identified loci associated with aridity, suggested the use of the current approach for identification of adaptations at the molecular level (Westengen et al. 2012). Characterization of genetic structure has also been used for identification of genetic gaps in breeding material.

FUTURE WORK

Because monthly data are more of a coarse type of climate site data and thus more likely to be out of phase in relation to critical stages of crop development (Coops et al. 2001), we will be using in the near future fine-resolution daily data. This is even more important as the out phase is being amplified by changing climate conditions.

Daily datasets will be used to align the data to eliminate out-of-phase variation due to the difference in the onset date of growing among the sites where wild wheat was originally sampled. For this monthly data will be used to generate daily data based on the early work by Stern and Coe (1984) followed by the work of McCaskill (1990) and later Epstein (1991). The latter represented the climate data as a mathematical equation on the form of a set of functional building blocks $\emptyset_k$, $k = 1, \ldots, K$ known as basis functions (Ramsay et al. 2010):

$$x(t) = \sum_{k=1}^{K} c_k \emptyset(t)$$

The daily data and the onset data were derived from models involving the proposed model by Epstein (1991) in the form of a sum of harmonic components.

There is also ongoing work to generate accurate continuous surfaces for daily climatic data by downscaling the 0.25° (25 km^2) European Center for Medium-Range Weather Forecasts reanalysis data to 1 km^2 grid sizes. We expect that daily data will allow us to calculate more accurately the timing of crop development phases in relation to climatic conditions. This, in turn, will allow the detection of environment-trait relationships with greater resolution. This will be carried out with the application of functional data analysis where records will be considered as functions rather than observations.

CONCLUSION

Recent studies have shown that rapid adaptation can occur based on rapid rates of evolution taking place within species, which in turn has implications on the abundance and distribution of these species. Evidence on adaptation even in response to contemporary climate change was found to take place over a time frame of tens of years. However, these responses will only occur provided there is readily available genetic variability in populations. If enough genetic diversity for adaptive evolution could be protected in situ for a species, it will not only help in adaptation by the species, but might go well beyond ensuring the persistence of the species, by serving as insurance for resilience in response to climate extremes (Sgrò et al. 2011).

REFERENCES

Acharjee, A., R. Finkers, R.G.F. Visser, and C. Maliepaard. 2013. Comparison of regularized regression methods for ~omics data. *Metabolomics* 3(3):126. doi: 10.4172/2153-0769.1000126.

Araus, J.L., G.A. Slafer, C. Royo et al. 2008. Breeding for yield potential and stress adaptation in cereals. *Crit Rev Plant Sci* 27:377–412.

Barboni, D., S.P. Harrison, P.J. Bartlein et al. 2009. Relationships between plant traits and climate in the Mediterranean region: A pollen data analysis. *J Veg Sci* 635–646. doi: 10.1111/j.1654-1103.2004.tb02305.x.

Breiman, L. 2004. *Consistency for a simple model of random forests*. Technical Report 670. Department of Statistics, University of California, Berkeley, CA.

Coops, N., A. Loughhead, P. Ryan, and R. Hutton. 2001. Development of daily spatial heat unit mapping from monthly climatic surfaces for the Australian continent. *Int J Geogr Inf Sci* 15:345–361.

Curtis, T. and N.G. Halford. 2014. Food security: The challenge of increasing wheat yield and the importance of not compromising food safety. *Ann Appl Biol* 164:354–372. doi:10.1111/aab.12108.

Cutler, D.R., T.C. Edwards Jr., K.H. Beard et al. 2007. Random forests for classification in ecology. *Ecology* 88(11):2783–2792.

Damania, A.B. 2008. History, achievements, and current status of genetic resources conservation. *Agron J* 100(1):9–21.

Davis, A.P., T.W. Gole, S. Baena, and J. Moat. 2012. The impact of climate change on indigenous Arabica coffee (*Coffea arabica*): Predicting future trends and identifying priorities. *PLoS ONE* 7(11):e47981. doi:10.1371/journal.pone.0047981.

De Pauw, E. 2008. Climatic and soil datasets for the ICARDA wheat genetic resource collections of the Eurasia Region Explanatory Notes. Technical Note, ICARDA GIS Unit.

Duckett, P.E., P.D. Wilson, and A.J. Stow. 2013. Keeping up with the neighbours: Using a genetic measurement of dispersal and species distribution modelling to assess the impact of climate change on an Australian arid zone gecko (*Gehyra variegata*). *Divers Distribut* 19:964–976. doi:10.1111/ddi.12071.

Ehdaie, B. and J.G. Waines. 1992. Heat resistance in wild *Triticum* and *Aegilops*. *J Genet Breed* 46:221–228.

Epstein, E.S. 1991. On obtaining daily climatological values from monthly means. *J Climate* 4:365–368.

Erskine, W., A. Sarker, and S. Kumar. 2011. Crops that feed the world 3: Investing in lentil improvement toward a food secure world. *Food Sec* 3:127–139.

Farooq, M., H. Bramley, J.A. Palta, and K.H.M. Siddique. 2011. Heat stress in wheat during reproductive and grain-filling phases. *Crit Rev Plant Sci* 30:6, 491–507. doi:10.1080/07352689.2011.615687.

Fawcett, T. 2006. An introduction to ROC analysis. *Pattern Recogn Lett* 27:861–874.

Feelders, A.J. 1999. Statistical concepts. In *Intelligent data analysis: An introduction*, eds. M. Berthold and D.J. Hand, 15–66. Berlin, Germany: Springer-Verlag.

Girvetz, E.H., C. Zganjar, G.T. Raber, E.P. Maurer, P. Kareiva, and J.J. Lawler. 2009. Applied climate-change analysis: The climate wizard tool. *PLoS ONE* 4:e8320.

Gobin, A. 2012. Impact of heat and drought stress on arable crop production in Belgium. *Nat Hazards Earth Syst Sci* 12:1911–1922. doi:10.5194/nhess-12-1911-2012.

Gollin, D., M. Smale, and B. Skovmand. 2000. Searching an ex situ collection of wheat genetic resources. *Am J Agric Econ* 82:812–827.

Hatfield, J., K. Boote, P. Fay et al. 2008. Agriculture. In *The effects of climate change on agriculture, land resources, water resources, and biodiversity in the United States*, 21–74. A Report by the U.S. Climate Change Science Program and the Subcommittee on Global Change Research. Washington, DC: CCSP.

Henry, R.J. and E. Nevo. 2014. Exploring natural selection to guide breeding for agriculture. *Plant Biotech J* 12:655–662. doi:10.1111/pbi.12215.

Herring, S.C., M.P. Hoerling, T.C. Peterson, and P.A. Stott. 2014. Explaining extreme events of 2013 from a climate perspective. *Bull Am Meteor Soc* 95(9):S1–S96.

Hothorn, T., K. Hornik, and A. Zeileis. 2006. Unbiased recursive partitioning: A conditional inference framework. *J Comput Graph Stat* 15:651–674.

Jones, H., N. Gosman, R. Horsnell et al. 2013. Strategy for exploiting exotic germplasm using genetic, morphological, and environmental diversity: The *Aegilops tauschii* Coss. example. *Theor Appl Genet* 126:1793–1808. doi:10.1007/s00122-013-2093-x.

Keilwagen, J., B. Kilian, H. Ozkan et al. 2014. Separating the wheat from the chaff—A strategy to utilize plant genetic resources from ex-situ genebanks. *Sci Rep* 4:5231. doi:10.1038/srep05231.

Khanna-Chopra, R. and C. Viswanathan. 1999. Evaluation of heat stress tolerance in irrigated environment of *T. aestivum* and related species. I. Stability in yield and yield components. *Euphytica* 106:169–180.

Khazaei, H., K. Street, A. Bari, M. Mackay, and F.L. Stoddard. 2013. The FIGS (focused identification of germplasm strategy) approach identifies traits related to drought adaptation in *Vicia faba* genetic resources. *PLoS ONE* 8(5):e63107. doi:10.1371/journal.pone.0063107.

Koo, B. and B.D. Wright. 2000. The optimal timing of evaluation of genebank accessions and the effects of biotechnology. *Am J Agric Econ* 82:797–811.

Landis, J.R. and G.G. Koch. 1977. The measurement of observer agreement for categorical data. *Biometrics* 33(1):159–174.

Liaw, A. and M. Wiener. 2002. Classification and regression by randomForest. *R News* 2(3):18–22.

McCaskill, M.R. 1990. An efficient method for generation of full climatological records from daily rainfall. *Aust J Agric Res* 41:595–602.

Mendelsohn, R. and A. Dinar. 2009. *Climate change and agriculture: An economic analysis of global impacts, adaptation, and distributional effects*. Cheltenham: Edward Elgar Publishing.

Mittelbach, G.G., D.W. Schemske, H.V. Cornell et al. 2007. Evolution and the latitudinal diversity gradient: Speciation, extinction and biogeography. *Ecol Lett* 10:315–331. doi:10.1111/j.1461-0248.2007.01020.x.

Mizuno, N., M. Yamasaki, Y. Matsouka, T. Kawahara, and S. Takumi. 2010. Population structure of wild wheat D-genome progenitor *Aegilops tauschii* Coss.: Implications for intraspecific lineage diversification and evolution of common wheat. *Mol Ecol* 19(5):999–1013.

Pohlmann, J.T. and D.W. Leitner. 2003. A comparison of ordinary least squares and logistic regression. *Ohio J Sci* 103(5):118–125.

Prasad, P.V.V., S.R. Pisipati, I. Momčilović, and Z. Ristic. 2011. Independent and combined effects of high temperature and drought stress during grain filling on plant yield and chloroplast EF-Tu expression in spring wheat. *J Agron Crop Sci* 197:430–441. doi:10.1111/j.1439-037X.2011.00477.x.

Ramsay, J., G. Hooker, and G. Spencer. 2010. *Functional data analysis with R and Matlab*. New York: Springer.

Read, Q.D., L.C. Moorhead, N.G. Swenson, J.K. Bailey, and N.J. Sanders. 2014. Convergent effects of elevation on functional leaf traits within and among species. *Funct Ecol* 28:37–45. doi:10.1111/1365-2435.12162.

Schiffers, K., E.C. Bourne, S. Lavergne, W. Thuiller, and J.M.J. Travis. 2013. Limited evolutionary rescue of locally adapted populations facing climate change. *Philos T R Soc B* 368:20120083. doi:10.1098/rstb.2012.0083.

Scott, J.M., P.J. Heglund, and M.L. Morrison. 2002. *Predicting species occurrences: Issues of accuracy and scale*. Covelo, CL: Island Press.

Semenov, M.A. and N.G. Halford. 2009. Identifying target traits and molecular mechanisms for wheat breeding under a changing climate. *J Exp Bot* 60(10):2791–2804.

Semenov, M.A. and P.R. Shewry. 2011. Modelling predicts that heat stress, not drought, will increase vulnerability of wheat in Europe. *Sci Rep* 1:66. doi:10.1038/srep00066.

Sgrò, C.M., A.J. Lowe, and A.A. Hoffmann. 2011. Building evolutionary resilience for conserving biodiversity under climate change. *Evol Appl* 4:326–337. doi:10.1111/j.1752-4571.2010.00157.x.

Shaw, R.G. and J.R. Etterson. 2012. Rapid climate change and the rate of adaptation: Insight from experimental quantitative genetics. *New Phytol* 195:752–765.

Stern, R.D. and R. Coe. 1984. A model fitting analysis of daily rainfall data. *J Roy Stat Soc A Sta* 147:1–3.

Strobl, C., A.L. Boulesteix, T. Kneib, T. Augustin, and A. Zeileis. 2008. Conditional variable importance for random forests. *BMC Bioinform* 9:307.

Strobl, C., J. Malley, and G. Tutz. 2009. An introduction to recursive partitioning: Rationale, application, and characteristics of classification and regression trees, bagging, and random forests. *Psychol Methods* 14(4):323.

Swets, J.A., R.M. Dawes, and J. Monahan. 2000. Better decisions through science. *Sci Am* 283:82–87.

Thompson, J., A. Charpentier, G. Bouguet et al. 2013. Evolution of a genetic polymorphism with climate change in a Mediterranean landscape. *Proc Natl Acad Sci USA* 110: 2893–2897.

van Zonneveld, M., X. Scheldeman, P. Escribano et al. 2012. Mapping genetic diversity of cherimoya (*Annona cherimola* Mill.): Application of spatial analysis for conservation and use of plant genetic resources. *PLoS ONE* 7(1):e29845. doi:10.1371/journal.pone.0029845.

Vasudevan, K., C.M. Vera Cruz, W. Gruissem, and N.K. Bhullar. 2014. Large scale germplasm screening for identification of novel rice blast resistance sources. *Front Plant Sci* 5:505. doi:10.3389/fpls.2014.00505.

Westengen, O.T., P.R. Berg, M.P. Kent, and A.K. Brysting. 2012. Spatial structure and climatic adaptation in African maize revealed by surveying SNP diversity in relation to global breeding and landrace panels. *PLoS ONE* 7(10):e47832. doi:10.1371/journal.pone.0047832.

Zink, R.M. 2014. Homage to Hutchinson, and the role of ecology in lineage divergence and speciation. *J Biogeogr* 41:999–1006. doi:10.1111/jbi.12252.

15 Plant Genetic Diversity
Statistical Methods for Analyzing Distribution and Diversity of Species

M. Singh, A.B. Damania, and Y.P. Chaubey

CONTENTS

Availability of plant genetic diversity is fundamental to the existence of the living planet. Conservation of biodiversity has been a practice of all concerned professions, including farmers, since ancient times. However, the changes in environmental conditions and pressures from population and technological change have resulted in genetic modification, including replacement of landraces and erosion of genetic diversity.

The relationship between loss of diversity and climate change has been well recognized with unprecedented higher levels of species extinction (Hooper et al. 2012). In order to minimize genetic erosion and capture diversity, a need for collection and conservation of biodiversity has been stressed in the Convention on Biological Diversity (Articles 8 and 9, CBD 1992), Agenda 21 (Chapters 14 and 15, UNCED 1992), and the Global Biodiversity Strategy (WRI et al. 1992), using several mechanisms such as in-situ, ex-situ, and in-vitro conservation. A number of references dealing with various aspects can be found in Guarino et al. (1995). Technical guidelines for germplasm exploration and collection including planning, methods, and procedures illustrated with real germplasm collection missions are given in Engels et al. (1995), whereas examples of planning and execution of a genetic resource collection have been given in Bennett (1970), Chang (1985), Damania (1987), and Kameswara Rao and Bramel (2000). This chapter discusses, in brief, statistical features of collection and analysis of data in this context.

SAMPLING FOR SPECIES AND GENETIC DIVERSITY

One practical way for preserving the species and genetic diversity is to collect and conserve (and regenerate) samples with maximum diversity in species and genomic information in relation to the environment/region. The sampling strategy will depend on the population structure, distribution of the traits or genes, and the statistical measure of the diversity captured in the sample as well as the precision required. The sample (sampling fraction) should be as large as possible with maximum information, on one hand, to serve the principle, but should be small enough to be collected and maintained within the limited time and resources available in practice on the other. An optimum sampling strategy for genetic conservation of crop plants under threat of extinction has been discussed by Marshall and Brown (1975), Weir (1990), and Brown and Marshall (1995), among others. This depends on the genetic variation in the set of populations under investigation in terms of genotype and allele frequencies or allelic richness, gene diversity, heterozygosity levels, and disequilibrium coefficients, and so on. Often in practice, for a single population, allelic richness is measured as the average number of alleles for a large number of markers. In case of sampling from several populations, the sampling strategy depends on the extent of genetic divergence among populations (e.g., in terms of number of alleles that attain appreciable frequencies in individual populations) and the level of genetic variation (e.g., in the distribution of number of alleles per locus).

SAMPLE SIZE

From the neutral theory of Kimura and Crow (1964), the approximate number of neutral alleles (k) in a sample of S random gametes, from a population of size N in equilibrium, at a locus with mutation rate u is given as (Brown and Briggs 1991, Brown and Marshall 1995).

$$k \approx \theta \ln\left[\frac{(S+\theta)}{\theta}\right] + 0.6, \text{ where } \theta = 4Nu > 0.1 \text{ and } S > 10$$

A basic sampling strategy should take into account the number and location of sampling units, the number of individual plants sampled at a site, the choice of individuals, and the number and type of propagules per plant. This strategy needs refinement in view of the information available on the genetic structure of the target populations. Thus, modifications are required to the basic sampling strategy for different species to address spatio-temporal distribution, life history, genetic system (mating structure), and mixture of the populations. Modifications in sampling strategy are also required when sampling is for specific goals (e.g., collecting for additional genes for a resistance to a specific disease). Sedcole (1977) gave expressions for the sample size (S) to recover, with 95% confidence, a minimum number r of plants with a trait that occurs in population with frequency p:

$$S \approx \frac{\left[r + 1.645r^{0.5} + 0.5\right]}{p}$$

In cases where partial information is available from a previous collection on the target species, re-sampling from the region could be done to improve the information content on the genetic diversity. A discussion of the advantages and disadvantages from various angles of covering unstudied areas, returning to the areas with high genetic diversity, or collecting for specific ecotypes are presented in Nabhan (1990). The gain due to re-sampling could be obtained in terms of change in diversity measure. A statistical test for significance of additional information from recollection is available in Rao (1973).

Brown (1989, 1992) made recommendations on the sample sizes: sample about 50 populations in an eco-geographical area or on a specific mission; the size of a sample field should be 50 individuals per site to capture locally common alleles with $p > .95$; the total collection size should be 30,000 individuals per species to include widespread rare alleles present at mutation rate; the core collection size should be 3,000 individuals per species to give the expected number of alleles equal to number of alleles in the species with a frequency $>10^{-4}$; the minimum number of sites for endangered species should be five (collecting 10 individuals per site) to ensure survival of worthwhile genotypes.

Methods of Sampling

General sampling procedures have been covered in standard texts (Cochran 1977, Sukhatme et al. 1984, among others). Application of any sampling technique requires preparation of a sampling frame (the list of all sampling units) of the population, or, of all the subpopulations (in the case of stratified sampling, where the population under study is divided into strata or subpopulations). In a simple random sampling, the sampling units are selected with equal probability. In systematic sampling, the sampling units are aligned on a rectangular grid (rows and columns). One of the rows (or columns, also called *clusters*) is selected using simple random sampling. It may be noted that while systematic sampling may be operationally very convenient, it does not provide an unbiased estimate of the population variance. Applications of various sampling methods in the context of wheat germplasm collection in the West Asia and Northern Africa region are discussed in Damania (1987), Valkoun and Damania (1990), and Porceddu and Damania (1991). When the samples are collected with geo-reference coordinates, spatial models should be used to analyze such data (Cressie 1993). The distance-sampling approach, used to collect the information on the spatial pattern of species or genetic diversity, comprises data on distances from a randomly placed line or point to the object of interest (Buckland et al. 1993).

DATA ANALYSIS METHODS

Several statistical methods are available to analyze data collected for various objectives relevant to plant genetic resources. We discuss a few of these methods used for specific purposes.

Measurement of Diversity

Diversity is widely used to judge the suitability of a habitat for conservation (Magurran 1988). It has two components: (1) variety/richness in terms of entities such as alleles, genes, varieties, populations, species, and genus and (2) relative abundance of the entities. The diversity measures have been defined by combining these two components in various ways. The most popular indices are Margalef's diversity index (Clifford and Stephenson 1975): $D_{\mathrm{Mg}} = (S-1)/\ln N$ and Menhinick's index (Whittaker 1977): $D_{\mathrm{Mn}} = S/\sqrt{N}$, where S is the number of species (groups/clusters) recorded and N is the total number of individuals (samples) summed over all the species (groups/clusters).

Shannon and Simpson Indices

Shannon and Simpson indices are heterogeneity measures, and they include both richness and abundance in one single value. The Shannon index is $H' = -\sum p_i \times \ln(p_i)$ and the Simpson index is $D = E(p_i)^2$, where p_i is estimated as n_i/N, where n_i is the species record (specimens, records from the flora, and germplasm data) and N is the total number of individual records of all the species.

The Shannon index has been corrected for bias and is given as

$$H_1 = -\sum p_i \times \ln(p_i) - \left[\frac{(S-1)}{N}\right] + \left[\frac{\left(1 - \sum p_i^{-1}\right)}{\left(12N^2\right)}\right] + \sum\left[\frac{\left(p_i^{-1} - p_i^{-2}\right)}{\left(12N^3\right)}\right]$$

with variance

$$\operatorname{var}(H_1) = \left\{\frac{\sum p_i \times \left[\ln(p_i)\right]^2 - \left[\sum p_i \times \ln(p_i)^2\right]}{N}\right\} - \left[\frac{S-1}{2N^2}\right]$$

See Hutcheson (1970) and Bowman et al. (1971).

A study of plant diversity may throw light on describing abundance of species, estimating the number of species in a given region, species diversity via indices, comparing diversities across regions, association between species and geographical region, spatial modeling of abundance of species, and so on. Various indices, such as those given above, may be computed using a number of tools in Excel, R-package (R Development Core Team 2009), Genstat statistical software (Payne 2014), and others. To illustrate, consider a hypothetical dataset with observed abundances (53, 33, 26, 16, 8, 2, and 1) for seven species. The following Genstat procedure

```
ECDIVERSITY [PRINT=index,estimate;
INDEX=hshannon,jshannon,simpson,isimpson; BMETHOD=bootstrap;\
CIPROBABILITY=0.95; NBOOT=100; SEED=12431]!(53,33,26,16,8,2,1)
```

would yield the output of Table 15.1.

The standard errors of the diversity indices have been evaluated using bootstraps of 100 replications. An open source package *BiodiversityR* in the R-software application (R Development Core Team 2009) may also be used to compute these indices (Kindt and Coe 2005).

Distributional Behavior of Abundance of Species

A number of theoretical models have been studied for their goodness of fit to the observed abundances of species (frequencies) (Engen and Taillie 1979). Let us denote the probability or relative abundance of the ith species by p_i, where $i = 1,\ldots, s$, and s is the number of species in the population (in the region under consideration). The most frequently used models include the following:

1. Uniform or completely even model

$$p_i = \frac{1}{s}, \quad (i = 1,\ldots,s)$$

TABLE 15.1
Values of Diversity Indices, Bootstrap Diversity Statistics, and Confidence Intervals Produced by Genstat from a Hypothetical Dataset (See Text)

	Shannon–Wiener H	Shannon–Wiener J	Simpson 1 − *D*	Simpson 1/*D*
Index	1.532	0.7874	0.7519	4.030
Bootstrap estimate	1.510	0.8099	0.7469	3.970
Bootstrap s.e.	0.052	0.0399	0.0177	0.271
95% bootstrap confidence interval	(1.404, 1.607)	(0.7333, 0.9088)	(0.7059, 0.7777)	(3.400, 4.499)

s.e. is the standard error.

2. Broken stick model

$$p_i = \left(\frac{1}{s}\right)\left(\frac{1}{i}+\frac{1}{(i+1)}+\cdots+\frac{1}{s}\right),\ (i = 1,\ldots,s)$$

3. Geometric series model
   A function to describe the distribution of such frequencies is the geometric frequency distribution function (May 1975), which is given by

$$n_i = NC(k)k(1-k)^{i-1},\quad i=1,\ldots,s,\ 0<k<1$$

   where:
   - k is the unknown parameter representing the proportion of available niche space or resource that each species occupies
   - n_i is the number of individuals in the ith species
   - N is the total number of individuals
   - s is the total number of species
   - $C(k) = [(1-(1-k)^s]^{i-1}$

   An example of the geometric distribution fitted satisfactorily to data on wild wheat (species of *Aegilops*) with the estimates of the parameter $k = 0.1966$ with a standard error of 0.0213 is provided by Bari and Singh (1998).
4. Infinite geometric model

$$n_i = Nk(1-k)^{i-1},\quad i=1,\ldots,s,\ 0<k<1$$

   As an example, to fit a geometric model to the hypothetical data used above—observed abundances (53, 33, 26, 16, 8, 2, and 1) for seven species—we may use the following command in the Genstat software:

```
ECFIT [PRINT=summary,estimates; MODELTYPE=geometric]
!(53,33,26,16,8,2,1)
```

   This results in the following output (Table 15.2) and the plot indicating the goodness of fit (Figure 15.1).
5. Gamma model

In order to account for variation in abundances over time and space, a number of continuous probability models have been found suitable. The gamma model, with index k and mean k/θ, has the following probability density function:

$$f(p) = \left(\frac{\theta^k p^{k-1} e^{-\theta p}}{\Gamma(k)}\right),\ \theta>0$$

where $\Gamma(k)$ is Gamma function of k.

TABLE 15.2
Summary of Genestat Output for the Geometric Model with a Hypothetical Dataset

Parameter	Value
Deviance	5.89 on 6 d.f.
Number of individuals	139
Number of species	7
Estimate of k	0.4006
Standard error of estimate	0.03487

d.f. is the degrees of freedom.

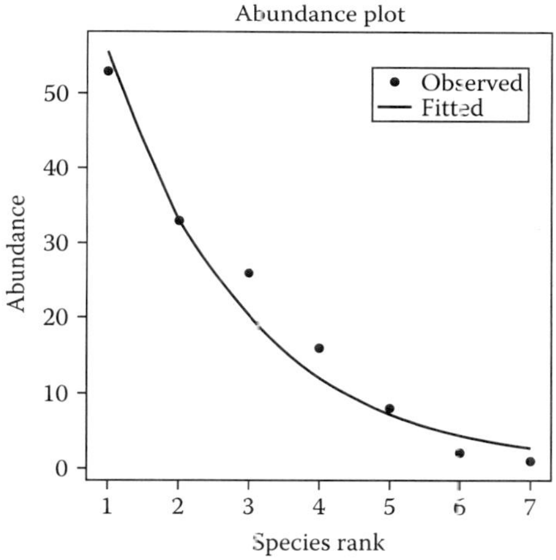

FIGURE 15.1 Geometric model fitted to the hypothetical data on abundances.

Expected Number of Species in a Region

Based on an observed abundance data, one may be interested in estimating the species richness in terms of the number of species that may be expected from a given sample size. Across various regions of collection, sample sizes are not always equal. In such cases, rarefaction is a way to counter this problem. This can be done by using the rarefaction technique of Sanders and modified by Hurlbert (1971):

$$E(S) = \Sigma \left[1 - \frac{\binom{N-Ni}{n}}{\binom{N}{n}} \right]$$

where:
$E(S)$ is the expected number of species
n is a standardized sample size (number of individual in the smallest sample)
N is the total number of individuals recorded
N_i is the number of individual in the ith species

In order to compute a confidence interval for the number of species in the region, it would be worthwhile to obtain an expression for variance of S. In its absence, one may use either a bootstrap method (Efron and Tibshirani 1993) or standard error from a number of independent estimates, for example, by dividing the region randomly into a number of groups (Bari and Singh 1998).

Core Collections

The idea of a core collection was put forth by Frankel (1984). Frankel and Brown (1984) and Brown (1989) described the essential feature as a *limited set of accessions representing the maximum diversity present in the population.* The advantages of having a core collection includes serving as a standard for including new accessions, efficiency of conservation, characterization, evaluation, enhancement, and distribution (Brown 1995).

The basic issues in forming a core collection include determining the (optimum) size and quality of the accessions selected in terms of representing the diversity (Brown 1995). The first step is to begin with the passport and characterization data for the available collection, followed by grouping of the accessions, selection of entries from each group, and evaluating the core accessions thus selected. Yonezawa et al. (1995) concluded that an optimal stratification sampling strategy is a proportional allocation to the number of accessions in the group and they gave procedures to retain both the pattern and the range of genetic diversity in the whole collection.

M- and *H*-Strategies

Schoen and Brown (1995) presented two core-collection strategies, termed *M-strategy* and *H-strategy*, for maximizing genetic diversity in core collection, using stratified sampling and marker-gene data for selecting allele-rich accessions from different regions. The *H*-strategy (after Nei's diversity index) allots the number of accessions in proportion to the sum, over all loci, of the θ coefficient ($=h/(1-h)$), a function of Nei's diversity index (h). The *M*-strategy (maximization strategy) pinpoints the individual accessions from each geographic group to be selected in the core collection to maximize the allelic diversity (using genetic-marker loci) expected to maximize the target allelic diversity. *M*-strategy uses a linear programming approach for minimizing the loss of marker alleles with respect to those present in the entire collection, subject to the conditions based on the number of accession in each group and on the total number of accessions in the core.

Coefficient of Variation

The coefficient of variation (CV) of a quantitative character used in defining the core collection can indicate the genetic diversity/variability of the collection and allow an estimate of how the inclusion or deletion of an accession can influence the diversity of the core.

Multivariate Methods

We discuss a number of multivariate analyses that have been used in diversity studies from a number of perspectives (Digby and Kempton 1987). The data at the DNA level are most appropriate for quantifying genetic diversity, while morphological and agronomic traits indicate the influence of environmental factors.

Classification Methods

Multivariate information has generally been used to show similarity of the entries using grouping/classification and ordination methods. Genetic diversity can be depicted by classifying/grouping the accessions into genetic diversity trees or dendrograms. The structure of genetic diversity in the form of the tree can then guide selection of accessions to form the core. The classification methods are of two types: hierarchical (often known as *cluster analyses*) and nonhierarchical (known as *K-means cluster analysis*).

In hierarchical analysis, the accessions will be arranged into groups with similar properties and the number of groups, at a given level of similarity, is unknown in advance. This method thus reflects a more natural interrelation, or closeness, or diversity in the accessions. An algorithm for a hierarchical method requires the following:

1. A measure of similarity between two items (e.g., two accessions). Depending on the nature of the traits observed, several measures have been suggested: Euclidean distance, cityblock (also known as *Manhattan*), simple matching, ecological distance, and Jaccard's measure.
2. A concept of treating initially all the accessions as separate clusters and fusing the two most nearby clusters into one at each of the subsequent stages (agglomerative method) or treating initially all the accessions as a single cluster and dividing it into two groups at subsequent stages (divisive methods).
3. A method of deriving similarity between two groups or clusters of items. Often-used procedures are single linkage or nearest neighbor, complete linkage or furthest neighbor, average linkage, centroid method, or a group average (Cormack 1971). The Genstat routine HCLUSTER can be used to produce the required information.

In a nonhierarchical method, accessions are divided into a given number (determined in advance) of disjoint groups such that the groups are reasonably homogeneous within and different between. Again, a number of criteria influence the algorithm for grouping. The commonly used ones are based on maximizing the between-group sum of squares; minimizing the determinant of the pooled within-class dispersion matrix; maximizing the total Mahalanobis squared distance between the groups; and choosing the maximal predictive classification (for which the Genstat routine CLUSTER could be used).

Comparison of Classifications

When there are several methods of classification, then one may wonder whether there is any natural order in the items. One may either work out some other independent grouping based on environmental and or morphological traits, or try examining whether two or more classifications give the same groupings. Comparison of two grouping methods could be done using chi-square contingency tests for independence of the two classifications produced by the two methods.

Ordination Techniques

Ordination techniques are used to order a group of objects, for example, accessions, populations in multidimensions (e.g., based on the response from analyses with multivariables or multimarker information) with a view to finding if there is any interrelationship among the set of objects and their association with other factors such as site or environmental factors. The ordination can be done using direct gradients of environmental factors (e.g., abundance versus soil pH) and weighted scores for objects over environments. In absence of environmental factor information, indirect methods can be used, such as principal component analysis (PCA), bi-plots, correspondence analysis, canonical variate analysis (CVA), and principal coordinate analysis (PCO), which are based on data in the form of two-way tables or matrices of objects and environments.

Principal Component Analysis

Let the $X = (x_{ij})$ be the data matrix (e.g., abundance for the ith object [species] from the jth variate [site], where $i = 1,\ldots, n$ and $j = 1,\ldots, p$). The values of the variates are standardized into matrix Y to have mean of zero and unit variance. The singular value decomposition of matrix Y is given as follows:

$$Y = USV'$$

where:

U is the $n \times p$ orthonormal matrix
S is the diagonal matrix of order p
V is the transpose of the orthogonal matrix V $(p \times p)$

The diagonal elements of S are arranged in descending order to give an order of approximation to Y in terms of the matrix $A = US$, called *scores*. Another PCA approach is to find a new set of transformed variates (linear combinations of observed variates), which account more effectively for the variation among the individuals. For this, one finds the use of spectral decompositions of the sum of squares, then the product matrix in Y, that is, $\Sigma = Y'Y$, yields the eigenvalues and eigenvectors (loadings) of Σ, arranged in descending order. Such an ordering assigns the associated vectors (scores as linear combinations of Y values and loadings) as principal components (PCs) and the associated eigenvalue gives the variation attributable to that PC. PCs can be used to display the objects in two dimensions (e.g., the first PC vs. the average abundance of species). The Genstat software directive for PCA is PCP.

Bi-Plot

PCs can be worked out for environment or sites as well, following the above procedure. The two sets of PCs, one for the species and the other for the environments, will give bi-plots. Such bi-plots can be used to determine if a specific set of species are associated with certain sites. The Genstat library procedure for this is BIPLOT.

Correspondence Analysis

Considering the species-by-site data as row-columns, one can obtain iteratively the scores for species and sites using the direct gradient method. The transformed data $Y (= y_{ij})$ are obtained from the observed frequency data $X (= x_{ij})$ after correcting for the proportional model. Thus, $y_{ij} = x_{ij} - x_{i.}x_{.j}/x_{..}$ (where the dots indicate totals over that suffix set), which can be used to calculate the scores for species and for sites:

$$a_i = \left(\frac{\rho^{-1} \sum y_{ij} b_j}{x_{i.}} \right) \text{and } b_j = \left(\frac{\rho^{-1} \sum y_{ij} a_i}{x_{.j}} \right)$$

where ρ is a constant to keep the score within range.

This is called *reciprocal averaging* (Digby and Kempton 1987). The correspondence analysis generalizes the reciprocal averaging approach in two dimensions, using matrix algebra tools and the results of spectral decomposition. The correspondence analysis was also approached using concepts in mechanics (Benzecri 1973, Greenacre 1984) leading to similar results. The Genstat library procedure is CORRESP.

Principal Coordinate Analysis

PCO, unlike PCA and other methods, uses an $n \times n$ (e.g., number of species) symmetric matrix of associations, similarity, or distances (Gower 1966). Then PCO is employed on such a matrix to give principal coordinate scores. The Genstat software directive is PCO.

Canonical Variate Analysis

When the species or sites are grouped (or have some structure, perhaps indicating similarity), CVA could be used to validate the existing groupings or to assign the membership to a new species in one of the groups. Using the $n \times p$ data from n species and p sites (variates), one can obtain within-group and between-group sums of squares and product matrices. CVA, also known as *linear discriminant analysis*, provides linear functions to maximize the ratio of between-group to within-group variation. The Genstat software directive for this analysis is CVA.

Canonical Correlation Analysis

Consider the case where a set of variables can be divided into two groups (e.g., abundance of species as one set of variables and environmental factors as the second set, observed over sites). Canonical correlation analysis provides a linear combination of variables in the first set (species abundance) and another linear combination of variables in the second set (environmental factors) such that the correlation between the two variables generated by the two linear combinations is maximal. Similarly, a second pair of linear combinations could be generated to give the next maximal correlation. Thus, the linear combinations of the variables could be used for prediction. The GENSTAT library procedure CANCOR can be used to generate these linear combinations.

Methods for Comparing Ordinations

Procrustes rotation was named after the innkeeper in Greek mythology who used to match the guest to the bed by adjusting the limbs of the guests. If there are more than one ordination of the same set of objects, the Procrustes rotation is used to determine the consistency between them. For example, one may be interested in comparing two ordinations for the same set of sites, perhaps one based on species abundance and the other based on environmental factors. Here, one of the two sets of coordinates of the n points in r-dimension is treated as a fixed configuration (the X-matrix), while the other configuration (the Y-matrix) is shifted and rotated to best match with X-matrix. A measure of goodness of fit is produced as the residual sum of squares. Its generalization to more than two ordinations is called the generalized Procrustes rotation (Gower 1975, 1985). The Genstat software directive ROTATE and library procedure GENPROC can be used.

Spatial Pattern of the Species

A feature of species may be that their abundance is associated with specific locations. Spatial distribution of the abundance of species with site references can be used to predict the abundance at a site where data were not collected. Spatial models use the stochastic behavior of the variable over space (Cressie 1993), in contrast to the classical approaches. The observations of the variable being modeled are assumed to be independent and randomly distributed, and predictions are made on means irrespective of the location. The modeling requires the concept of variation with distance, a variogram, and its parameters such as nugget (microscale variation), sill (the maximum variation between any two points), and range (distance within which there is variation in the variogram) beyond which variance does not depend on distance. Geostatistical programs are used. The Genstat software directive FVARIOGRAM and library procedures MVARIOGRAM and KRIGE could be used.

GERMPLASM DATABASES, BIOINFORMATICS, AND SOFTWARE

Information on collection missions, environment, sites of collection, and GIS maps/spatial tables of environmental variables, accession-specific data, and morphological and molecular data are valuable resources (e.g., the genetic resource database at

ICARDA). With advances in biotechnology, the data at the molecular levels are being electronically stored with volume of the database growing with time. Bioinformatics, a system to store, retrieve, manipulate, and interpret genomic data, is now being recognized as a very useful and active discipline.

REFERENCES

Bari, A. and M. Singh. 1998. Analysis of plant genetic resources data. Lecture notes for a training course on *Documentation and information management of plant genetic resources*, 22 Nov–3 Dec 1998. Aleppo, Syria: ICARDA.

Bennett, E. 1970. Tactics of plant exploration. In *Genetic resources in plants—Their exploration and conservation*, eds. O.H. Frankel and E. Bennett, 157–179. Oxford: Blackwell.

Benzecri, P.J. 1973. *L'analyse des donnees, Vol 2. L'analyse des correspondances.* Paris, France: Dunod.

Bowman, K.O., K. Hutcheson, E.P. Odum, and L.R. Shenton. 1971. Comments on the distribution of indices of diversity. In *Statistical ecology, Vol. 3. Many species populations, ecosystems, and systems analysis*, eds. G.P. Patil, E.C. Pielou, and W.E. Walters, 315–336. University Park, PA: Pennsylvania State University Press.

Brown, A.H.D. 1989. Core collections: A practical approach to genetic resources management. *Genome* 31:818–824.

Brown, A.H.D. 1992. Human impact on plant gene pools and sampling for their conservation. *Oikos* 63:109–118.

Brown, A.H.D. 1995. The core collection at the crossroads. In *Core collections of plant genetic resources*, eds. T. Hodgkin, A.H.D. Brown, Th.J.L. van Hintum, and E.A.V. Morales, 3–19. New York: John Wiley & Sons.

Brown, A.H.D. and J.D. Briggs. 1991. Sampling strategies for genetic variation in ex situ collections of endangered plant species. In *Genetics and conservation of rare plants*, eds. D.A. Falk and K.E. Holsinger, 99–122 New York: Oxford University Press.

Brown, A.H.D. and D.R. Marshall 1995. A basic sampling strategy: Theory and practice. In *Collecting plant genetic diversity: Technical guidelines*, eds. L. Guarino, V. Ramanatha Rao, and R. Reid, 76–91. Wallingford, CT: CAB International.

Buckland, S.T., D.R. Anderson, K.P. Burnham, and J.L. Laake. 1993. *Distance sampling: Estimating abundance of biological populations.* London: Chapman & Hall.

CBD. 1992. *Convention on biological diversity.* https://www.cbd.int/doc/legal/cbd-en.pdf.

Chang, T.T. 1985. Collection of crop germplasm. *Iowa State J Res* 59:349–364.

Clifford, H.T. and W. Stephenson. 1975. An introduction to numerical classification. New York: Academic Press.

Cochran, W.G. 1977. *Sampling techniques*, 3rd ed. New York: John Wiley & Sons.

Cormack, R.M. 1971. A review of classification. *J Roy Stat Soc A Sta* 134:321–367.

Cressie, N.A.C. 1993. *Statistics for spatial data*, Rev. ed. New York: John Wiley & Sons.

Damania, A.B. 1987. Sampling cereal diversity in Morocco. *Plant Genetic Resources Newsletter* 72:29–30.

Digby, P.G.N. and R.A. Kempton. 1987. *Multivariate analysis of ecological communities.* London: Chapman & Hall.

Efron, B. and R.J. Tibshirani. 1993. *An introduction to the bootstrap.* New York: Chapman & Hall.

Engels, J.M.M., R.K. Arora, and L. Guarino. 1995. An introduction to plant germplasm exploration and collecting: Planning, methods and procedures, follow-up. In *Collecting plant genetic diversity: Technical guidelines*, eds. L. Guarino, V. Ramanatha Rao, and R. Reid, 31–63. Wallingford, CT: CAB International.

Engen, S. and C. Taillie. 1979. A basic development of abundance models: Community description. In *Statistical distributions in ecological work*, eds. J.K. Ord, G.P. Patil, and C. Taillie, 289–311. Fairland, MD: International Co-operative Publishing House.

Frankel, O.H. 1984. Genetic perspectives of germplasm conservation. In *Genetic manipulation: Impact on man and society*, eds. W. Arber, K. Illmensee, W.J. Peacock, and P. Starlinger, 161–169. Cambridge: Cambridge University Press.

Frankel, O.H. and A.H.D. Brown. 1984. Current plant genetic resources: A critical appraisal. In *Genetics: New frontiers, Vol. 4. Applied genetics. Proceedings of the International Congress of Genetics, 15th*, New Delhi, India. December 12–21, 1983, eds. V.L. Chopra, B.C. Joshi, R.P. Sharma, and H.C. Bansal, 1–11. New Delhi, India: Oxford and IBH.

Gower, J.C. 1966. Some distance properties of latent root and vector methods used in multivariate analysis. *Biometrika* 53:325–328.

Gower, J.C. 1975. General procrustes analysis. *Psychometrika* 40:33–51.

Gower, J.C. 1985. Measures of similarity, dissimilarity and distance. In *Encyclopaedia of Statistics, Vol. 5*, eds. N.L. Johnson, S. Kotz, and C.B. Read, 397–405. New York: John Wiley & Sons.

Greenacre, M.J. 1984. *Theory and applications of correspondence analysis*. London: Academic Press.

Guarino, L., V. Ramanatha Rao, and R. Reid (eds.). 1995. *Collecting plant genetic diversity: Technical guidelines*. Wallingford, CT: CAB International.

Hooper, D.U., E.C. Adair, B.J. Cardinale et al. 2012. A global synthesis reveals biodiversity loss as a major driver of ecosystem change. *Nature* 486:105–108.

Hurlbert, S.H. 1971. The nonconcept of species diversity: A critique and alternative parameters. *Ecology* 52(4):577–586.

Hutcheson, K. 1970. A test for comparing diversities based on the Shannon formula. *J Theor Biol* 29:151–154.

Kameswara Rao, N. and P. J. Bramel (eds.). 2000. *Manual of genebank operations and procedures*. Technical Manual no. 6. India: ICRISAT.

Kimura, M. and J.F. Crow. 1964. The number of alleles that can be maintained in a finite population. *Genetics* 49:725–738.

Kindt, R. and R. Coe. 2005. *Tree diversity analysis: A manual and software for common statistical methods for ecological and biodiversity studies*. http://www.worldagroforestry.org/resources/databases/tree-diversity-analysis.

Magurran, A.E. 1988. *Ecological diversity and its measurement*. Kent: Croom Helm.

Marshall, D.R. and A.H.D. Brown. 1975. Optimum sampling strategies in genetic conservation. In *Crop genetic resources for today and tomorrow*, eds. O.H. Frankel and J.G. Hawkes, 53–80. Cambridge: Cambridge University Press.

May, R.M. 1975. Patterns of species abundance and diversity. In *Ecology and evolution of communities*, eds. M.L. Cody and J.M. Diamond. 81–120. Cambridge, MA: Harvard University Press.

Nabhan, G.P. 1990. *Wild Phaseolus Ecogeography in the Sierra Madre Occidental, Mexico: Aerographic Techniques for Targetting and Conserving Species Diversity*, Systematic and Ecogeographic Studies on Crop Genepools 5. Rome, Italy: IBPGR and FAO/UN.

Payne, R.W. (ed.). 2014. *The guide to GenStat® Release 17. Part 2: Statistics*. Hemel Hempstead: VSN International.

Porceddu, E. and A.B. Damania. 1991. *Sampling strategies for conserving variability of crop genetic resources in seed crops*. Technical Manual No. 17. Aleppo, Syria: ICARDA.

R Development Core Team. 2009. *R: A language and environment for statistical computing*. Vienna, Austria: R Foundation for Statistical Computing. http://www.R-project.org.

Rao, C.R. 1973. *Linear statistical inference and its applications*. New York: John Wiley & Sons.

Schoen, D.J. and A.H.D. Brown. 1995. Maximizing genetic diversity in core collections of wild relatives of crop species. In *Core collections of plant genetic resources*, eds. T. Hodgkin, A.H.D. Brown, Th.J.L. van Hintum, and E.A.V. Morales, 55–76. New York: John Wiley & Sons.

Sedcole, J.R. 1977. Number of plants necessary to recover a trait. *Crop Sci* 17:667–668.

Sukhatme, P.V., B.V. Sukhatme, S. Sukhatme, and C. Asok. 1984. *Sampling theory of surveys with applications*, 3rd ed. Ames, IA: Iowa State University Press.

UNCED. 1992. *Agenda 21*. United Nations Conference on Environment and Development, June 3–14, 1992. http://www.unep.org/Documents.Multilingual/Default.asp?documentid=52.

Valkoun, J. and A.B. Damania. 1990. *Report on the germplasm collecting mission to Tibet (Autonomous Region of China)* (unpublished). Aleppo, Syria: ICARDA.

Weir, B.S. 1990. *Genetic data analysis*. Sunderland, MA: Sinauer Associate.

Whittaker, R.H. 1977. Evolution of species diversity in land communities. *Evol Biol* 10:1–67.

WRI, IUCN, and UNEP. 1992. *Global biodiversity strategy: Guidelines for action to save, study, and use Earth's biotic wealth sustainably and equitably*. Washington, DC: WRI, IUCN and UNEP. http://pdf.wri.org/globalbiodiversitystrategy_bw.pdf.

Yonezawa, K., T. Nomura, and H. Morishima. 1995. Sampling strategies for use in stratified germplasm collections. In *Core collections of plant genetic resources*, eds. T. Hodgkin, A.H.D. Brown, Th.J.L. van Hintum, and E.A.V. Morales, 35–53. New York: John Wiley & Sons.

Section IV

Applied Omics Technologies

16 Exploiting Germplasm Resources for Climate-Change Adaptation in Faba Bean

F.L. Stoddard, H. Khazaei, and K.Y. Belachew

CONTENTS

The environmental benefits of grain legumes in crop rotations are well established (Bues et al. 2013), but because of their relatively minor role in agriculture in comparison with cereals, breeding of legumes for stress resistance is poorly developed. Technologies for germplasm identification, genomic analysis, gene identification, and rapid breeding have greatly advanced, so it is now possible for legume breeding to catch up with cereal and oilseed breeding.

Faba bean (*Vicia faba* L.) is chosen for this study for the following reasons. It has a higher protein concentration than other starchy cool-season legumes, and often the biomass production is high with high protein yield. It is grown from 63°N to 45°S on all inhabited continents, so there is a wealth of diversity in the ~20,000 accessions in the germplasm banks. It is of increasing importance in the European Union following the latest reform of the Common Agricultural Policy, and it is a key food crop in West Asia and North Africa. It has a reputation for being more tolerant to waterlogging and less tolerant to water deficit than other grain legumes (Stoddard et al. 2006, Khan et al. 2010), so there are traits that need strengthening and others that can be more widely exploited. Since it has a very large genome, 13 Gbp, its genomic analysis lags behind that of some other crops, but alternatives lie in new techniques such as genotyping-by-sequencing (GBS), where the complete expressed genome is sequenced rather than the complete genome (Elshire et al. 2011), transcriptomic studies, and population sequencing (POPSEQ), where the genome fragments are joined together and ordered (Mascher et al. 2013). Genomic information from faba bean, as it is gathered, can be referred to the extensively (though still incompletely) annotated complete genome of the model legume, *Medicago truncatula* (Webb et al. 2015), allowing identification of potential candidate genes in quantitative trait loci (QTLs) for stress response (Khazaei et al. 2014b). Methods for single-seed descent have been developed that allow up to seven generations to be advanced in a single year (Mobini et al. 2015).

The first step in the process is the identification of germplasm carrying beneficial traits. The focused identification of germplasm strategy (FIGS; Mackay and Street 2004) is a tool for selecting material most likely to carry those traits, based on the provenance of the accession. Passport data are often incomplete, however, and alternative tools such as core collections have a role. The set of germplasm then needs to be narrowed down by appropriate phenotyping to a handful of accessions that can be used in a breeding program. The source of the stress-resistance trait is crossed with an otherwise adapted and high-yielding breeding line or cultivar, and may be backcrossed to the adapted parent in order to reduce linkage drag of undesirable alleles from the donor parent. At this stage, modern technologies for marker-assisted selection become highly effective, as the desirable alleles may be recessive or quantitatively expressed, and thus hidden by the other alleles. The closer the marker to the gene in question, the better, and ideally the marker is part of the gene itself. The identification of an appropriate marker is greatly accelerated when there is a genome sequence available, hence part of the importance of genomics.

SCREENING GERMPLASM

Although it is grown in many drought-prone regions around the world, faba bean is often considered deficient in drought tolerance and avoidance (Khan et al. 2010). Transient drought is expected to increase in frequency and severity as climate changes in the coming decades (Dai 2013), so resilient cultivars are needed for all parts of the world. For a study on resistance to water deficit, we contacted the International Centre for Agricultural Research in the Dry Areas (ICARDA) for germplasm and discussed the application of FIGS to the question. Sufficient glasshouse space and student time were available to screen about 400 accessions, so 200 were selected from moisture-limited regions and 200 from moisture-adequate regions, according to FIGS (Khazaei et al. 2013a). These were grown in a climate-controlled glasshouse and various aspects of stomatal morphology and physiology were evaluated (Khazaei et al. 2013a,b). The means of most traits showed significant differences between the two germplasm sets, but there was considerable overlap. The breadth of scatter was such that sets of 100 would clearly have been adequate to distinguish the two groups, and perhaps even 50 would have sufficed.

Root traits were not systematically investigated in that study, and considerable variation in root length, root length distribution, and root biomass remains to be quantified (Figure 16.1). The ability of roots to access water in drying soils is an important part of drought avoidance. Maintenance of root growth during early phases of water deficit helps to achieve this, but measurement of this trait will require the development of new techniques. Further examination of the sets may lead to the identification of other traits contributing to the maintenance of productivity in conditions of transient water deficit. Since modeling indicates that stomata optimize the uptake of carbon per unit of water that is transpired to the atmosphere (Duursma et al. 2013), genotypic differences in the model parameter g_1, relating to the marginal cost of water to the plant, could be exploited in breeding for drought resilience.

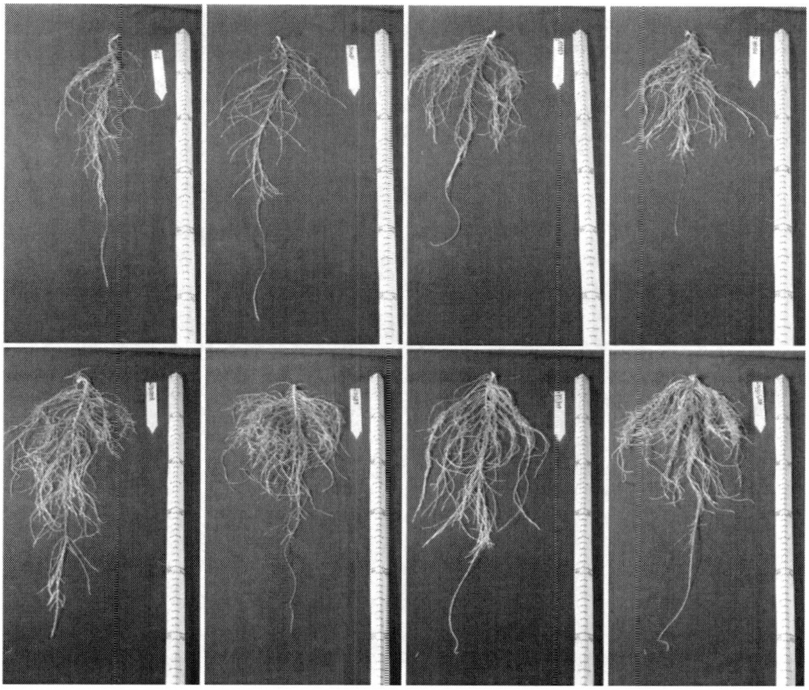

FIGURE 16.1 Root systems of faba bean accessions from wet regions (*top row*) and dry regions (*bottom row*).

Attention to other traits during screening may produce some useful outcomes. During the screening for stomatal response, data were gathered on flowering date. Several sources of extremely early flowering were identified (Khazaei et al. 2012), and some of them are also earlier to mature than the Finnish cultivar *Kontu*, which is considered the world's earliest-maturing faba bean cultivar (Figure 16.2) (Stoddard and Hämäläinen 2011). These have already been used in crossing for both experimental and commercial purposes.

Acidity and aluminum toxicity have been highlighted as major challenges for future sustainable cropping systems (Witcombe et al. 2008). It is widely recommended that faba bean be sown into soils with a nearly neutral pH, from 6.5 to 8.0 (Hebblethwaite et al. 1983). Lower soil pH is associated with paler leaves, reduced seed weight per plant, and greater incidence of chocolate spot infection (Elliott and Whittington 1978). While liming is well understood in many countries, in others, it is not understood or lime is not available (Tesema et al. 2012). Phosphorus availability is also low in acid soils (Rengel 2011), and the farmer may not be able to afford phosphorus fertilizer. Soil acidification in Ethiopia has thus led farmers to abandon faba bean in favor of white lupin, a less-liked food. The effects of climate change, particularly changed rainfall patterns, on soil acidification have not been modeled, so it is not yet known whether the problem

FIGURE 16.2 Faba bean cultivar *Kontu* (*left*) and earliness accession *IG 132238* (*right*).

will spread significantly, but it is already severe in regions as far apart as parts of Australia and Finland, and Rengel (2011) expected it to expand. Aluminum is present in its ionic form Al^{3+} at low pH values, and this is considered the most yield-limiting factor in many acid soils (Rengel 2011). Hence, improved tolerance to soil acidity and particularly to Al^{3+} is desirable in faba bean germplasm. A set of 20 accessions has been collected from acid-soil regions of Ethiopia, and 10 cultivars have been gathered from cool-climate regions where beans are usually grown in neutral to alkaline soils. Tests are based on hematoxylin binding to $AlPO_4$ formed at the root surface during 24 h of exposure to Al^{3+}, and on the regrowth of roots after exposure (Hede et al. 2001), and the first trials show a wide range of responses in this germplasm set.

Crop-quality traits are seldom subject to the same selection pressures as adaptation traits, unless they are culturally important. Finding a low vicine–convicine allele required screening of 919 germplasm accessions, until *vc-* was found in a Greek landrace (Duc et al. 1989). Finding a low or null lipoxygenase allele that would improve food quality may be even harder, and require mutagenesis rather than simple screening of available material. Null alleles for three lipoxygenase genes have been identified and recombined in soybean, leading to a null-lipoxygenase phenotype that is suitable for many food uses (Lenis et al. 2010). Since soybean is an ancient tetraploid with some silencing and neofunctionalization of the extra copies, we may expect to seek two lipoxygenase genes for silencing in faba bean.

UNDERSTANDING AND USING THE VARIATION

Having demonstrated that there is useful variation in a trait, one then has to capture it into a useful form. Historically, this would be done by crossing the source plant to an adapted cultivar and backcrossing it two to three times to recover as much of the adapted parent as possible, while testing for the desired phenotype from the source. This process is made faster and more reliable when there is more information about the genetic loci involved in the trait.

We used inbred lines derived from *Aurora*, a drought-sensitive Swedish cultivar; *Mélodie*, a drought-avoiding French cultivar; and *ILB938*, a drought-tolerant Ecuadorian landrace, for crossing in all possible combinations and advanced the families by single-seed descent. The F_5 generation of Mélodie/2 × ILB938/2, comprising 211 recombinant inbred lines, was subjected to SNP genotyping at 188 loci and was phenotyped in the glasshouse (Khazaei et al. 2014b). The stomatal conductance of Mélodie was much greater than that of ILB938 in well-watered conditions, but dropped below that of ILB938 when soil moisture was reduced to one-tenth of field capacity. In the progeny, QTLs for stomatal density, stomatal length, and both stomatal conductance and canopy temperature in well-watered conditions, co-located on chromosome II, with increased conductance and cooler leaves associated with the Mélodie allele. At the other end of the same chromosome, about 130 cm away, was another locus affecting canopy temperature in water-deficit conditions, where the allele conferring cooler leaves came from ILB938 (Khazaei et al. 2014b). We now have the possibility to test the four possible combinations of these alleles for their effects on water use in well-watered and water-limited conditions, in an otherwise random genetic background, and determine whether there is an advantage conferred by any particular combination.

The same population segregated for vicine–convicine content (Khazaei et al. 2015), stipule spot pigmentation (Khazaei et al. 2014a), green seed-coat color, and two loci affecting seed size (Khazaei et al. 2014b). In most of these cases, synteny with the *Medicago truncatula* genome has allowed us to nominate candidates for the genes affecting these traits. Root traits need to be quantified in this population, as clear differences in root biomass and distribution were visible when the pots were emptied at the end of the experiment. Water-use efficiency and water productivity (or *efficient use of water*) can also be determined, and QTLs mapped. The population (now at F_7) is about to be subjected to GBS to provide a dense array of markers and extensive genomic information. It has been shared with international colleagues for further phenotyping, so that we can accumulate information about both overall drought response and individual components of that response from a single population, with the aim of developing an understanding of how the components interact and which are more important in a given circumstance.

Our project on aluminum tolerance is at an earlier stage. Treatment with aluminum led to responses in many hundreds of genes in soybean (Duressa et al. 2010) and *M. truncatula* (Chandran et al. 2008), and in the latter species, a citrate-exuding multidrug and toxin efflux (MATE) transporter was identified as key. The current version of the *M. truncatula* genome lists three MATE efflux genes, two aluminum-activated carboxylate transporters, and three other aluminum-response genes, so there is a small number of candidates on which to base hypotheses that can be tested in faba bean.

The region in *M. truncatula* enclosing the syntenic region to the QTL for vicine–convicine concentration includes no genes identifiably associated with pyrimidine metabolism (Khazaei et al. 2015), but has seven transcription factors, one of which could be the candidate for downregulation in the *vc-* mutation. Candidates for neofunctionalization of pyrimidine metabolism lie elsewhere in the genome.

CLIMATE CHANGE AND THE PROTEIN DEFICIT

The European Union's self-sufficiency in plant-based protein-rich feed materials was only 31% in 2013 (Bouxin 2014), so the figure of 70% dependence on imports is widely quoted. About two-thirds of this import is soybean meal. While some of the imported protein feed is used for ruminants (particularly dairy animals), most of it goes into poultry and pig production (Bues et al. 2013) on account of its high content of ileal-digestible lysine, which is key for these animals. Crop production in many European countries has become increasingly cereal-based over the last 50 years, partly because Europe's climates and soils are so well suited to the filling of wheat and barley grains. In France, for example, wheat yields have risen from 1.5 times those of soya bean in 1971 to 3 times in 2011, whereas in the United States, yields of the two crop species have remained equal to each other (Eurostat and US Department of Agriculture data, reported by Bues et al. 2013). At the same time, pastures and silage crops have become increasingly grass based rather than including forage legumes. When the nitrogen fertilizer for these cereals and grasses is synthetic, it is manufactured with the use of energy from fossil fuel, and whether it is synthetic or manure, some of its nitrate leaches into groundwater and denitrification releases N_2O, a powerful greenhouse gas. European agriculture thus contributes to global change: its residents consume too much meat, there is too much cereal or grass monoculture, and too much pollution, whereas feed producers and feed consumers have become geographically disconnected. High and unstable prices for fossil fuel and soybean meal have led to concerns about food and feed security in the European Union. Production of grain legumes (including soybean) in the European Union is desirable to solve some of these problems, and it is achievable if certain impediments are overcome.

One of those impediments is that legumes yield less in tons and euros per hectare than cereals, although the beneficial pre-crop effects of the legume on the following cereal help to make up the financial gap. In order to fix the yield deficit, modern technologies need to be applied, which requires public sector funding to public sector organizations such as universities and research institutes that can collaborate across national and geographic boundaries. Since the stresses that reduce legume yields occur globally, research funded or undertaken in one country can benefit many others. Developing cultivars of faba bean that are less sensitive to water deficit (or heat, waterlogging, cold, salinity, or acidity) will increase seed yield per hectare and reduce several of the causes of yield instability, enabling more farmers in developed and developing countries to grow the crop profitably. Cultivars with improved quality will open up new food and feed markets, relieving another impediment, more common in developed than in developing countries, the lack of markets for the harvested seed. Achieving these aims requires focused exploitation of genetic resources, in order to identify sources of desirable alleles, combined with the application of modern, marker-assisted breeding methods, in order to combine the stress resistances with quality, adaptation, and yield factors into broadly adapted cultivars.

REFERENCES

Bouxin, A. 2014. *Food and feed statistical yearbook 2013*. Brussels, Belgium: FEFAC, European Feed Manufacturers' Association.

Bues, A., S. Preissel, M. Reckling et al. 2013. *The environmental role of protein crops in the new Common Agricultural Policy, Study*. European Parliament, Directorate General for Internal Policies, Policy Department B: Structural and Cohesion Policies, Agricultural and Rural Development IP/B/AGRI/IC/2012-067. doi:10.2861/27627. http://www.europarl.europa.eu/RegData/etudes/etudes/join/2013/495856/IPOL-AGRI_ET(2013)495856_EN.pdf.

Chandran, D., N. Sharopova, K.A. VandenBosch et al. 2008. Physiological and molecular characterization of aluminum resistance in *Medicago truncatula*. *BMC Plant Biol* 8:89. doi:10.1186/1471-2229-8-89.

Dai, A. 2013. Increasing drought under global warming in observations and models. *Nature Climate Change* 3:52–58. doi:10.1038/nclimate1633.

Duc, G., G. Sixdenier, M. Lila, and V. Furstoss. 1989. Search of genetic variability for vicine and convicine content in *Vicia faba* L. A first report of a gene which codes for nearly zero-vicine and zero-convicine contents. In *Recent advances of research in antinutritional factors in legume seeds*, eds. J. Huismann, T.F.B. van der Poel, and I.E. Liener, 305–313. Wageningen, The Netherlands: Pudoc.

Duressa, D., K. Soliman, and D. Chen. 2010. Identification of aluminum responsive genes in Al-tolerant soybean line PI 416937. *Int J Genomics* 2010:ID 164862. doi:10.1155/2010/164862.

Duursma, R.A., P. Payton, M.P. Bange et al. 2013. Near-optimal response of instantaneous transpiration efficiency to vapour pressure deficit, temperature and [CO_2] in cotton (*Gossypium hirsutum* L.). *Agr Forest Meteorol* 168:168–176. doi:10.1016/j.agrformet.2012.09.005.

Elliott, J.E.M. and W.J. Whittington. 1978. The effect of soil pH on the severity of chocolate spot infection on field bean varieties. *J Agr Sci Cambridge* 91:563–567. doi:10.1017/S0021859600059943.

Elshire, R.J., J.C. Glaubitz, Q. Sun et al. 2011. A robust, simple genotyping-by-sequencing (GBS) approach for high diversity species. *PLoS ONE* 6:e19379. doi:10.1371/journal.pone.0019379.

Hebblethwaite, P.D., G.C. Hawtin, and P.J.W. Lutman. 1983. The husbandry of establishment and maintenance. Chapter 13. In *The faba bean (*Vicia faba *L.), a basis for improvement*, ed. P.D. Hebblethwaite, 271–312. London: Butterworth.

Hede, A.R., B. Skovmand, and J. López-Cesati. 2001. Acid soils and aluminum toxicity. Chapter 15. In *Application of physiology in wheat breeding*, eds. M.P. Reynolds, J.I. Ortiz-Monasterio, and A. McNab, 172–182. Mexico, DF: CIMMYT.

Khan, H.R., J.G. Paull, K.H.M. Siddique, and F.L. Stoddard. 2010. Faba bean breeding for drought-affected environments: A physiological and agronomic perspective. *Field Crop Res* 115:279–286. doi:10.1016/j.fcr.2009.09.003.

Khazaei, H., D.M. O'Sullivan, H. Jones et al. 2015. Flanking SNP markers for vicine-convicine concentration in faba bean (*Vicia faba* L.). *Mol Breed* 35:38. doi:10.1007/s11032-015-0214-8.

Khazaei, H., D.M. O'Sullivan, M.J. Sillanpää, and F.L. Stoddard. 2014a. Genetic analysis reveals a novel locus in *Vicia faba* decoupling pigmentation in the flower from that in the extra-floral nectaries. *Mol Breed* 34:1507–1513. doi:10.1007/s11032-014-0100-9.

Khazaei, H., D.M. O'Sullivan, M.J. Sillanpää, and F.L. Stoddard. 2014b. Use of synteny to identify candidate genes underlying QTL controlling stomatal traits in faba bean (*Vicia faba* L.). *Theor Appl Genet* 127:2371–2385. doi:10.1007/s00122-014-2383-y.

Khazaei, H., F.L. Stoddard, C. Lizarazo, and K. Street. 2012. New sources of earliness for Finnish faba bean breeding. Paper presented at Maataloustieteenpäivät 2012, Viikki, Helsinki, Finland, January 10–11, 2012, ed. N. Schulman, 147. http://www.smts.fi/Kasvintuotanto_jalostuu/Khazaei_New%20sources.pdf.

Khazaei, H., K. Street, A. Bari, M. Mackay, and F.L. Stoddard. 2013a. The FIGS (Focused Identification of Germplasm Strategy) approach identifies traits related to drought adaptation in *Vicia faba* genetic resources. *PLoS ONE* 8:e63107. doi:10.1371/journal.pone.0063107.

Khazaei, H., K. Street, A. Santanen et al. 2013b. Do faba bean (*Vicia faba* L.) accessions from environments with contrasting seasonal moisture availabilities differ in stomatal characteristics and related traits? *Genet Resour Crop Ev* 60:2343–2357. doi:10.1007/s10722-013-0002-4.

Lenis, J.M., J.D. Gillman, J.D. Lee, J.G. Shannon, and K.D. Bilyeu. 2010. Soybean seed lipoxygenase genes: Molecular characterization and development of molecular marker assays. *Theor Appl Genet* 120:1139–1149. doi:10.1007/s00122-009-1241-9.

Mackay, M. and K. Street. 2004. Focused identification of germplasm strategy—FIGS. In *Proceedings of the 54th Australian cereal chemistry conference and the 11th wheat breeders' assembly*, September 21–24, 2004, Canberra, ACT, Australia, eds. C.K. Black, J.F. Panozzo, and G.J. Rebetzke, 138–141. Melbourne, Australia: Cereal Chemistry Division, Royal Australian Chemical Institute (RACI).

Mascher, M., G.J. Muehlbauer, D.S. Rokhsar et al. 2013. Anchoring and ordering NGS contig assemblies by population sequencing (POPSEQ). *Plant J* 76:718–727. doi:10.1111/tpj.12319.

Mobini, S.H., M. Luisdorf, T.D. Warkentin, and A. Vandenberg. 2015. Plant growth regulators improve in vitro flowering and rapid generation advancement in lentil and faba bean. *In vitro Cell Dev Biol Plant* 51:71–79. doi:10.1007/s11627-014-9647-8.

Rengel, Z. 2011. Soil pH, soil health and climate change. Chapter 4. In *Soil health and climate change, soil biology*, eds. B.P. Singh, A.L. Cowie, and K.Y. Chan, 69–86. Berlin, Germany: Springer-Verlag.

Stoddard, F.L., C. Balko, W. Erskine, H.R. Khan, W. Link, and A. Sarker. 2006. Screening techniques and sources of resistance to abiotic stresses in cool season food legumes. *Euphytica* 147:167–186. doi:10.1007/s10681-006-4723-8.

Stoddard, F.L. and K. Hämäläinen. 2011. Towards the world's earliest maturing faba beans. *Grain Legumes* 56:9–10.

Tesema, G., M. Argaw, and E. Adgo. 2012. Farmers soil management practices and their perceptions to soil acidity, at Ankesha District of Awi Zone, Northwestern Ethiopia. *Libyan Agr Res Cent J Int* 3:64–72.

Webb, A., A. Cottage, T. Wood et al. 2015. A SNP-based consensus genetic map for synteny-based trait targeting in faba bean (*Vicia faba* L.). *Plant Biotech J* (preprint version). doi:10.1111/pbi.12371.

Witcombe, J.R., P.A. Hollington, C.J. Howarth, S. Reader, and K.A. Steele. 2008. Breeding for abiotic stresses for sustainable agriculture. *Philos T Roy Soc B* 363:703–716. doi:10.1098/rstb.2007.2179.

17 Developing Climate-Change Adaptive Crops to Sustain Agriculture in Dryland Systems through Applied Mathematics and Genomics

S. Dayanandan and A. Bari

CONTENTS

Increasing crop production to feed the rapidly growing human population in a sustainable manner while maintaining healthy ecosystems under changing climatic conditions remains one of the greatest challenges of our time (Godfray et al. 2010, Wheeler and von Braun 2013). Crop cultivars with increased tolerance to heat, drought, and salinity, and resistance to emerging pests and diseases are urgently needed as a means to adapt and sustain agricultural productivity under changing climatic conditions and mitigate the negative effects of climate change. Although predictions of global climate models differ substantially (IPCC 2013), they largely converge in projections of increase in temperature and frequency of droughts in drylands, one of the most climate-change-sensitive land systems of the world. Drylands cover over 40% of the land surface and are home to over 2.5 billion (mostly poor) people. Improvement of crops to meet climate-change-imposed challenges in drylands will depend largely on the identification and effective utilization of genes related to climate-change-adaptive traits in plant genomes. Most genetic variants related to these traits are likely present in germplasm in genebanks or in wild relatives of crop plants. Discovering new and useful alleles of these genes in over seven million accessions of crop germplasm stored in more than 1700 national and international genebanks remains a challenge.

Plants have evolved adapting to various environmental conditions ranging from dry deserts and wet tropics to cold arctic and alpine environments, showing an exceptional ability to adapt to their environment. The historical environmental changes that accompanied the ice ages (Ruddiman et al. 1989, Webb and Bartlein 1992, Lisiecki and Raymo 2005) during the Quaternary period (2.4 mya to the present) have played important roles in the genetic diversification of plants and animals (Hewitt 2000, 2004, Willis and Niklas 2004). During the Pleistocene glacial period, most tropical plants adapted to wet climatic conditions may have been confined to moist patchy areas or *islands* embedded in seasonally dry areas (Whitmore 1990). In contrast to tropics, ice-free areas embedded in or adjacent to large ice sheets served as refugia in temperate regions (Pielou 1991). During the interglacial period, individuals may have migrated from these refugia, expanding their distribution ranges (Ashton and Gunatilleke 1987), and diversified (Waser and Campbell 2004, Zhao et al. 2013, Thomson et al. 2015) through selection induced by environmental factors such as soil type (McNeilly and Antonovics 1968), pollinators (Grant 1949), floral traits (Johnston 1991), or through neutral (genetic drift) genetic processes (Knowles and Richards 2005).

In recent history, crop domestication, which began around 10,000 years ago in the Fertile Crescent region and subsequently in various other regions of the world, led to intensive human-mediated selection of crop plants targeted for various traits. During domestication, individual plants with desired traits have been selected from wild progenitors, resulting in reduced genetic variation in domesticated crops (Tanksley and McCouch 1997, Doebley et al. 2006). These crop plants have undergone further selection in response to local environmental conditions. Thus, there is an overall reduction in genetic diversity during crop domestication, with highest standing diversity in wild progenitors followed by traditionally cultivated landraces and agronomically improved crop cultivars (Choudhury et al. 2013). Therefore, exploring genetic diversity in progenitors of modern crops (including accessions stored in genebanks) and integrating the genetic diversity into elite lines through breeding are crucial strategies for maintaining food security under changing climatic conditions.

In this chapter, we highlight the use of applied mathematics, environmental data, and landscape genomic approaches for targeted identification of germplasm with traits relevant for climate-change adaptation, and propose a platform for developing climate-change adaptive crops. The focus however is more on landscape genomics where inference involves association analyses of several layers of genomics and geographical data to develop tools as surrogates for collection site georeferences. The use of landscape genetics concepts and associated challenges are discussed in Segelbacher et al. (2010).

There is a potential of using landscape genomics to identify main factors driving patterns of selection by modeling the response of multiple genes involving gradients of multiple environmental variables, where the patterns of variation in adaptive genes (AG) can be considered as functions of environmental selection gradients (Schwartz et al. 2009).

The approach of landscape genetics will rely on the relationship that may exist between mostly spatial and genetic distances using metrics such as the Mantel test, which is a product term:

$$z = \sum_{i=1}^{n} \sum_{j=1}^{n} x_{ij} y_{ij}$$

where x_{ij} refers to the dissimilarity matrix over the spatial measurements corresponding to n sites whereas y_{ij} refers to that over the genetic measurements.

The equation is used in its normalized form as

$$r = \sum_{i=1}^{n} \sum_{j=1}^{n} \frac{(x_{ij} - \bar{x})}{\text{SD}_x} \frac{(y_{ij} - x\bar{y}}{\text{SD}_y}$$

where:

x and y refer to variables measured at locations i and j, respectively

n is the number of elements in the distance matrices

The terms SD_x and SD_y refer to the standard deviations for each of the two variable x and y

The Mantel test is used for searching patterns using matrices of association coefficients for departure from randomness against components of environmental variability (Burgman 1987, Mantel 1967).

In terms of spatial relationships, based on Mantel test results, the hypothesis of *un-relatedness* of the two matrices, genetic distance and geographical distance, can be either accepted or rejected. It is thus expected that the differences among pairs of sites that are closer to each other will be lower when compared to the differences among pairs of sites that are far from each other. Sites that are environmentally similar are also expected to be similar in terms of trait-states descriptions.

APPLIED MATHEMATICS AND FOCUSED IDENTIFICATION OF GERMPLASM WITH CLIMATE-CHANGE ADAPTIVE TRAITS

Plants are well known for their exceptional ability to adapt to local environmental conditions through natural selection. Thus, the likelihood of finding plants with traits adaptive to a given environment could be predicted based on the climate data of the geographical region where a given plant normally occurs. Therefore, choosing germplasm accessions from collection sites with selection pressure for the sought-after trait maximizes the chance of finding a germplasm with the desired trait (Mackay and Street 2004). The analyses of germplasm accession data (passport data), trait data, and environmental data employing sophisticated mathematical models provide a means to minimize the number of accessions to be screened to identify accessions with desirable traits that could be used for crop improvement through breeding. Through this approach, commonly known as focused identification of germplasm strategy (FIGS), El Bouhssini et al. (2009) identified one durum and eight bread wheat accessions with resistance to Sunn pest by screening only approximately 500 accessions, where previous screening of almost 2000 accessions failed to identify any accessions resistant to Sunn pest. Endresen (2010) used a multilinear data-modeling method to show predictive association between ecogeography and morphological traits in Nordic barley landraces. Endresen et al. (2011) evaluated stem rust in wheat and net blotch in barley using four different mathematical classification methods (LDA, PLS-DA, kNN, and SIMCA) and showed that ecogeographic distribution of both stem rust in wheat and net blotch in barley are linked to climatic factors. In a more recent study, Bari et al. (2012) used five modeling techniques (PCLR, GPLS, RF, NN, and SVM) and showed a strong predictive power in association between stem rust in wheat and agro-ecoclimatic factors of the collection sites. Overall, the FIGS approach has improved over the years, incorporating advanced mathematical modeling techniques, and has become an invaluable tool for ranking a large number of crop germplasm accessions to choose a manageable subset for trait evaluation through field trials.

PHYLOGEOGRAPHY, LANDSCAPE GENOMICS, AND IDENTIFICATION OF CLIMATE-CHANGE ADAPTIVE GENES

Phylogeography, or the analysis of evolutionary relatedness of organisms in a geographical framework (Avise et al. 1987, Avise 2000), has advanced over time as a science of integrative comparative study (Hickerson et al. 2010) leading to landscape genomics, which analyzes genome-wide genetic variation in a geographical context (Holderegger et al. 2006, Manel et al. 2010). These approaches provide an opportunity to examine migration, selection, and local adaptation of plants (Sork et al. 2013). A detailed understanding of the current geographical distribution patterns of genetic variability of crops in relation to putative refugia and centers of domestication provides a means to understand the role of past climate changes on evolution of biodiversity (Holder et al. 1999), and to gain insight into how species may respond and adapt to ongoing changes in the climate (Reusch and Wood 2007). The results of phylogeographic and landscape genomic analyses provide a framework for comparative genomic studies to develop taxon- or germplasm-diagnostic molecular markers to ascertain identity and purity of germplasm in breeding programs; determine evolutionary genetic distance among crop cultivars and choose appropriate germplasm for breeding for maximizing genetic diversity; and identify genomic regions associated with traits adaptive for given environmental conditions, providing opportunities to develop climate-change adaptive crops.

The approaches described here are proposed as alternatives to the methods requiring geographical data or georeferencing prior to use. *Landscape genetics* was first coined by Manel et al. (2003) and is a process to quantify the relationship between genetic variation, including genetic structure and the landscape variable (Storfer et al. 2007). This approach (Manel et al. 2003, Wagner et al. 2005) can be used where geographical information is lacking. Relationships between landscape features and genetic variation between and within plant populations can be explored and used as predictors for those accessions lacking both *ex ante* evaluation and geographical location information (Podolsky and Holtsford 1995, Bari et al. 2012).

Thus landscape-genetic associations, involving different layers of available omics data could be used as a surrogate for collection site georeferences.

The mechanisms through which climate change may directly or indirectly influence organisms will be manifested through AG, which are involved in controlling traits adaptive for a given environmental condition. Identifying AGs has been a tedious process and geographically explicit approaches, including comparison of genetic differentiation statistics across loci and geographic locations to detect genomic elements under selection, are gaining popularity (Handley et al. 2007, Manel et al. 2010, Neff and Fraser 2010). Recent advances in molecular techniques, particularly high-throughput next-generation DNA sequencing (NGS) technologies (Metzker 2010, Stapley et al. 2010, Ekblom and Galindo 2011), along with theoretical advancements (Siol et al. 2010), provide an unprecedented opportunity for discovering AGs in a broad range of organisms, including nonmodel plants. Selection in response to the environment acts in a locus-specific manner, leading to reduced genetic diversity in the target locus with no effect in other areas of the genome (Strasburg et al. 2012), and serving as a signature of selection. For example, in sunflower, quantitative trait loci (QTLs) and microsatellite-based analysis (Burke et al. 2005) revealed reduced diversity in parts of the genome due to selective sweeps during the post-domestication era.

AGs are generally identified through either top-down or bottom-up approaches (Ehrenreich and Purugganan 2006, Ross-Ibarra et al. 2007). The top-down approach uses the phenotype-associated genomic regions identified through QTL and linkage mapping to identify candidate genes. Although this approach has been successfully used to identify many candidate genes with adaptive importance, mapping to localize genomic regions requires carefully designed mapping populations and extensive work. In contrast, the genome-wide scanning of genetic polymorphism using high-throughput genotyping and NGS technologies (Davey et al. 2011) provides a novel approach to pinpoint AG and to identify genes or genomic regions associated with traits adapted to various environmental conditions. Recent reviews of the use of natural variations in genome-wide association studies (GWAS) in crop plants (Huang and Han 2014) and genomic insights into crop domestication (Olsen and Wendel 2013) outline a variety of genomic approaches for the identification of climate-adaptive traits in crops. GWAS have been successfully used to identify climate-sensitive QTLs (Li et al. 2010) and candidate loci for local adaptation with geographic and climatic signatures (Fournier-Level et al. 2011, Hancock et al. 2011) The genomic regions associated with climate-adaptive and agronomic traits have been identified through GWAS in several crops, including sorghum (Morris et al. 2013), barley (Pasam et al. 2012), rice (Huang et al. 2012), maize (Buckler et al. 2009, Kump et al. 2011, Poland et al. 2011, Tian et al. 2011), and millet (Jia et al. 2013). Although only a small fraction of the phenotypic variation is explained by genetic markers identified through GWAS in humans (Ott et al. 2011), a large proportion of phenotypic variation is associated with genetic markers identified through GWAS in plants (Atwell et al. 2010). Controlling for genetic relatedness and population structure (Brachi et al. 2011, Pasam et al. 2012) in GWAS is crucial to improve results by minimizing false positives and false negatives. Phylogeographic or landscape genomics approaches provide a means to control for genetic relatedness and population structure in GWAS analyses.

Among several commercially available NGS platforms (Metzker 2010, Glenn 2011), the Illumina NGS platform serves as one of the most cost-effective means for population genomic analyses. Although sequence data of the entire genome are ideal for genome-wide scanning, sequencing whole genomes of multiple individuals for population-level studies is often cost prohibitive. Therefore, reduced representation sequencing of enriched targets using solid phase oligonucleotide microarrays (Singh-Gasson et al. 1999), liquid phase biotinylated RNA capture (Gnirke et al. 2009), or restriction-site associated DNA sequencing approaches are often used. Reduced representation sequencing with barcoded, restriction-site associated DNA (RAD) (Baird et al. 2008, Ekblom and Galindo 2011) is one of the most cost-effective means to obtain genome sequence data for single nucleotide polymorphism (SNP) discovery and genotyping for AG identification through GWAS. Typically, equal amounts of DNA from multiple individuals of each population are barcoded and then pooled before sequencing, so that haplotypes from each population may be identified in the pool of sequences. The methods of library construction and inference of SNPs are described in Hohenlohe et al. (2010) and Emerson et al. (2010). In brief, each population sample comprising an equal quantity of DNA from multiple individuals is digested with a restriction enzyme, barcoded with a unique DNA sequence (about 5 bp), pooled, and then sequenced using the NGS platform. The resulting sequences are sorted based on the barcode and aligned de novo using bioinformatics tools such as Stacks and RADtools (Catchen et al. 2011) software packages. The pairwise population differentiation (F_{ST}) based on each locus is estimated using FDIST2 and BAYESCAN software (Narum and Hess 2011), and the loci that show significant deviation (F_{ST} outliers) are analyzed further to identify adaptive genomic regions or AGs. The identified AGs are annotated using NCBI-GenBank databases and used for studying the mechanisms of adaptation of plants to specific environments. The SNP and insertion/deletion (indel) diversity along the genome is scanned and compared across ecotypes. Regions with lower diversity compared to rest of the genome and F_{ST} outliers are inferred to be associated with a trait of interest. The association of allele frequencies across the genome with ecotypes and environmental parameters may be analyzed following Turner et al. (2010) and Namroud et al. (2008), and the predicted functions of such genes may be further investigated through biotechnological methods.

PLATFORM TO DEVELOP CLIMATE-CHANGE ADAPTIVE CROPS

Integrating mathematical modeling and landscape genomics along with traditional breeding provides a unique opportunity for mining genebanks and wild relatives of crop plants for rapid identification of germplasm with climate-change adaptive traits (Figure 17.1). Based upon the needs of stakeholders (farmers, agricultural research centers, universities, and NGOs) for climate-change adaptive traits in the crop plant of interest, passport data of georeferenced germplasm can be analyzed through the FIGS approach to identify subsets of germplasm with a high likelihood of containing the target trait. This germplasm can be subject to phenotyping to determine the specific accessions with the desired trait. The accessions with confirmed phenotypes can be subject to high-throughput DNA sequencing approaches as described

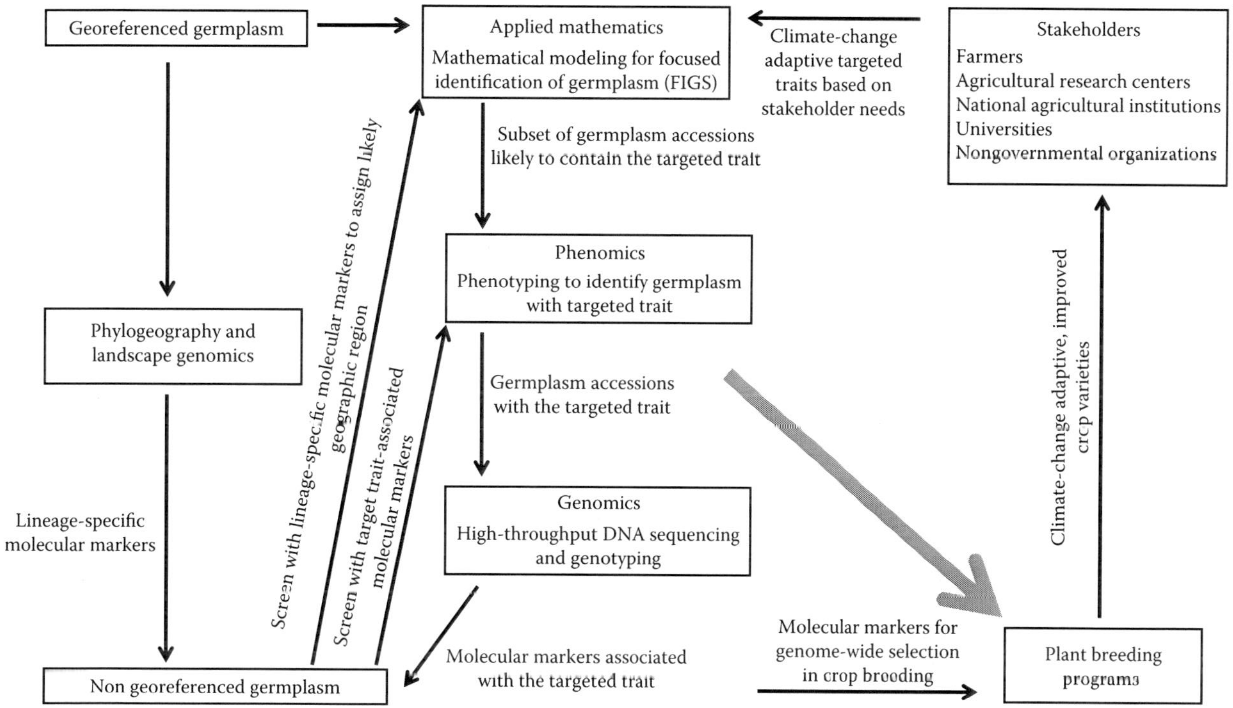

FIGURE 17.1 Platform to develop climate-change adaptive crop plants. The passport data of georeferenced germplasm are used for selecting subsets likely to contain targeted traits through mathematical modeling, phenotyped to identify the germplasm with the targeted trait and used for characterizing trait associated molecular markers. Phylogeography and landscape genomic analyses of georeferenced germplasm are used for characterizing lineage-specific molecular markers, which will be used for assessing genetic relatedness of nongeoreferenced germplasm and explore accessions with target traits in germplasm with limited passport data.

above, and the resulting genomic sequence data can be used for characterization of molecular markers associated with traits of interest. The DNA sequence data of the georeferenced germplasm can be used for phylogeographic and landscape genomics analyses. The resulting molecular phylogenetic trees are used for controlling the population structure in GWAS analyses, and characterizing lineage, lineage-specific molecular markers, or geographical source-specific DNA barcodes are to be used for assigning germplasm lacking passport data to corresponding genetic lineages. The DNA barcode-based assignment of nongeoreferenced germplasm to corresponding lineages with well-characterized germplasm provides a means to identify the likely geographical source of the germplasm lacking passport data. Germplasm with predicted geographic sources may be used in FIGS analyses and screened through phenomics to confirm the phenotype. This characterized germplasm could then be used by plant breeders to develop climate-change adaptive crops to maintain food security under changing climatic conditions.

CONCLUSION

Crop cultivars with traits adaptive to changing climatic conditions need to be developed to increase and sustain agricultural productivity under changing climatic conditions. Discovering new and useful alleles of climate-change AG in wild relatives of crop plants and germplasm in genebanks can be accomplished through detailed analyses of passport, trait, and environmental data from collection sites of crop accessions through advanced mathematical modeling followed by landscape genomics approaches. A multidisciplinary platform integrating mathematical modeling, genomic analyses, and breeding is needed to develop climate-change adaptive crops to sustain agricultural productivity under changing climatic conditions.

REFERENCES

Ashton, P.S. and C.V.S. Gunatilleke. 1987. New light on the plant geography of Ceylon. I. Historical plant geography. *J Biogeogr* 14:249–285.

Atwell, S., Y.S. Huang, B.J. Vilhjálmsson et al. 2010. Genome-wide association study of 107 phenotypes in a common set of *Arabidopsis thaliana* inbred lines. *Nature* 465:627–631.

Avise, J.C. 2000. *Phylogeography: The history and formation of species*. Cambridge: Harvard University Press.

Avise, J.C., J. Arnold, R.M. Ball et al. 1987. Intraspecific phylogeography: The mitochondrial DNA bridge between population genetics and systematics. *Annu Rev Ecol Systemat* 18:489–522.

Baird, N.A., P.D. Etter, T.S. Atwood et al. 2008. Rapid SNP discovery and genetic mapping using sequenced RAD markers. *PLoS One* 3:e3376. doi:10.1371/journal.pone.0003376.

Bari, A., K. Street, M. Mackay, D.T.F. Endresen, E. De Pauw, and A. Amri. 2012. Focused identification of germplasm strategy (FIGS) detects wheat stem rust resistance linked to environmental variables. *Genet Resour Crop Ev* 59:1465–1481.

Brachi, B., G.P. Morris, and J.O. Borevitz. 2011. Genome-wide association studies in plants: The missing heritability is in the field. *Genome Biol* 12:232. doi:10.1186/gb-2011-12-10-232.

Buckler, E.S., J.B. Holland, P.J. Bradbury et al. 2009. The genetic architecture of maize flowering time. *Science* 325:714–718.

Burgman, M.A. 1987. An analysis of the distribution of plants on granite outcrops in southern Western Australia using Mantel tests. *Vegetation* 71:79–86.

Burke, J.M., S.J. Knapp, and L.H. Rieseberg. 2005. Genetic consequences of selection during the evolution of cultivated sunflower. *Genetics* 171:1933–1940.

Catchen, J.M., A. Amores, P.A. Hohenlohe, W.A. Cresko, and J.H. Postlethwait. 2011. Stacks: Building and genotyping loci *de novo* from short-read sequences. *G3* 1:171–182.

Choudhury, B., M.L. Khan, and S. Dayanandan. 2013. Genetic structure and diversity of indigenous rice (*Oryza sativa*) varieties in the Eastern Himalayan region of Northeast India. *SpringerPlus* 2:228. doi:10.1186/2193-1801-2-228.

Davey, J.W., P.A. Hohenlohe, P.D. Etter, J.Q. Boone, J.M. Catchen, and M.L. Blaxter. 2011. Genome-wide genetic marker discovery and genotyping using next-generation sequencing. *Nat Rev Genet* 12:499–510.

Doebley, J.F., B.S. Gaut, and B.D. Smith. 2006. The molecular genetics of crop domestication. *Cell* 127:1309–1321.

Ehrenreich, I.M. and M.D. Purugganan. 2006. The molecular genetic basis of plant adaptation. *Am J Bot* 93:953–962.

Ekblom, R. and J. Galindo. 2011. Applications of next generation sequencing in molecular ecology of non-model organisms. *Heredity* 107:1–15.

El Bouhssini, M., K. Street, A. Joubi, Z. Ibrahim, and F. Rihawi. 2009. Sources of wheat resistance to Sunn pest, *Eurygaster integriceps* Puton, in Syria. *Genet Resour Crop Ev* 56:1065–1069.

Emerson, K.J., C.R. Merz, J.M. Catchen et al. 2010. Resolving postglacial phylogeography using high-throughput sequencing. *Proc Natl Acad Sci USA* 107 16196–16200.

Endresen, D.T.F. 2010. Predictive association between trait data and ecogeographic data for Nordic barley landraces. *Crop Sci* 50:2418–2430.

Endresen, D.T.F., K. Street, M. Mackay, A. Bari, and E. De Pauw. 2011. Predictive association between biotic stress traits and eco-geographic data for wheat and barley landraces. *Crop Sci* 51:2036–2055.

Fournier-Level, A., A. Korte, M.D. Cooper, M. Nordborg, J. Schmitt, and A.M. Wilczek. 2011. A map of local adaptation in *Arabidopsis thaliana*. *Science* 334:86–89.

Glenn, T.C. 2011. Field guide to next-generation DNA sequencers. *Mol Ecol Resources* 11:759–769.

Gnirke, A., A. Melnikov, J. Maguire et al. 2009. Solution hybrid selection with ultra-long oligonucleotides for massively parallel targeted sequencing. *Nat Biotechnol* 27:182–189.

Godfray, H.C.J., J.R. Beddington, I.R. Crute et al. 2010. Food security: The challenge of feeding 9 billion people. *Science* 327:812–818.

Grant, V. 1949. Pollination systems as isolating mechanisms in angiosperms. *Evolution* 3:82–97.

Hancock, A.M., B. Brachi, N. Faure et al. 2011. Adaptation to climate across the *Arabidopsis thaliana* genome. *Science* 334:83–86.

Handley, L.J.L., A. Manica, J. Goudet, and F. Balloux. 2007. Going the distance: Human population genetics in a clinical world. *Trends Genet* 23:432–439.

Hewitt, G. 2000. The genetic legacy of the Quaternary ice ages. *Nature* 405:907–913.

Hewitt, G.M. 2004. Genetic consequences of climatic oscillations in the Quaternary. *Philos T Roy Soc B* 359:183–195.

Hickerson, M.J., B.C. Carstens, J. Cavender-Bares et al. 2010. Phylogeography's past, present, and future: 10 years after Avise, 2000. *Mol Phylogenet Evol* 54:291–301.

Hohenlohe, P.A., S. Bassham, P.D. Etter, N. Stiffler, E.A. Johnson, and W.A. Cresko. 2010. Population genomics of parallel adaptation in threespine stickleback using sequenced RAD tags. *PLoS Genetics* 6:e1000862. doi:10.1371/journal.pgen.1000862.

Holder, K., R. Montgomerie, and V.L. Friesen. 1999. A test of the glacial refugium hypothesis using patterns of mitochondrial and nuclear DNA sequence variation in rock ptarmigan (*Lagopus mutus*). *Evolution* 53:1936–1950.

Holderegger, R., U. Kamm, and F. Gugerli. 2006. Adaptive vs. neutral genetic diversity: Implications for landscape genetics. *Landscape Ecol* 21:797–807.

Huang, X. and B. Han. 2014. Natural variations and genome-wide association studies in crop plants. *Annu Rev Plant Biol* 65:531–551.

Huang, X., N. Kurata, X. Wei et al. 2012. A map of rice genome variation reveals the origin of cultivated rice. *Nature* 490:497–501.

IPCC. 2013. *Climate Change 2013: The Physical Science Basis. Contribution of Working Group I to the Fifth Assessment Report of the Intergovernmental Panel on Climate Change*, eds. T.F. Stocker, D. Qin, G.-K. Plattner, M. Tignor, S.K. Allen, J. Boschung, A. Nauels, Y. Xia, V. Bex, and P.M. Midgley. Cambridge: Cambridge University Press. doi:10.1017/CBO9781107415324. https://www.ipcc.ch/report/ar5/.

Jia, G., X. Huang, H. Zhi et al. 2013. A haplotype map of genomic variations and genome-wide association studies of agronomic traits in foxtail millet (*Setaria italica*). *Nat Genet* 45:957–961.

Johnston, M.O. 1991. Natural selection on floral traits in two species of *Lobelia* with different pollinators. *Evolution*. 45:1468–1479.

Knowles, L.L. and C.L. Richards. 2005. Importance of genetic drift during Pleistocene divergence as revealed by analyses of genomic variation. *Mol Ecol* 14:4023–4032.

Kump, K.L., P.J. Bradbury, R.J. Wisser et al. 2011. Genome-wide association study of quantitative resistance to southern leaf blight in the maize nested association mapping population. *Nat Genet* 43:163–168.

Li, Y., Y. Huang, J. Bergelson, M. Nordborg, and J.O. Borevitz. 2010. Association mapping of local climate-sensitive quantitative trait loci in *Arabidopsis thaliana*. *Proc Natl Acad Sci USA* 107:21199–21204.

Lisiecki, L.E. and M.E. Raymo. 2005. A Pliocene-Pleistocene stack of 57 globally distributed benthic δ18O records. *Paleoceanography* 20:PA1003. doi:10.1029/2004PA001071.

Mackay, M. and K. Street. 2004. Focused identification of germplasm strategy—FIGS. In *Proceedings of the 54th Australian Cereal Chemistry Conference and the 11th Wheat Breeders' Assembly*, September 21–24, 2004, Canberra, ACT, Australia, eds. C.K. Black, J.F. Panozzo, and G.J. Rebetzke, 138–141. Melbourne, Australia: Cereal Chemistry Division, Royal Australian Chemical Institute (RACI).

Manel, S., S. Joost, B.K. Epperson et al. 2010. Perspectives on the use of landscape genetics to detect genetic adaptive variation in the field. *Mol Ecol* 19:3760–3772.

Manel, S., M.K. Schwartz, G. Luikart, and P. Taberlet. 2003. Landscape genetics: Combining landscape ecology and population genetics. *Trends Ecol Evol* 18:189–197.

Mantel, N. 1967. The detection of disease clustering and a generalized regression approach. *Cancer Research* 27:209–220.

McNeilly, T. and J. Antonovics. 1968. Evolution in closely adjacent plant populations IV. Barriers to gene flow. *Heredity* 23:205–218.

Metzker, M.L. 2010. Sequencing technologies—The next generation. *Nat Rev Genet* 11:31–46.

Morris, G.P., P. Ramu, S.P. Deshpande et al. 2013. Population genomic and genome-wide association studies of agroclimatic traits in sorghum. *Proc Natl Acad Sci USA* 110:453–458.

Namroud, M.-C., J. Beaulieu, N. Juge, J. Laroche, and J. Bousquet. 2008. Scanning the genome for gene single nucleotide polymorphisms involved in adaptive population differentiation in white spruce. *Mol Ecol* 17:3599–3613.

Narum, S.R. and J.E. Hess. 2011. Comparison of F_{ST} outlier tests for SNP loci under selection. *Mol Ecol Resour* 11:184–194.

Neff, B.D. and B.A. Fraser. 2010. A program to compare genetic differentiation statistics across loci using resampling of individuals and loci. *Mol Ecol Resources* 10:546–550.

Olsen, K.M. and J.F. Wendel. 2013. A bountiful harvest: Genomic insights into crop domestication phenotypes. *Annu Rev Plant Biol* 64:47–70.

Ott, J., Y. Kamatani, and M. Lathrop. 2011. Family-based designs for genome-wide association studies. *Nat Rev Genet* 12:465–474.

Pasam, R.K., R. Sharma, M. Malosetti et al. 2012. Genome-wide association studies for agronomical traits in a world wide spring barley collection. *BMC Plant Biol* 12:16. doi:10.1186/1471-2229-12-16.

Pielou, E.C. 1991. *After the Ice Age: The return of life to glaciated North America*. Chicago, IL: University of Chicago Press.

Podolsky, R.H. and T.P. Holtsford. 1995. Population structure of morphological traits in *Clarkia dudleyana*. I. Comparison of FST between allozymes and morphological traits. *Genetics* 140:733–744.

Poland, J.A., P.J. Bradbury, E.S. Buckler, and R.J. Nelson. 2011. Genome-wide nested association mapping of quantitative resistance to northern leaf blight in maize. *Proc Natl Acad Sci USA* 108:6893–6898.

Reusch, T.B.H. and T.E. Wood. 2007. Molecular ecology of global change. *Mol Ecol* 16:3973–3992.

Ross-Ibarra, J., P.L. Morrell, and B.S. Gaut. 2007. Plant domestication, a unique opportunity to identify the genetic basis of adaptation. *Proc Natl Acad Sci USA* 104:8641–8648.

Ruddiman, W.F., M.E. Raymo, D.G. Martinson, B.M. Clement, and J. Backman. 1989. Pleistocene evolution: Northern hemisphere ice sheets and North Atlantic Ocean. *Paleoceanography* 4:353–412.

Schwartz, M.K., G. Luikart, K.S. McKelvey, and S.A. Cushman. 2009. Landscape genomics: A brief perspective. In *Spatial complexity, informatics, and wildlife conservation*, eds. S.A. Cushman and F. Huettmann, 165–174, Tokyo, Japan: Springer-Verlag. doi:10.1007/978-4-431-87771-4_9.

Segelbacher, G., S.A. Cushman, B.K. Epperson et al. 2010. Applications of landscape genetics in conservation biology: Concepts and challenges. *Conserv Genet* 11:375–385. doi:10.1007/s10592-009-0044-5.

Singh-Gasson, S., R.D. Green, Y. Yue et al. 1999. Maskless fabrication of light-directed oligonucleotide microarrays using a digital micromirror array. *Nat Biotechnol* 17:974–978.

Siol, M., S.I. Wright, and S.C.H. Barrett. 2010. The population genomics of plant adaptation. *New Phytol* 188:313–332.

Sork, V.L., S.N. Aitken, R.J. Dyer, A.J. Eckert, P. Legendre, and D.B. Neale. 2013. Putting the landscape into the genomics of trees: Approaches for understanding local adaptation and population responses to changing climate. *Tree Genet Genomes* 9:901–911.

Stapley, J., J. Reger, P.G.D. Feulner et al. 2010. Adaptive genomics: The next generation. *Trends Ecol Evol* 25:705–712.

Storfer, A., M.A. Murphy, J.S. Evans et al. 2007. Putting the 'landscape' in landscape genetics. *Heredity* 98:128–142.

Strasburg, J.L., N.A. Sherman, K.M. Wright, L.C. Moyle, J.H. Willis, and L.H. Rieseberg. 2012. What can patterns of differentiation across plant genomes tell us about adaptation and speciation? *Philos T Roy Soc B* 367:364–373.

Tanksley, S.D. and S.R. McCouch. 1997. Seed banks and molecular maps: Unlocking genetic potential from the wild. *Science* 277:1063–1066.

Thomson, A.M., C.W. Dick, and S. Dayanandan. 2015. A similar phylogeographical structure among sympatric North American birches (*Betula*) is better explained by introgression than by shared biogeographical history. *Biogeogr* 42:339–350.

Tian, F., P.J. Bradbury, P.J. Brown et al. 2011. Genome-wide association study of leaf architecture in the maize nested association mapping population. *Nat Genet* 43:159–162.

Turner, T.L., E.C. Bourne, E.J. von Wettberg, T.T. Hu, and S.V. Nuzhdin. 2010. Population resequencing reveals local adaptation of *Arabidopsis lyrata* to serpentine soils. *Nat Genet* 42:260–263.

Wagner, H.H., R. Holderegger, S. Werth, F. Gugerli, S.E. Hoebee, and C. Scheidegger. 2005. Variogram analysis of the spatial genetic structure of continuous populations using multilocus microsatellite data. *Genetics* 169:1739–1752.

Waser, N.M. and D.R. Campbell. 2004. Ecological speciation in flowering plants. In *Adaptive speciation*, eds. U. Dieckmann, M. Doebeli, J.A.J. Metz, and D. Tautz, 264–277. Cambridge: Cambridge University Press.

Webb, T. and P.J. Bartlein. 1992. Global changes during the last 3 million years: Climatic controls and biotic responses. *Annu Rev Ecol Evol Systemat* 23:141–173.

Wheeler, T. and J. von Braun. 2013. Climate change impacts on global food security. *Science* 341:508–513.

Whitmore, T.C. 1990. *An introduction to tropical rain forests*. New York: The Clarendon Press.

Willis, K.J. and K.J. Niklas. 2004. The role of Quaternary environmental change in plant macroevolution: The exception or the rule? *Philos T Roy Soc B* 359:159–172.

Zhao, J.-L., L. Zhang, S. Dayanandan, S. Nagaraju, D.-M. Liu, and Q.-M. Li. 2013. Tertiary origin and Pleistocene diversivication of dragon blood tree (*Dracaena cambodiana*-Asparagaceae) populations in the Asian tropical forests. *PLoS One* 8:e60102.

18 Toward the Rapid Domestication of Perennial Grains

Developing Genetic and Genomic Resources for Intermediate Wheatgrass

T. Kantarski, L. DeHaan, J. Poland, and J. Borevitz

CONTENTS

FOOD SECURITY

As the global population continues to grow, the amount of arable land decreases, and climate change leads to less-favorable production conditions, humanity will face new food-security challenges (Glover et al. 2010, Godfray et al. 2010). The Green Revolution of the 1930s–1960s modernized agriculture, resulting in high-yielding varieties of cereals that saved hundreds of millions of people from starvation (Woodward et al. 2009). However, one in eight people (842 million) still suffer from chronic hunger (FAO 2013), grain-yield improvement in major annual crops has slowed (Godfray et al. 2010), and most of the land suitable for production is in use and suffering from degradation (Pimentel 2006, Monfreda et al. 2008). While food insecurity is a multidimensional issue, the development of new and improved crops optimized for alternative sustainable farming practices could address many of those components and is urgently needed.

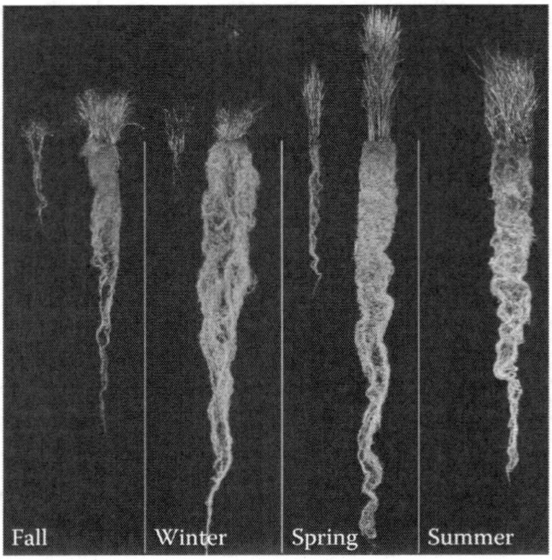

FIGURE 18.1 Seasonal comparisons of root growth in fall, winter, and spring seasons between wheat (left) in each of the first three panels, and intermediate wheatgrass (right); for the summer panel, only intermediate wheatgrass is shown. (Adapted from Glover, J.D. et al., *Science* 328[5986], 1638–1639, 2010 by permission of publisher.)

Perennial plants dominate the natural landscape, but are currently of minor importance in modern grain production. Perennials for grain production would offer many natural benefits over annuals, including permanent ground cover to reduce soil erosion, greater interception of light energy due to a longer growing season, reduced passes across the field for seeding and tillage operations (reducing erosion and conserving energy), and a large root system that reduces nutrient leaching, gives greater access to water and nutrients, and increases carbon fixation (Cox et al. 2006, 2010) (Figure 18.1). However, perennial grain species have not experienced improvement through domestication or intensive modern breeding for grain yield and quality. The development of perennial versions of major grain crops could provide farmers with the options necessary for producing food in challenging conditions (Cox et al. 2006, 2010, Glover et al. 2010).

LACK OF PERENNIAL GRAINS IN AGRICULTURE

Herbaceous perennial plants have leaves and nonwoody stems that die back at the end of each growing season; however, the plants can be long lived, growing back from parts of the plant that remain alive below or close to ground level. Herbaceous annual plants die completely at the end of their respective growing season, possessing only one opportunity to pass their genetic information to the next generation. Therefore, it is assumed that in a given growing season, selection has optimized a schedule of carbon allocation so that final seed set is maximized (Chiariello and Roughgarden 1984). In comparison, perennials vary considerably in their allocation of reproductive effort to seeds, not only between species but also between years within a given species (Primack 1979, Stewart and Thompson 1982). While perennial species with high reproductive allocation and annual species with low reproductive allocation have been observed, comparisons between congeneric annual and perennial species support the prediction of higher average reproductive allocation in annuals than perennials (Bazzaz et al. 1987). The question remains, though, whether a perennial's evolutionary life history is a set trade-off or whether it is alterable via artificial selection, that is, domestication. Although a herbaceous perennial may not have experienced selection pressure in nature to consistently maximize seed set, this does not mean it cannot be artificially bred to do so. Much effort has been directed toward answering this question, from theoretical perspectives and through active breeding programs (Cox et al. 2010, Van Tassel et al. 2010, DeHaan et al. 2014, DeHaan and Van Tassel 2014).

The domestication of plants began approximately 10,000 years ago, but only annual species have prevailed as easier candidates for selection and domestication. The initial stages of plant domestication arguably involved the Darwinian process of natural selection and the coevolution of plants and humans (Cox 2009, Van Tassel et al. 2010). Early farmers are likely to have unconsciously selected for reduced shattering, reduced dormancy, and large seeds, followed by conscious selection for these and additional traits (Cox 2009). Both modes of selection were effective, as they increased the individual fitness of the plants being selected and the farmers' benefit from the harvested seeds (Van Tassel et al. 2010). The early agricultural environment was disturbed by human activity, resulting in the prevalence of those species that could best thrive in those areas, generally annuals. Differences in annual and perennial species' life history strategies provide the basis for five hypotheses as to why perennials were not as likely to be domesticated (Van Tassel et al. 2010). These hypotheses can be divided into two groups, each based on one assumption; either it is—or it is not—physiologically possible to have a high grain-yielding herbaceous perennial species.

The first group, based on the assumption that a high grain-yielding herbaceous perennial is not possible, contains only one hypothesis, which simply states that there is an evolutionary constraint to achieving a high-yielding herbaceous perennial species in any environment along any evolutionary trajectory. Hypotheses two through five assume that a high grain-yielding herbaceous perennial is possible, and has perhaps arisen, but that the combination of high grain yield and longevity is not stable over evolutionary time. Hypothesis two assumes incomplete sampling: there has not been enough time for all trait combinations to arise. Hypothesis three proposes that a certain combination of traits may not be possible because the species is developmentally canalized, that is, the traits are robust to environmental perturbations and/or genetic changes. Hypothesis four proposes that the combination of traits has arisen repeatedly but conferred low fitness each time, eventually leading to the extinction of those genotypes. Hypothesis five also allows that the combination of traits has arisen repeatedly but never spread; while the combination was novel, perhaps there was no novel adaptive function (Wainwright 2007, Van Tassel et al. 2010).

Van Tassel et al. (2010) thoroughly examined the literature pertaining to these five hypotheses and made two major conclusions. First, hypothesis one is unlikely to explain the lack of high grain-yielding herbaceous perennial species. The existence of larger, slower growing, longer lived, and higher yielding organisms than perennial herbs (i.e., tree crops) implies that size and longevity are not necessarily in opposition to reproductive effort. Next, explanation by hypothesis four of the lack of high grain-yielding herbaceous perennials is the most highly supported. There are several possible reasons for a perennial plant's reduced fitness in the early agricultural environment. Propagation by seed, frequent disturbance from replanting, and tillage for weed and nitrogen management were more favorable for annual species' weedy tendencies than for a perennial species' longevity strategy (Cox 2009, Van Tassel et al. 2010). Additionally, longevity is associated with lower seed fertility, as lethal recessive mutations accumulate in perennials from cycle to cycle (DeHaan and Van Tassel 2014). This accrual is referred to as a *genetic load* and occurs because a large number of mitotic cell divisions occur between meiotic cell divisions. In plants, these somatic (accrued) mutations can be passed along to the offspring. Somatic mutations generate heterozygosity in perennials, which can be good for maintaining diversity in a population, but deleterious recessive alleles are also maintained in the population (Morgan 2001). These species tend to be allogamic (cross-fertilizing); hence, recombination not only disrupts associations built and favored by selection (Agrawal 2006) but also inefficiently purges deleterious alleles (Morgan 2001).

Understanding of the early agricultural environment and the characteristics that tend to accompany perennials has helped explain why herbaceous perennials did not stand out to early farmers as promising grain producers. However, this knowledge, combined with modern plant-breeding approaches, opens possibilities for designing a strategy for domesticating perennial grain crops (DeHaan and Van Tassel 2014). Progress is being made toward perennial rice, perennial wheat, perennial sorghum, perennial maize, perennial sunflowers, perennial chickpeas, and toward the direct domestication of new perennial candidates for agriculture (Cox et al. 2010, Batello et al. 2014).

MOLECULAR MARKERS

Genetic markers correspond to differences in DNA sequences that can be used to distinguish individuals, populations, or species. Each genetic marker has a specific location in the genome of interest, and the differences in genotypes that each marker represents can be tested for association with a phenotypic trait of interest (such as plant height or seed size). Genetic markers can be used to divide traits into their major components as quantitative trait loci (QTLs), while accounting for the polygenic contribution from shared ancestry. This information can be used to better understand the inheritance and gene action for traits of interest and to predict the adult phenotype from the genotype of the seed or seedling.

The use of genetic markers to guide plant-breeding efforts is known as *marker-assisted selection*, and its implementation accelerates the plant-selection process. Marker-assisted selection is especially advantageous when traits are difficult and/or expensive to measure and tends to be utilized to screen populations at the seedling stage, as plants with desirable trait combinations can be identified early and those with undesirable traits can be eliminated (Collard and Mackill 2008).

A single nucleotide polymorphism (SNP) is a type of genetic marker and is the difference in a nucleotide at a particular genomic location between individuals.

SNPs are becoming the marker of choice in the scientific community because they are digital and can be read out in the millions (Nordborg and Weigel 2008, Ganal et al. 2009). Tens or hundreds of thousands of markers per sample can now be assayed at a relatively low cost. High-throughput approaches may involve reduced representation sequencing, targeting sites across the genome of interest with a restriction-enzyme or a combination of enzymes. Methods for simultaneous marker discovery and genotyping across multiplexed individuals continue to evolve. Recently, genotyping-by-sequencing (GBS) methods have greatly improved upon previous marker-development platforms (Poland and Rife 2012).

One of the most important advantages of GBS is the elimination of ascertainment bias arising from genotyping one population with a panel of SNPs developed from another population. Most recently, a two-enzyme GBS approach was developed and used to construct high-density genetic maps for barley and wheat (Poland et al. 2012b). This approach utilizes a common-cutting restriction enzyme and a rare-cutting restriction enzyme and is improved from the original GBS protocol, as only fragments with both a rare cut-site and a common cut-site amplify in the PCR (Poland et al. 2012b). All amplified fragments include a barcoded forward adapter with the rare-cutter and a common reverse (Y) adapter with the common-cutter, which avoids excess amplification of common fragments (Poland et al. 2012b). With a unique barcode sequence to identify reads from each individual, up to 384 samples can be pooled and sequenced as one sample. The appropriate plex level depends upon the size of the genome and how much marker coverage is needed. The development and use of genetic markers within and between species have been valuable for isolating genes of interest via map-based cloning, mapping the genetic architecture underlying traits of interest, and improving understanding of genome structure, function, and evolution (Nordborg and Weigel 2008).

GENETIC LINKAGE MAPPING OF MARKERS

A genetic linkage map is composed of genetic markers that are grouped according to how often they are inherited together. The concept of linkage refers to the idea that the closer two markers or genes are on a chromosome, the more likely they are to be inherited together because they are less likely to be separated during recombination. Given enough markers, each group represents a chromosome, with the relative positions of markers calculated by a mapping function. A mapping function is a mathematical adjustment to the genetic distance between two markers that relates the calculated recombination frequency to map distance, which is represented by centimorgans (cMs). One cM is equivalent to the distance between two markers in which a 1% recombination rate is observed between them. When there is little distance between genetic markers, the recombination frequency closely estimates map distance; however, accuracy is reduced for loci that are not tightly linked, as the likelihood of multiple crossovers increases.

The two most commonly used mapping functions are the Haldane and Kosambi functions. Haldane discovered that as the recombination frequency approaches 10%, the relationship between recombination frequency and map distance deviates from linearity. The chance of multiple crossovers increases as the distance between loci increases; additionally, even numbers of crossovers may go unnoticed, resulting in an underestimation of map distance. With the increase in marker density due to the increased ease of marker discovery, missing a double crossover is no longer an issue. Haldane's mapping function operates with the assumption that crossovers occur randomly and independently over a given chromosome. Alternatively, Kosambi's mapping function allows for interference, based on the observation that the occurrence of a crossover reduces the likelihood of another crossover in adjacent loci. Both mapping functions have been used extensively and give similar estimates of map distance when the distance between markers is small.

The development of high-density genetic linkage maps for species with little or no genetic resources has become possible with the ability to discover and type new markers through sequencing. The ability to generate a high-density genetic linkage map for a species of interest facilitates additional studies, including quantitative trait mapping, marker-assisted selection, gene-cloning experiments, and evolutionary studies. These advances have provided opportunities for increasing selection efficiency in plant breeding and improved the understanding of the evolutionary history of crop domestication.

EVOLUTION OF CEREAL CROPS

Although angiosperms (plants with seeds enclosed by an ovary) diverged from gymnosperms (plants with seeds not enclosed by an ovary) over 200 million years ago, the grass family, Poaceae, which contains all cereal crops, diverged from a common ancestor relatively recently, 50–70 million years ago (Kellogg 1998). Poaceae is organized into 12 subfamilies, with each of the following groups of crops belonging to different subfamilies: wheat, barley, oats, and rye in Pooideae; rice in Ehrhartoideae; and maize, sorghum, millet, and sugarcane in Panicoideae. Genome size, ploidy level (number of chromosome sets), and chromosome number vary widely among domestic grasses, from rice's 466 Mb genome (Yu et al. 2002) ($2n = 24$) to wheat's allohexaploid 17 Gb genome (IWGSC 2014) ($2n = 6x = 42$).

Comparative analyses from the genome sequences of *Brachypodium*, rice, and maize, representing the three lineages previously mentioned, respectively, revealed a common feature that is likely shared among all grasses. After a whole genome duplication 70 million years ago, most duplicated gene sets lost one copy before the major grass lineages began to diverge closer to 50 million years ago (Paterson et al. 2004, Tang et al. 2010). Since this discovery, more evidence has accumulated for at least one additional whole genome duplication 130 million years ago (Zhang et al. 2005, Jaillon et al. 2007, Salse et al. 2008). Therefore, while the common ancestor to all grasses experienced at least two whole genome duplications, subsequent diploidization rendered it a functional diploid with conflicting models as to haploid chromosome number.

Early comparative mapping studies revealed extensive collinearity among the genomes of rice, wheat, barley, rye, oat, maize, sorghum, and others (Ahn et al. 1993, Moore et al. 1995, Devos and Gale 2000). Although the extent of microcollinearity (sequence level) varies in certain genomic regions, the extent of macrocollinearity (chromosome level) is mostly conserved across cereals. For example, comparisons of genomic regions of wheat and rice have revealed rearrangements in gene order and content (Bennetzen 2000, Feuillet and Keller 2002, Li and Gill 2002, Sorrells et al. 2003, Lu and Faris 2006), as well as conservation of syntenic (conserved gene order) regions (Yan et al. 2003, Distelfeld et al. 2004, Valárik et al. 2006, Kuraparthy et al. 2008). Rice sequence data from collinear regions of rice and wheat have been used to identify candidate genes for traits of interest and develop markers for fine-mapping genes of interest in wheat (Distelfeld et al. 2004, Valárik et al. 2006). More recently, two wild *Brachypodium* species, *B. sylvaticum* and *B. distachyon*, have been proposed as genomic models for the Triticeae crops due to their small genomes and more recent divergence than rice (Bossolini et al. 2007). Results from comparative studies between wheat and *Brachypodium* have been similar to those of rice, some demonstrating perfect gene-level conservation (Griffiths et al. 2006) and others reaching the conclusion that rice has better conservation in some regions than *Brachypodium* (Faris et al. 2008). While genomic sequences from species with smaller genomes contribute comparative knowledge to genomic studies involving cereals with large genomes, such as wheat and intermediate wheatgrass, it cannot substitute for genomic resources for the crop of interest.

GENETIC ARCHITECTURE OF DOMESTICATION

Darwin often used examples of domesticated plants and animals to better understand and illustrate the importance of variation and develop his theory of evolution via natural selection (Darwin 2009). With an improved understanding of evolutionary principles, such as convergent evolution, promising domestication *hotspots* could be targeted for desirable phenotypic changes, especially those known to control flowering time and determinate growth (Lenser and Theißen 2013). The most important traits for agriculture are often rare in nature, as they tend not to provide any evolutionary advantage. These traits are collectively referred to as the *domestication syndrome*, because they are often shared across domesticated crops and include examples such as reduced shattering, increased seed number and size, reduced dormancy, and determinate growth. Although the domestication of each major cereal was independent, mutations in homologous genes may have occurred. Recent molecular analyses have demonstrated that the causal mutations for convergent phenotypic changes are often located in particular genes (Lenser and Theißen 2013). This finding could be used to define candidate genes for strategically improving new plant species during the domestication process, simultaneously increasing crop diversity and food security.

The domesticated morphologies of the major crops are often strikingly diverged from their wild ancestors. A more complete picture of how this transition occurred has come with the ability to study the genetic and genomic differences between major crops and their progenitors. The genetic architecture underlying the drastic morphological changes that occur during domestication tends to consist of few genes or genomic regions with major-effect QTLs, with QTL clustering within a given genomic region (Hartman et al. 2013). While there are exceptions to these trends, studies of genes involved in domestication phenotypes have revealed that a change in one or few of these genes can result in dramatic morphological changes (Gross and Olsen 2010, Olsen and Wendel 2013).

Examples of one or few genes underlying domestication phenotypes have been found across most major crops, including maize, rice, wheat, and tomato (Figure 18.2). Although evidence of selection has been observed for thousands of genes (Tian et al. 2009), changes in approximately five genomic regions are responsible for the major phenotypic transition from the progenitor teosinte to domesticated maize (Doebley 1990, Tian et al. 2009). Specifically, *teosinte branched1* (*tb1*) was found to be largely

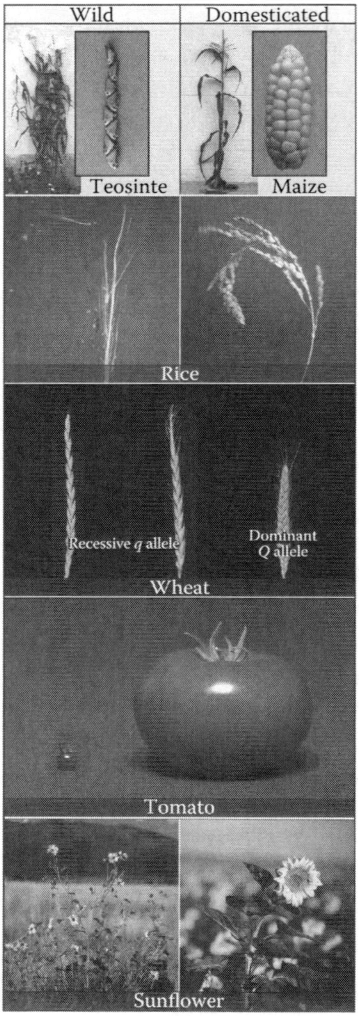

FIGURE 18.2 Examples of domestication-related phenotypes from maize (reduced branching and naked seed), rice (reduced shattering), wheat (compact head and free-threshing seed), tomato (increased fruit size), and sunflower (reduced branching/apical dominance); these phenotypic transitions in each crop, except sunflower, involved few loci of major effect. (Adapted from Doebley, J.F. et al., *Cell* 127, 1309–1321, 2006 by permission of publisher.)

responsible for reduced branching (Doebley et al. 1997), and the QTL *teosinte glume architecture1* (*tga1*) was found to control the formation of the casing that surrounds teosinte kernels (Wang et al. 2005) (Figure 18.2). A few genes of large effect were also involved in the domestication of wheat, barley, and rice, although each transition did not result in such a drastic morphological change as achieved in maize. Shatter reduction was often achieved by a key mutation in a gene or locus of major effect, including a mutation in homoeologous *Br* genes in wheat and possibly the homologous *Btr* gene in barley (Nalam et al. 2006), while changes in *sh4* and *qSH1* each gave similar effects in rice (Li et al. 2006, Xiong et al. 1999) (Figure 18.2). The super domestication gene, *Q*, is arguably the most important factor in the worldwide spread of wheat cultivation, pleiotropically affecting several traits, including rachis fragility, glume shape and toughness, spike architecture, flowering time, and plant height (Simons et al. 2006) (Figure 18.2). Likewise, the large effect QTL *fruitweight2.2* (*fw2.2*) controls 30% of the fruit mass difference between wild and domesticated tomato (Frary et al. 2000) (Figure 18.2). Although major morphological changes were also observed between wild and domesticated sunflower *Helianthus annuus* L., many genes of only small effect have been identified (Wills and Burke 2007) (Figure 18.2).

NEW DOMESTICATION

With the increased understanding of the genetics of domestication, combined with recent advances in sequencing technology and bioinformatics, new species can be strategically developed for agricultural use. Intermediate wheatgrass, *Thinopyrum intermedium* (Host) Barkworth and D. Dewey ($2n = 6x = 42$; genomes JJJ^sJ^sSS), is a perennial grass within the Triticeae tribe and shows great potential as a dual grain and biofuel crop, providing an economically and ecologically viable crop for food, feed, and biofuel without threatening soil quality, as occurs with annuals (Glover et al. 2010). However, its large and understudied genome has made genetic studies and genomic-assisted breeding intractable. The species is an allohexaploid with an estimated 12.95 Gb genome (Vogel et al. 1999), no reference genome, and previously minimal marker development.

To address the lack of genomic resources for this complex genome, we applied new genotyping approaches that use next-generation sequencing technology to characterize the intermediate wheatgrass genome and populations. GBS combines molecular marker discovery and genotyping and has been successfully utilized in species with large, complex genomes, such as barley and wheat (Elshire et al. 2011, Poland et al. 2012a).

Reduced height, free-threshing, increased seed size, and reduced shattering are important traits for the domestication of *Th. intermedium* (which is being marketed under the trademark name Kernza™). Multiple beneficial alleles were associated with these traits and are promising candidates for marker-assisted selection (Kantarski 2015). With the increased genetic complexity due to the vast allelic diversity and heterozygosity in this outcrossing polyploid, it would be difficult to stack and fix traits in the population. With these newly developed genetic and genomic resources, marker-assisted and genomic selection can be used to rapidly fix beneficial alleles in the breeding population.

It took thousands of years to domesticate the major annual crops, ultimately selecting for and stacking useful alleles underlying the domestication syndrome. This suite of traits generally included characteristics such as increased seed/fruit size, free-threshing, and yield, and reduced height and shattering. These traits have been selected for an intermediate wheatgrass breeding program, which had an effective population size of only 14 individuals. It took only three cycles of selection to achieve a plant with larger, free-threshing seeds, but such a plant has been difficult to reproduce (especially one with reduced height). The difficulty in consistently achieving a plant stacked with all the desirable traits could be one reason why an outcrossing perennial polyploid was never domesticated.

With modern genomics-assisted breeding, the speed at which an economically viable first perennial grain crop can be achieved may finally be reduced to within a lifetime. The implementation of marker-assisted selection will lead to earlier selection of plants with a suite of important domestication-related traits, thereby increasing the efficiency of the breeding program and decreasing the time to the first perennial grain crop. These results have far-reaching implications as the first case study of an on-going domestication of a perennial grain crop.

FUTURE WORK

GBS was utilized to discover SNPs that were used to construct a genetic map and to map important agronomic traits in *Th. intermedium*. These resources will be utilized to develop markers for marker-assisted selection for the intermediate wheatgrass-breeding program, fine map the gene(s) underlying free-threshing and other important domestication-related traits, and compare the structure and evolution of the *Q* gene across major cereals. Marker-assisted selection (MAS) and genomic selection become valuable tools for plant breeding when the time and cost of genotyping becomes less than that of phenotyping. With decreasing genotyping costs and development of high-throughput assays, this is now the case for most breeding programs and especially true for perennial crops (Myles et al. 2009). Advanced cycle breeding typically takes years to complete phenotypic evaluation and select the most productive individuals as parents for the next cycle of selection. This is particularly true for grain yield, which should be measured across years and locations to minimize environmental effects, and for perenniality traits such as longevity. The cycle length is even more profound for the breeding of perennials, which should be evaluated for several years to determine longevity and sustained production.

Genetic information in the form of whole-genome markers enables prediction of these complex traits with DNA from a single seedling or even a single seed (Meuwissen et al. 2001). GBS was the genotyping method of choice for wheat's challenging genome and gave higher prediction accuracies than established marker platforms for important agronomic traits (Poland et al. 2012b). Rapid selection for favorable alleles at known genes can then be accomplished using MAS (Lande and Thompson 1990, Singh et al. 2001). Future populations of many thousands of individuals can be genotyped, keeping only a fraction with the favorable alleles and the highest breeding values for phenotyping and continued improvement. GBS was successfully used for marker discovery and genotype calling in *Th. intermedium* and will be an efficient approach for genomics-assisted breeding.

The development of molecular markers for traits of interest will be an important contribution to the intermediate wheatgrass-breeding program and for conferring disease resistance to wheat. *Th. intermedium* has demonstrated resistance to many wheat pests and diseases, including leaf and stem rust (Larkin et al. 1995, Turner et al. 2013), wheat streak mosaic (Friebe et al. 1996, Cox et al. 2002), barley yellow dwarf (Larkin et al. 1995), *Cephalosporium* stripe (Cox et al. 2002), eyespot (Cox et al. 2002), and *Fusarium* head blight (Turner et al. 2013), which cause massive yield losses in wheat. Induced homoeologous pairing and recombination is commonly used in wheat to transfer a gene conferring a trait of interest from a closely related species or wild relative. Genes known to affect homoeologous pairing are well known and are often used to promote genetic exchange between chromosomes between two species. Ideally, molecular markers would then be used to determine whether the desired chromosomal segment (or gene) is present in each recombinant. While the genomic resources required for marker development are available for major crops, there are generally none available for the wild/related donor. Though considerable effort has been devoted to developing PCR-based simple-sequence-repeat markers that can be utilized for related species, there has been limited success and it remains a time-consuming process (Yu et al. 2004, Mullan et al. 2005, Peng and Lapitan 2005). The genetic and genomic resources generated for intermediate wheatgrass will alleviate this constraint for identifying *Th. intermedium* introgressions in wheat.

There has been considerable research to understand the domestication syndrome of cultivated species, but this work has primarily focused on understanding ancestral changes of our current crops. With the contemporary domestication of intermediate wheatgrass, mutations conferring key phenotypes for domestication can be targeted to both understand their functional effect in the population and select them for rapid improvement. This research is unique in that it will not only further our understanding of the genetic architecture of genes underlying domestication but also *apply* this understanding to the improvement of a perennial grain crop. The continuation of this work will use and improve upon the foundation of genetic and genomic resources recently built to accelerate the development of intermediate wheatgrass, give molecular insight into the domestication syndrome of cereals, and aid the transfer of disease resistance to wheat.

The United Nations Food and Agriculture Organization (FAO) research program on Climate Change, Agriculture, and Food Security has listed "Climate-Smart Agriculture" as one of its research priorities. Climate-smart agriculture is defined as "agriculture that sustainably increases productivity, resilience (adaptation), reduces/removes GHGs [green house gases] (mitigation), and enhances achievement of national food security and development goals" (FAO 2010, p. ii). Perennials are unmatched in their capacity for remediation and adaptation, but currently do not produce a sufficient amount of human-edible grain that is easily harvested, stored, and transported. The perennial grains currently being developed by research programs around the world could be improved to contribute to the food supply, facilitating the sustainable intensification of agricultural systems.

The spread of agriculture into marginal lands, climate disruption, and a growing global population contribute to the rising concern over food security. When prioritizing agronomic research, short-term demands, such as food security, profit, production, and resistance to new or evolving diseases, can outweigh long-term concerns, including sustainability, soil loss, and land degradation. This research continues to have both a short- and long-term focus through the development of genetic and genomic resources from *Th. intermedium* that would assist in both averting immediate food insecurity by improving bread wheat with disease resistance and improving intermediate wheatgrass as a sustainable source of grain. Principles from quantitative genetics, evolutionary biology, and applied breeding will continue to be utilized to understand and to accelerate the selection process for intermediate wheatgrass while providing new genomic resources to transfer disease resistance to wheat. By better understanding the genetics of the domestication syndrome in intermediate wheatgrass, we will be able to not only improve its breeding program but also provide a case study for the domestication of other perennial species, moving our agriculture systems toward sustainable intensification.

REFERENCES

Agrawal, A.F. 2006. Evolution of sex: Why do organisms shuffle their genotypes? *Curr Biol* 16:R696–R704.

Ahn, S., J.A. Anderson, M.E. Sorrells, and S.D. Tanksley. 1993. Homoeologous relationships of rice, wheat and maize chromosomes. *Mol Gen Genet* 241(5–6):483–490.

Batello, C., L. Wade, S. Cox et al., eds. 2014. *Perennial crops for food security*. Proceedings of the FAO Expert Workshop. Rome, Italy: Food and Agriculture Organization of the United Nations.

Bazzaz, F.A., N.R. Chiariello, P.D. Coley, and L.F. Pitelka. 1987. Allocating resources to reproduction and defense. *BioScience* 37(1):58–67.

Bennetzen, J.L. 2000. Comparative sequence analysis of plant nuclear genomes: Microcolinearity and its many exceptions. *Plant Cell* 12(7):1021–1029.

Bossolini, E., T. Wicker, P.A. Knobel, and B. Keller. 2007. Comparison of orthologous loci from small grass genomes *Brachypodium* and rice: Implications for wheat genomics and grass genome annotation. *Plant J* 49(4):704–717.

Chiariello, N. and J. Roughgarden. 1984. Storage allocation in seasonal races of an annual plant: Optimal versus actual allocation. *Ecology* 65(4):1290–1301.

Collard, B.C.Y. and D.J. Mackill. 2008. Marker-assisted selection: An approach for precision plant breeding in the twenty-first century. *Philos T Roy Soc B* 363(1491):557–572.

Cox, C.M., T.D. Murray, and S.S. Jones. 2002. Perennial wheat germ plasm lines resistant to eyespot, Cephalosporium stripe, and wheat streak mosaic. *Plant Dis* 86(9):1043–1048.

Cox, T., D. Van Tassel, C. Cox, and L. DeHaan. 2010. Progress in breeding perennial grains. *Crop Pasture Sci* 61(7):513–521.

Cox, T.S. 2009. Crop domestication and the first plant breeders. Chapter 1. In *Plant breeding and farmer participation*, eds. S. Ceccarelli, E.P. Guimarães, and E. Weltzien, 1–26. Rome, Italy: Food and Agriculture Organization of the United Nations.

Cox, T.S., J.D. Glover, D.L. Van Tassel, C.M. Cox, and L.R. DeHaan. 2006. Prospects for developing perennial-grain crops. *BioScience* 56:649–659.

Darwin, C. 2009. *The origin of species by means of natural selection: Or, the preservation of favored races in the struggle for life*, (edited and with an introduction by W.F. Bynum). New York: Penguin Classics.

DeHaan, L.R. and D.L. Van Tassel. 2014. Useful insights from evolutionary biology for developing perennial grain crops. *Am J Bot* 101(10):1801–1819.

DeHaan, L.R., S. Wang, S.R. Larson, D.J. Cattani, X. Zhang, and T.R. Kantarski. 2014. Current efforts to develop perennial wheat and domesticate thinopyrum intermedium as a perennial grain. In *Perennial crops for food security*. Proceedings of the FAO Expert Workshop. eds. C. Batello, L. Wade, S. Cox, N. Pogna, A. Bozzini, and J. Choptiany, 72–89. Rome, Italy: Food and Agriculture Organization of the United Nations.

Devos, K.M. and M.D. Gale. 2000. Genome relationships The grass model in current research. *Plant Cell* 12(5):637–646.

Distelfeld, A., C. Uauy, S. Olmos, A. Schlatter, J. Dubcovsky, and T. Fahima. 2004. Microcolinearity between a 2-cM region encompassing the grain protein content locus Gpc-6B1 on wheat chromosome 6B and a 350-kb region on rice chromosome 2. *Funct Integr Genomics* 4(1):59–66.

Doebley, J. 1990. Molecular evidence and the evolution of maize. *Econ Bot.* 44(3):6–27.

Doebley, J., A. Stec, and L. Hubbard. 1997. The evolution of apical dominance in maize. *Nature* 386(6624):485–488.

Doebley, J.F., B.S. Gaut, and B.D. Smith. 2006. The molecular genetics of crop domestication. *Cell* 127(7):1309–1321.

Elshire, R.J., J.C. Glaubitz, Q. Sun et al. 2011. A robust, simple genotyping-by-sequencing (GBS) approach for high diversity species. *PLoS One* 6(5):e19379.

FAO. 2010. *"Climate-smart" agriculture. Policies, practices and financing for food security, adaptation and mitigation*. Rome, Italy: Food and Agrigulture Organization of the United Nations.

FAO. 2013. *The state of food insecurity in the world. 2013. The multiple dimensions of of food security*. Rome, Italy: Food and Agriculture Organization of the United Nations.

Faris, J., Z. Zhang, J. Fellers, and B. Gill. 2008. Micro-colinearity between rice, *Brachypodium*, and *Triticum monococcum* at the wheat domestication locus Q. *Funct Integr Genomics* 8(2):149–164.

Feuillet, C. and B. Keller. 2002. Comparative genomics in the grass family: Molecular characterization of grass genome structure and evolution. *Ann Bot-London* 89(1):3–10.

Frary, A., T. C. Nesbitt, A. Frary et al. 2000. fw2.2: A quantitative trait locus key to the evolution of tomato fruit size. *Science* 289(5476):85–88.

Friebe, B., K.S. Gill, N.A. Tuleen, and B.S. Gill. 1996. Transfer of wheat streak mosaic virus resistance from *Agropyron intermedium* into wheat. *Crop Sci* 36:857–861.

Ganal, M.W., T. Altmann, and M.S. Röder. 2009. SNP identification in crop plants. *Curr Opin Plant Biol* 12(2):211–217.

Glover, J.D., J.P. Reganold, L.W. Bell et al. 2010. Increased food and ecosystem security via perennial grains. *Science* 328(5986):1638–1639.

Godfray, H.C.J., J.R. Beddington, I.R. Crute et al. 2010. Food security: The challenge of feeding 9 billion people. *Science* 327(5967):812–818.

Griffiths, S., R. Sharp, T.N. Foote et al. 2006. Molecular characterization of *Ph1* as a major chromosome pairing locus in polyploid wheat. *Nature* 439(7077):749–752.

Gross, B.L. and K.M. Olsen. 2010. Genetic perspectives on crop domestication. *Trends Plant Sci* 15(9):529–537.

Hartman, Y., D.P. Hooftman, M.E. Schranz, and P. van Tienderen. 2013. QTL analysis reveals the genetic architecture of domestication traits in Crisphead lettuce. *Genet Resour Crop Ev* 60(4):1487–1500.

IWGSC (International Wheat Genome Sequencing Consortium). 2014. A chromosome-based draft sequence of the hexaploid bread wheat (*Triticum aestivum*) genome. *Science* 345(6194):1251788.

Jaillon, O., J.M. Aury, B. Noel et al. 2007. The grapevine genome sequence suggests ancestral hexaploidization in major angiosperm phyla. *Nature* 449(7161):463–467.

Kantarski, T. R. 2015. The genetics of the domestication syndrome in the perennial intermediate wheatgrass. PhD dissertation, The University of Chicago, Chicago, IL.

Kellogg, E.A. 1998. Relationships of cereal crops and other grasses. *Proc Natl Acad Sci USA* 95(5):2005–2010.

Kuraparthy, V., S. Sood, and B. Gill. 2008. Genomic targeting and mapping of tiller inhibition gene (*tin3*) of wheat using ESTs and synteny with rice. *Funct Integr Genomics* 8(1):33–42.

Lande, R. and R. Thompson. 1990. Efficiency of marker-assisted selection in the improvement of quantitative traits. *Genetics* 124(3):743–756.

Larkin, P., P. Banks, E. Lagudah et al. 1995. Disomic *Thinopyrum intermedium* addition lines in wheat with barley yellow dwarf virus resistance and with rust resistances. *Genome* 38(2):385–394.

Lenser, T. and G. Theißen. 2013. Molecular mechanisms involved in convergent crop domestication. *Trends Plant Sci* 18(12):704–714.

Li, C., A. Zhou, and T. Sang. 2006. Rice domestication by reducing shattering. *Science* 311(5769):1936–1939.

Li, W. and B.S. Gill. 2002. The colinearity of the Sh2/A1 orthologous region in rice, sorghum and maize is interrupted and accompanied by genome expansion in the Triticeae. *Genetics* 160(3):1153–1162.

Lu, H. and J. Faris. 2006. Macro- and microcolinearity between the genomic region of wheat chromosome 5B containing the *Tsn1* gene and the rice genome. *Funct Integr Genomics* 6(2):90–103.

Meuwissen, T.H.E., B.J. Hayes, and M.E. Goddard. 2001. Prediction of total genetic value using genome-wide dense marker maps. *Genetics* 157(4):1819–1829.

Monfreda, C., N. Ramankutty, and J.A. Foley. 2008. Farming the planet: 2. Geographic distribution of crop areas, yields, physiological types, and net primary production in the year 2000. *Global Biogeochem Cy* 22(1):GB1022.

Moore, G., K.M. Devos, Z. Wang, and M.D. Gale. 1995. Cereal genome evolution: Grasses, line up and form a circle. *Curr Biol* 5(7):737–739.

Morgan, M.T. 2001. Consequences of life history for inbreeding depression and mating system evolution in plants. *Philos T Roy Soc B* 268(1478):1817–1824.

Mullan, D.J., A. Platteter, N.L. Teakle et al. 2005. EST-derived SSR markers from defined regions of the wheat genome to identify *Lophopyrum elongatum* specific loci. *Genome* 48(5):811–822.

Myles, S., J. Peiffer, P.J. Brown et al. 2009. Association mapping: Critical considerations shift from genotyping to experimental design. *Plant Cell* 21(8):2194–2202.

Nalam, V., M.I. Vales, C.W. Watson, S. Kianian, and O. Riera-Lizarazu. 2006. Map-based analysis of genes affecting the brittle rachis character in tetraploid wheat (*Triticum turgidum* L.). *Theor Appl Genet* 112(2):373–381.

Nordborg, M. and D. Weigel. 2008. Next-generation genetics in plants. *Nature* 456:720–723.

Olsen, K.M. and J.F. Wendel. 2013. A bountiful harvest: Genomic insights into crop domestication phenotypes. *Annu Rev Plant Biol* 64(1):47–70.

Paterson, A.H., J.E. Bowers, and B.A. Chapman. 2004. Ancient polyploidization predating divergence of the cereals, and its consequences for comparative genomics. *Proc Natl Acad Sci USA* 101(26):9903–9908.

Peng, J.H. and N.V. Lapitan. 2005. Characterization of EST-derived microsatellites in the wheat genome and development of eSSR markers. *Funct Integr Genomics* 5(2):80–96.

Pimentel, D. 2006. Soil erosion: A food and environmental threat. *Environ Dev Sustain* 8(1):119–137.

Poland, J.A., J. Endelman, J. Dawson et al. 2012b. Genomic selection in wheat breeding using genotyping-by-sequencing. *Plant Genet* 5(3):103–113.

Poland, J.A., P.J. Brown, M.E. Sorrells, and J.-L. Jannink. 2012a. Development of high-density genetic maps for barley and wheat using a novel two-enzyme genotyping-by-sequencing approach. *PLoS One* 7(2):e32253.

Poland, J.A. and T.W. Rife. 2012. Genotyping-by-sequencing for plant breeding and genetics. *Plant Genet* 5(3):92–102.

Primack, R.B. 1979. Reproductive effort in annual and perennial species of *Plantago* (Plantaginaceae). *Am Nat* 114(1):51–62.

Salse, J., S. Bolot, M. Throude et al. 2008. Identification and characterization of shared duplications between rice and wheat provide new insight into grass genome evolution. *Plant Cell* 20(1):11–24.

Simons, K.J., J.P. Fellers, H.N. Trick et al. 2006. Molecular characterization of the major wheat domestication gene *Q*. *Genetics* 172(1):547–555.

Singh, S., J.S. Sidhu, N. Huang et al. 2001. Pyramiding three bacterial blight resistance genes (*xa5*, *xa13* and *Xa21*) using marker-assisted selection into indica rice cultivar PR106. *Theor Appl Genet* 102(6–7):1011–1015.

Sorrells, M.E., M. La Rota, C.E. Bermudez-Kandianis et al. 2003. Comparative DNA sequence analysis of wheat and rice genomes. *Genome Res* 13(8):1818–1827.

Stewart, A.J.A. and K. Thompson. 1982. Reproductive strategies of six herbaceous perennial species in relation to a successional sequence. *Oecologia* 52(2):269–272.

Tang, H., J.E. Bowers, X. Wang, and A.H. Paterson. 2010. Angiosperm genome comparisons reveal early polyploidy in the monocot lineage. *Proc Natl Acad Sci USA* 107(1):472–477.

Tian, F., N.M. Stevens, and E.S. Buckler. 2009. Tracking footprints of maize domestication and evidence for a massive selective sweep on chromosome 10. *Proc Natl Acad Sci USA* 106(Supplement 1):9979–9986.

Turner, M.K., L.R. DeHaan, Y. Jin, and J.A. Anderson. 2013. Wheatgrass–wheat partial amphiploids as a novel source of stem rust and fusarium head blight resistance. *Crop Sci.* 53(5):1994–2005.

Valárik, M., A.M. Linkiewicz, and J. Dubcovsky. 2006. A microcolinearity study at the earliness per se gene *Eps-A^m 1* region reveals an ancient duplication that preceded the wheat–rice divergence. *Theor Appl Genet* 112(5):945–957.

Van Tassel, D.L., L.R. DeHaan, and T.S. Cox. 2010. Missing domesticated plant forms: Can artificial selection fill the gap? *Evol Appl* 3(5–6):434–452.

Vogel, K.P., K. Arumuganathan, and K.B. Jensen. 1999. Nuclear DNA content of perennial grasses of the Triticeae. *Crop Sci* 39(3):661–667.

Wainwright, P.C. 2007. Functional versus morphological diversity in macroevolution. *Annu Rev Ecol Evol Systemat* 38(1):381–401.

Wang, H., T. Nussbaum-Wagler, B. Li et al. 2005. The origin of the naked grains of maize. *Nature* 436(7051):714–719.

Wills, D.M. and J.M. Burke. 2007. Quantitative trait locus analysis of the early domestication of sunflower. *Genetics* 176(4):2589–2599.

Woodward, B., J. Shurkin, and D. Gordon. 2009. Scientists greater than Einstein: The biggest lifesavers of the twentieth century. Fresno, CA: Linden Publishing.

Xiong, L.Z., K.D. Liu, X.K. Dai, C.G. Xu, and Q. Zhang. 1999. Identification of genetic factors controlling domestication-related traits of rice using an F2 population of a cross between *Oryza sativa* and *O. rufipogon*. *Theor Appl Genet* 98(2):243–251.

Yan, L., A. Loukoianov, G. Tranquilli, M. Helguera, T. Fahima, and J. Dubcovsky. 2003. Positional cloning of the wheat vernalization gene *VRN1*. *Proc Natl Acad Sci USA* 100(10):6263–6268.

Yu, J., S. Hu, J. Wang et al. 2002. A draft sequence of the rice genome (*Oryza sativa* L. ssp *indica*). *Science* 296(5565):79–92.

Yu, J.K., M. La Rota, R.V. Kantety, and M.E. Sorrells. 2004. EST derived SSR markers for comparative mapping in wheat and rice. *Mol Genet Genomics* 271(6):742–751.

Zhang, Y., G.-H. Xu, X.-Y. Guo, and L.-J. Fan. 2005. Two ancient rounds of polyploidy in rice genome. *J Zhejiang U Sci B* 6(2):87–90.

19 Traits for Testing, Screening, and Improving Salt Tolerance of Durum Wheat Genetic Resources

R. Chaabane, A. Saidi, S. Moufida, S. Chaabane, S. Sayouri, M. Roussi, A. Ben Naceur, M. Inagaki, H. Bchini, M. Ben Naceur, I. Ayadi, and A. Bari

CONTENTS

According to the United States Department of Agriculture (USDA), Tunisia's proportion of durum wheat production out of its total wheat harvest is always the highest of the Maghreb countries, comprising about 80% of Tunisia's total wheat harvest. The goals of the Tunisian government include technology exchange, increasing certified seed usage, and reaching 120,000 ha of irrigated cereals in the short term. The new Tunisian cultivars are often characterized by high grain yield and are more responsive to increased seeding rates, added nitrogen fertilizer use, and additional supplementary irrigation and receptive to appropriate weed control techniques (Rezgui et al. 2008). However, most of the available water resources are of medium to poor quality, and the saline content is often high (Mougou et al. 2011). In arid and semiarid regions of Tunisia, the use of low-quality water for irrigation is accompanied by risks leading to plant stress and soil salinization with associated harmful consequences on plant development and yield. Salt accumulation in arable soils is mainly derived from irrigation water that contains trace amounts of sodium chloride (NaCl) and from seawater (Deinlein et al. 2014). In many coastal areas of Tunisia, excessive groundwater pumping from coastal freshwater wells increases saltwater intrusion, which leads to land contamination and salinization. Currently, about 50% of the total irrigated areas in Tunisia are considered at high risk for salinization (Bouksila et al. 2011). In these regions, salinity is one of the serious environmental problems, causing osmotic stress and reduction in plant growth and crop productivity. Projected climate-change impacts, rising temperature, rise in potential evapotranspiration rates, and declining rainfall exacerbate this problem of salinity.

Increased salt tolerance in sensitive species such as durum wheat has great economic potential beyond the improvement of yield in these saline areas. Having wheat with increased salt tolerance would be much less expensive for poor farmers in developing countries than having to use other management practices (Qureshi and Barrett-Lennard 1998). To meet this challenge, it is important to understand the mechanisms of salt tolerance for further improving salt tolerance of crops, by either traditional breeding or gene manipulation (Huang et al. 2006). Plant adaptation or tolerance to salinity stress involves complex physiological traits, metabolic pathways, and molecular or gene networks (Gupta and Huang 2014).

Possibilities for introducing salt tolerance into durum wheat involve traditional breeding techniques coupled with using physiologically based phenotyping, marker-assisted selection, and transformation of genes known to improve Na^+ exclusion or tissue tolerance (Lindsay et al. 2004). Identifying traits involved in tolerance will help to speed up the breeding efforts to develop tolerant genotypes. While many traits have been reported in the literature (El-Hendawy et al. 2005, Asadi and Khiabani 2007, Chaabane et al. 2011, Gupta and Huang 2014), these must be considered with respect to the analyzed genotypes and type of environment for which these genotypes are targeted.

Recent research has identified various adaptive responses to salinity stress at molecular, cellular, metabolic, and physiological levels, although mechanisms underlying salinity tolerance are far from being completely understood (Hasegawa et al. 2000). Salt tolerance in bread wheat (*Triticum aestivum* L.) and many other species is associated with the ability to exclude Na^+, so that high Na^+ concentrations do not occur in leaves, particularly in the leaf blade (Munns 2005). Durum (pasta) wheat (*Triticum turgidum* L. ssp. *durum* (Desf.) Husn.) is more salt sensitive than bread wheat, probably because of its poorer ability to exclude Na^+ from the leaf blade (Gorham et al. 1990, Munns and James 2003). Durum wheat lacks the Na^+-excluding locus *Kna1* (mapped to the distal region of chromosome 4DL), which enables hexaploid wheat to maintain lower leaf-Na^+ and a greater K^+ to Na^+ ratio than durum wheat (Dubcovsky et al. 1996). A new source of Na^+ exclusion was found in durum wheat, *Line 149*, which had low Na^+ concentrations and high K^+ to Na^+ ratios in the leaf blade similar to bread wheat (Munns et al. 2000). The low Na^+ phenotype was found to be controlled by two dominant interacting genes of major effect (Munns et al. 2003). The loci, named *Nax1* and *Nax2*, had originated in the diploid wheat *T. monococcum* L. and had been serendipitously crossed into Line 149 by a breeder introgressing genes for rust resistance from *T. monococcum* into durum wheat (The 1973, James et al. 2006, 2012).

The *Nax1* gene removes Na^+ from the xylem in roots and the lower parts of leaves, the leaf sheaths, whereas *Nax2* removes Na^+ from the xylem, only in the roots (James et al. 2006). They are likely to be Na^+ transporters in subfamily 1 of the *HKT* (high-affinity K^+ transporter) gene family. A gene in the *Nax2* locus, *TmHKT1;5A* encodes a Na^+-selective transport located on plasma membrane of root cells surrounding xylem vessels, which is therefore ideally localized to withdraw Na^+ from the xylem and reduce transport of Na^+ to leaves (Munns et al. 2012). The presence of *TmHKT1;5A* significantly reduces leaf Na^+ and increases durum wheat yield by 25% compared to near isogenic lines without the *Nax2* locus (Munns et al. 2012). The expression pattern of the *TmHKT7-A2* gene was consistent with the physiological role of *Nax1* in reducing Na^+ concentration in leaf blades by retaining Na^+ in the sheaths. *TmHKT7-A2* could control Na^+ unloading from xylem in roots and sheaths (Huang et al. 2006). *Nax1* was mapped as a QTL to the long arm of chromosome 2A (Lindsay et al. 2004). Tightly linked flanking molecular markers *gwm312*, *wmc170*, and *gwm249* (Lindsay et al. 2004) indicate the presence of *Nax1* and could be used for marker-assisted selection of it. Marker *gwm312* has been successfully used to accelerate the transfer of this trait into commercial varieties of durum wheat (Lindsay et al. 2004). *Nax2* was located in the terminal 14% of chromosome 5AL (Byrt et al. 2007). A tightly linked marker, *cslinkNax2*, has been used for selection of lines containing *Nax2*.

The goals of this chapter are to (1) assess Tunisian and Australian germplasm (containing Nax genes) for salinity response; (2) identify traits for salinity screening and improvement of this germplasm; (3) understand the relationship among different traits and their contribution to salinity tolerance; and (4) propose a salt-tolerance improving scheme for elite Tunisian varieties.

MATERIALS AND METHODS

Agro-Physiological Analysis

Twenty-three varieties of durum wheat were tested under salinity treatment and control: 4 Australian lines containing salt-tolerance *Nax* genes from CSIRO; 18 Tunisian old and new cultivars (Azizi, Baidha, Benbechir, Beskri, Bidi, Chili, INRAT69, Kerim, Khiar, Maali, Maghrbi, Mahmoudi, Nasr, Omrabiaa, Razzek, Richi, Selim, and SwebaEljia); and 1 line (CBD27) selected as an out-group from the INRAT crossing block. The 23 lines were grown under semi-controlled conditions in a rain-out shelter during the 2012/2013 growing season in 10-l pots (4 plants/pot) filled by a loamy-sand soil collected from the soil zone, 0–15 cm from the soil surface, at the Ariana Experimental Station of INRAT. The pots were placed on carts, so they could be moved under the shelter when it rained. The soil was air-dried, ground, passed through a 5-mm mesh screen, and thoroughly mixed. The experiment was conducted in triplicate with a completely randomized design. Two treatments were used: a saline treatment (150 mM NaCl) and a control (no NaCl). The salinity treatment was initiated at the three-leaf stage. Agro-physiological measurements were conducted at different growth stages. The height of the main shoot of each plant was measured with a ruler at 60, 90, and 120 days after sowing (DAS). Chlorophyll (Chl) content of the flag leaves was measured at 60, 90, and 120 DAS. In this protocol, the rate of Chl was estimated per unit SPAD. Three different measurements were performed at the base, center, and apex of the leaf using a portable Minolta SPAD 502 meter. Tiller number was recorded at 150 DAS. After harvesting, shoots were oven-dried at 70°C for 48 h to determine dry weight (DW). The number of spikes/plant, number of spikelets/spike, grain number, grain weight/spike, and 1000-grain weight were also determined at final harvest (150 DAS). The data were also converted to a salt-tolerance index (STI) to allow comparisons among genotypes for salt sensitivity. STI was defined as the observation at salinity divided by the average of the controls (Zeng et al. 2002, as cited in El-Hendawy et al. 2005). Analysis of variance (ANOVA) was performed using Statistica 5.0 v. '98 Edition.

Molecular Analysis

DNA Extraction

Total DNA was extracted from young leaves of a single plant per genotype. The extraction buffer (pH 8) was composed of 20 mM EDTA, 100 mM Tris-HCl (pH 8.0), 1.44 mM NaCl, 3% CTAB (w/v), and 1% β-mercaptoethanol (v/v). All reagents were from Sigma-Aldrich (St. Louis, Missouri). DNA was purified by a treatment with RNase (10 mg/ml, Fermentas) at a final concentration of 10 μg/ml followed by a phenolic extraction (treatment with an equal volume of phenol:chloroform:isoamyl alcohol (25:24:1, v/v/v), followed by treatment with an equal volume of chloroform:isoamyl alcohol (24:1, v/v). DNA concentration was quantified by spectro-photometry and quality was checked by gel electrophoresis. The average DNA yield was 15 μg DNA/g of tissue.

SSR Analysis

A set of microsatellites primer pairs amplifying the tightly linked Nax loci were used. PCR reactions were carried out in a 25-μl reaction volume containing 1 U of *Taq* polymerase, 50 to 100 ng of template DNA, 0.25 μM of each primer, 0.2 mM of each dNTP, 2 mM of $MgCl_2$, and 1X PCR reaction buffer. Amplifications were performed in a DNA thermocycler (Biometra Thermocycler, Goettingen, Germany) programmed for one cycle of 95°C for 3 min and 35 consecutive cycles (of 1 min denaturing at 94°C, 1 min annealing at 55°C, and 2 min extension at 72°C) followed by 10 min at 72°C. Amplified PCR products were separated by electrophoresis using a 3% agarose 1X TBE gel, stained with 0.5 mg/ml ethidium bromide, visualized under UV light, and photographed by a gel documentation system (GDS). A 100-bp DNA ladder (Promega, Ariana, Tunisia) was used as the molecular size standard.

RESULTS

Salinity affected all of the growth and development agro-physiological parameters measured at different growth stages. The values for tiller number, shoot DW (SDW), plant height at different stages, heading date, and root DW (RDW) for the salinity treatment varied significantly (Table 19.1) from those of the control. The flowering date was not significantly affected by salinity. The STIs of tiller number, SDW, plant height (60 and 90 DAS), flowering date, heading date, and RDW were significantly affected by salinity.

TABLE 19.1
Variance Analysis of Growth and Development Traits

		Trait								
Source	**df**	**Tiller**	**SDW**	**H60**	**H90**	**H120**	**H150**	**Flower**	**Heading**	**RDW**
Genotype (G)	22	0.54***	48.01***	42.76***	136.1***	995.47*	2160.62	535.57***	1640.65***	0.72***
Salinity (S)	1	10.19***	569.46***	226.16***	6,074.08***	8,934.49***	19,637.49***	47.54	218.13***	8.55***
G × S	22	0.39***	21.32***	9.82	31.65	619.21	1135.88	69.11	17.8	0.61***
Error	92	0.15	4.41	9.03	28.95	559.11	1343.69	42.2	16.38	0.04***
STI										
Genotype (G)	22	0.1***	0.03**	0.01*	0.01*	0.11	0.7	0.008*	0.003***	3.02***
Error	46	0.02	0.01	0.006	0.006	0.09	0.63	0.003	0.0009	0.38***

Note: Tiller = tiller number; SDW = shoot dry weight (g); H60 through H150 = plant height (cm) at 60 through 150 DAS, respectively; flower = flowering date, heading = heading date; RDW = root dry weight (g/plant).

*, **, and *** indicate significance at 0.05, 0.01, and 0.001 levels, respectively.

Mean tiller number for the SwebaEljia cultivar in the salinity treatment exceeded that of the control. This trait did not vary in the saline treatment compared to control for the varieties Richi and Azizi. For the rest of the analyzed varieties, the tiller number in the salinity treatment was lower by an average of 29% compared to that of the control (Figure 19.1). The STIs of tiller number ranged from 0.46 (Khiar) to 1.22 (SwebaEljia). For tiller number, Khiar was most affected by salinity and SwebaEljia was least affected.

Mean SDW for all varieties in the salinity treatment was reduced by 49.6% compared to the control. The STI of SDW ranged from 0.32 (Omrabiaa) to 0.7 (Mahmoudi). For SDW, Omrabiaa was most affected by salinity and Mahmoudi was least affected (Figure 19.2).

Mean plant height in the salinity treatment was reduced by 6%, 20%, 18%, and 14%, respectively, at 60, 90, 120, and 150 DAS compared to the control. At 60 DAS, the STI of plant height ranged from 0.84 (Swebaa Eljia) to 1.11 (Nax242). At 90 DAS, the STI of plant height ranged from 0.63 (Bidi) to 0.9 (Nasr). At 120 DAS, the STI of plant height ranged from 0.44 (CBD) to 1.37 (Baidha). At 150 DAS, the STI of plant height ranged from 0.32 (Kerim) to 2.92 (Mahmoudi).

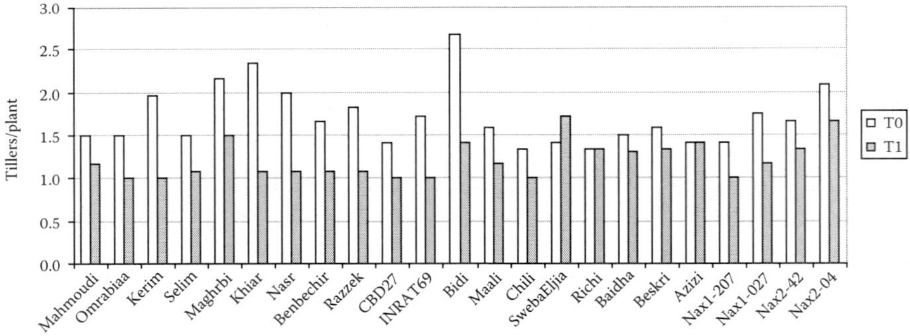

FIGURE 19.1 Effect of salinity treatment on tiller number (T1: salinity treatment, T0: control).

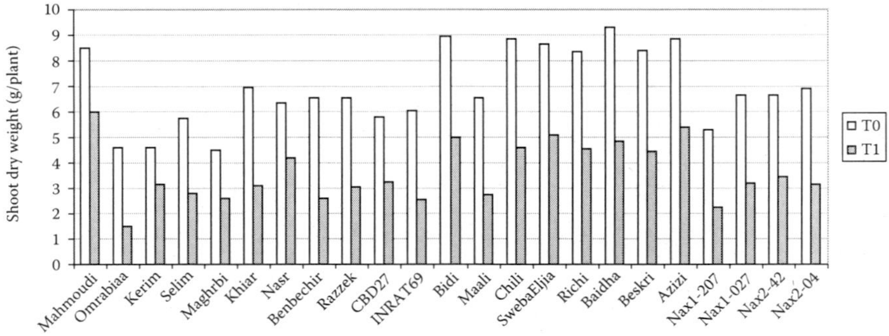

FIGURE 19.2 Shoot dry weight (g) variation induced by salt stress (T1: salinity treatment, T0: control).

The flowering date was in some cases earlier in the salinity treatment and later in other cases, compared to the control. The flowering date on average was earlier by a maximum of 9.5 days (Nax1-207) and was later by a maximum of 10 days (Khiar) in salinity treatment compared to the control.

The heading date was in some cases earlier in the salinity treatment and later in other cases compared to the control. The heading date on average was earlier by a maximum of 11.75 days (SwebaEljia) and was later by a maximum of 11 days (Bidi) in the salinity treatment compared to the control.

Mean RDW for the three Nax lines (Nax1-207, Nax1-027, and Nax2-04) in the salinity treatment exceeds that of the control. For the rest of the analyzed varieties the RDW in the salinity treatment was reduced by 78% compared to the control. The STI of RDW ranged from 0.06 (Khair) to 4.9 (Nax1-027). For RDW, Khair was the most affected by salinity, and the Nax genotypes were the least affected.

Except for spike length and spike weight, the rest of final harvest parameters in the salinity treatment varied significantly from those in the control (Table 19.2). Mean number of spikes per plant for all varieties in the salinity treatment was reduced by 28.73% compared to the control. The STI of the number of spikes per plant ranged from 0.26 (Bidi) to 1 (Azizi). As measured by number of spikes per plant, Bidi was the most affected by salinity and Azizi was the least affected. For some genotypes (Mahmoudi, Kerim, Selim, Nasr, INRAT69, Bidi, Bidha, Nax1-207, and Nax2-04), spike length was greater in the salinity treatment compared to the control. For these varieties, mean spike length was increased by 12% compared to the control. For the rest of analyzed varieties, spike length was reduced by 8.7% compared to the control. The STI of the spike length ranged from 0.82 (Razzek) to 1.29 (Nasr). The spike weight of the Nasr cultivar was higher in the salinity treatment compared to the control. For the rest of the analyzed varieties, the mean spike weight in the salinity treatment was reduced by 44.58% compared to the control. The STI of the spike weight ranged from 0.04 (Khair) to 1 01 (Nasr).

TABLE 19.2
Variance Analysis of Final Harvest Traits

		Trait					
Source	**df**	**Spikes/ Plant**	**Spike Length (cm)**	**Spike Weight (g)**	**Spikelets/ Spike**	**Grains/Spike**	**Grain Yield**
Genotype (G)	22	0.38**	3.07***	102.218	19.62***	139.9*	1.73***
Salinity (S)	1	9.69***	0.23	274.17	191.66***	1600.06***	65.10***
G × S	22	0.37**	0.81	98.22	2.3	48.06	0.94***
Error	92	0.15	0.81	99.54	2.67	72.8	0.34
STI							
Genotype (G)	22	0.06***	0.04**	0.23***	0.03***	0.13**	0.06***
Error	46	0.01	0.02	0.07	0.007	0.05	0.02

Note: Grain yield = total grain weight/plant (g).

*, **, and *** indicate significance at 0.05, 0.01, and 0.001 levels, respectively.

For the variety Selim, the number of spikelets per spike was higher in the salinity treatment compared to the control. For the rest of analyzed varieties, the mean spikelet number per spike in the salinity treatment was lower by 16.6% compared to the control. The STI of the spikelet number per spike ranged from 0.71 (Bidi) to 1.05 (Selim); thus, Bidi was the most affected by salinity, and Selim was the least affected for this trait.

For the varieties Kerim and Nasr, mean grain number per spike was higher by 11% and 16%, respectively, in the salinity treatment compared to the control. For the rest of the analyzed varieties, the global average of grain number in the salinity treatment was lower by 27.3% compared to the control. The STI of the grain number per spike ranged from 0.35 (Maghrbi) to 1.11 (Kerim); thus, Maghrbi was the most affected by salinity, and Kerim was the least affected for this trait.

Mean grain yield per plant for all varieties in the salinity treatment was lower by 58.6% compared to the control (Figure 19.3). The STI of the grain yield per plant ranged from 0.14 (Omrabiaa) to 0.74 (Mahmoudi); thus, Omrabiaa was the most affected by salinity, and Mahmoudi was the least affected for this trait. The average chlorophyll content of flag leaves (Chl content) varied over time. At 90 and 150 DAS, it varied significantly in the salinity treatment compared to the control (Table 19.3). The STI of Chl content varied significantly. Chl content increased slowly at early vegetative stages, reaching a maximum at advanced stages, and then fell quickly at senescence. Compared to the control, the average Chl content in salinity treatment increased by 8% and 1.97% at 90 and 120 DAS, respectively. At 150 DAS, the average Chl content of the analyzed genotypes decreased by 56% in the salinity treatment compared to the control (Figure 19.4). This reveals that senescence was advanced by salinity. These results appear to fully support our previous studies reported in Chaabane et al. (2011, 2012).

DISCUSSION

The various agro-physiological traits showed different responses to salinity. To simplify the reading of the various statistical tables, we reported the most important traits in Table 19.4, those significantly influenced by salinity and having an STI that

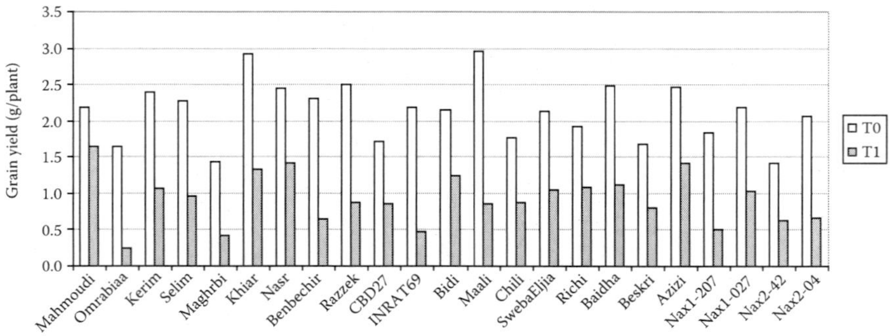

FIGURE 19.3 Grain yield (g/plant) variation induced by salt stress (T1: salinity treatment, T0: control).

TABLE 19.3
Variance Analysis of Physiological Parameter (Chlorophyll Content in SPAD Units) at 90, 120, and 150 DAS

		Chl Content (SPAD Units)		
Source	**df**	**90 DAS**	**120 DAS**	**150 DAS**
Genotype (G)	22	34.21*	93.53*	752.30***
Salinity (S)	1	393.27***	0.65	1117.50***
G × S	22	26.048	111.64**	92.92**
Error	92	18.896	51.019	39.78
STI				
Genotype (G)	22	0.04*	0.22*	0.83***
Error	46	0.01	0.06	0.04

*, **, and *** indicate significance at 0.05, 0.01, and 0.001 levels, respectively.

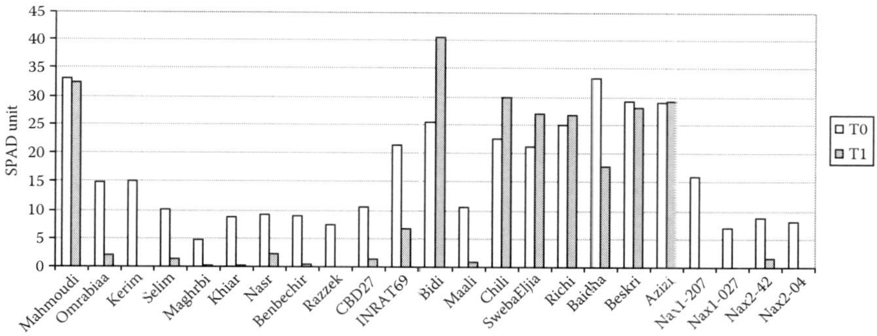

FIGURE 19.4 Effect of salinity on chlorophyll content (SPAD units) at 90 DAS (T1: salinity treatment, T0: control).

differed significantly from the corresponding STI of grain yield. In other words, these seven traits are significantly affected by salinity and their sensitivity to salinity affects grain yield significantly.

According to these results, the most important traits for salinity-tolerance characterization at vegetative stages are tillering, plant height, and chlorophyll content at 90 DAS. Crop yield is generally dependent upon various yield-contributing components, tillering, being the most important one (Jamel et al. 2011). Salinity affected tillering in all analyzed genotypes (Figure 19.1). Reduction in the tillering capacity of various varieties might be due to the toxic effects of salt on plant growth, particularly at an early growth stage (Goudarzi and Pakhiyat 2008).

These results are in accord with those obtained by several other authors. Goudarzi and Pakhiyat (2008) reported significant reduction in tillers per plant of various

TABLE 19.4
Pearson Correlation of the STIs of the Most Important Salinity-Tolerance Traits with the STI of Grain Yield

Trait	Correlation
Tillering	0.33*
Shoot dry weight	0.57*
Plant height 90 DAS	0.31*
Chl content (SPAD) 90 DAS	0.26*
Spikes per plant	0.39*
Grains per spike	0.42*
1000 grain weight	0.61*

* indicates significance at 0.01.

wheat cultivars due to salinity. El-Hendawy et al. (2005) reported that tiller number was significantly more affected by salinity than leaf number and leaf area at the vegetative stage. Shahzad et al. (2012) reported that the number of tillers/plant in wheat landraces was reduced significantly less than 250 mm NaCl stress. Nicolas et al. (1994) found that salt stress during tiller emergence can inhibit tiller formation and can cause their abortion at later stages. Jones and Kirby (1977) reported that breeding genotypes with fewer but less vulnerable tillers could substantially increase yields on salt-affected soils. This research may suggest that tiller number under salt-stressed conditions can be used as a simple and nondestructive measurement for the assessment of salt tolerance in wheat-breeding programs. Use of tiller number has been proposed as a morphological trait for the screening of salt-tolerant genotypes in rice (Alam et al. 2004, Singh 2007) and in barley-breeding programs (Bchini et al. 2010).

According to our results, the two other important traits for salinity screening at the vegetative stage are plant height and chlorophyll content at 90 DAS, which corresponds to the heading stage for the majority of genotypes. In some earlier studies, a positive correlation has been found between rate of photosynthesis and final yield. Ashraf and Bashir (2003) reported that the relatively higher photosynthetic capacity of Inqlab-91 (a high yielding cultivar of wheat) at the vegetative stage might have played a significant role in its higher grain yield. In the present study, the values obtained for chlorophyll content differ from one genotype to another. Chlorophyll content at different growth stages showed significant variation (Figure 19.4). These genotypes differed significantly in chlorophyll content variation under salinity (Table 19.3). At 90 DAS, a decrease in chlorophyll content in salinity treatments compared to the control was observed for Omrabiaa, Maghrbi, and Bidi, but increased chlorophyll content was observed in the rest of analyzed genotypes. At this stage, most all of the analyzed genotypes recorded higher chlorophyll content in the salinity treatment compared to the control. Increase in chlorophyll content under salt stress has already been reported (Ghogdi et al. 2012). After the 90-DAS stage, there was gradual reduction in total chlorophyll accumulation in the leaves for the salinity treatments as compared to the controls. At 150 DAS, there was a high reduction (56%) of chlorophyll content in the salinity treatments compared to the controls. The decrease in chlorophyll content under stress at advanced stages is a commonly reported phenomenon in various studies (Ashraf and Shahbaz 2003, Raza et al. 2006). This may be due to different reasons: (1) membrane deterioration (Ashraf and Bhatti 2000) or (2) interference of salt ions with the de novo synthesis of proteins, the structural component of chlorophyll, rather than the breakdown of chlorophyll (Jaleel et al. 2007). The large decrease in chlorophyll content in salinity treatments at 150 DAS reveals that senescence processes were promoted by salinity, which confirms our results in Chaabane et al. (2012). This decrease in chlorophyll content most closely parallels the change in photosynthesis occurring during the period of grain filling, which in turn affects grain weight and final yield. Reduction in grain weight can be due to a shortening of the grain-filling period resulting from accelerated senescence (Hochman 1982). The absence of a significant correlation between the STI of grain yield and the STI of chlorophyll content at 150 DAS could be partly due to the fact that the chlorophyll content was measured instantaneously, whereas grain yield is the result of many processes across the entire grain-filling period. Frequent measurements of chlorophyll content during the grain-filling period could give better results for salinity-tolerance characterization.

Salt stress induced significant variation in plant growth. Plant height decreased with salt stress. The reduction of plant height under salt-stress conditions was also reported by Khan et al. (2007). The most relevant variation in plant growth induced by salt stress was observed in the SDW trait. The SDW and grain yield showed large and significant variation (Table 19.2) in the salinity treatments in this set of 23 genotypes. SDW was among the most affected traits with a 49.6% reduction compared to the controls.

The STI of SDW was highly significantly correlated with the STI of grain yield (r = 0.57), showing that these two parameters are linearly, greatly related (R^2 = 0.67) (Figure 19.5). On the basis of these findings, it can be concluded that SDW is important for the assessment of salt tolerance in durum wheat. It was also previously reported that genotype ranking in terms of SDW leads to the same result as ranking by grain yield (El-Hendawy et al. 2011).

The evaluation of final grain yield and of growth parameters determining grain yield is critical to breeding programs (Turki et al. 2014). The final yield of wheat is determined by the number of spikes per plant and various yield components, such as grain number and 1000-grain weight. All the final yield traits were significantly affected by salinity, and their STI was highly significantly correlated with STI of grain yield (Table 19.2). ANOVA (Table 19.2) showed that salt stress had a significant effect on number of spikelets per spike, seed weight, seed number, and grain yield. These traits have remarkable variation among the analyzed genotypes; thus, they can be helpful as candidate traits for subsequent studies screening for salt tolerance with larger populations.

RDW was highly significantly affected by salinity but the STI of RDW was not significantly correlated with the STI of grain yield. For this reason, we have not considered this parameter as a good marker trait for salt-tolerance screening and improvement.

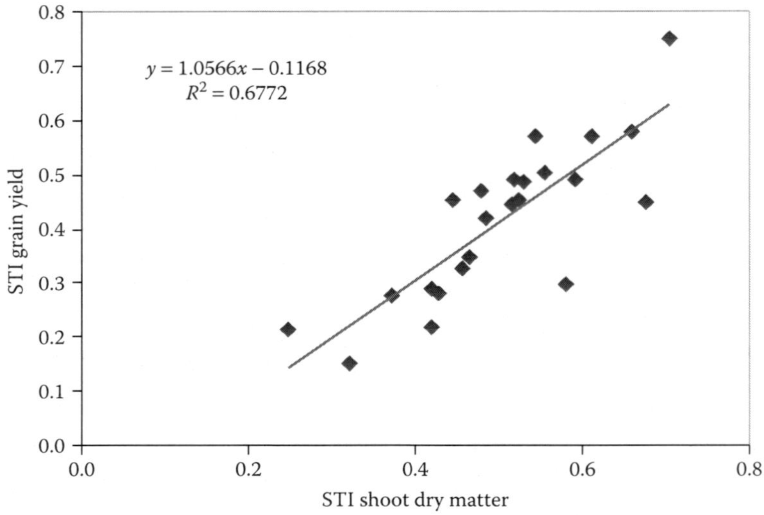

FIGURE 19.5 Relationship between STI of shoot dry matter and the STI of grain yield.

The complex nature of the effect of salt on plants makes it difficult for breeders to select superior genotypes based only on phenotypic evaluation. The development of DNA markers linked to salt-tolerant genes allows breeders to select individuals carrying target genes in a segregating population based on linked markers rather than on their phenotypes. Tightly linked flanking molecular markers to the salt-tolerance *Nax* genes could be used for marker-assisted selection of these loci. We used these markers in a backcross-assisted selection scheme to transfer these genes from Nax Australian lines to Tunisian elite varieties to improve their salt tolerance. A cross between an elite Tunisian variety (Selim) and Nax1 lines was made. The results of gel electrophoresis of the PCR products of the DNA extracted from the parents and F_1 offspring are shown in Figure 19.6. The PCR was generated by the SSR primer *wmc170*, which is linked to the *Nax1* gene.

The results of the electrophoresis show two alleles (Table 19.5): one allele with 199 base pairs (bp) belonged to the Nax lines, and another allele with 236 bp belonged to the Tunisian elite variety Selim. These results show that all the crosses succeeded because they have both parental alleles in their electrophoresis profile. This is an example of how to follow the transfer of the alleles from one generation to another. A series of backcrosses with marker-assisted selection will facilitate the transfer of the *Nax1* gene into the Tunisian variety Selim. In addition, we can also screen based on morpho-physiological traits in each backcross generation to better improve the salt-tolerance performances of the variety Selim.

Finally, we conclude that both molecular and agro-physiological approaches could be combined to select individuals carrying target genes and traits for salt-tolerance improvement.

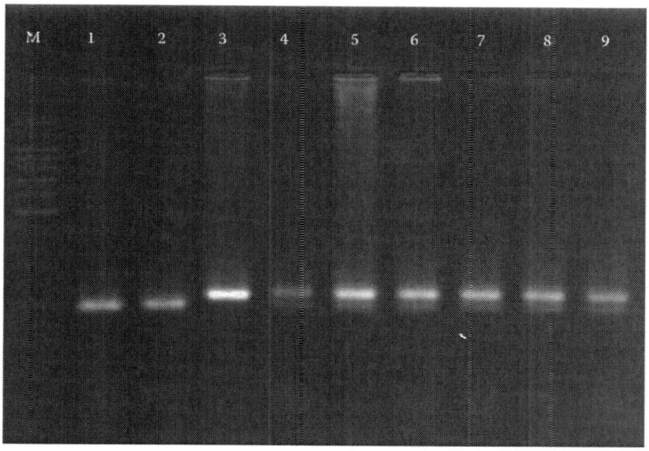

FIGURE 19.6 Agarose gel (2%) electrophoresis of PCR products generated by SSR primer *wmc170* from DNA of the Nax1 lines (lanes 1 and 2), the Selim parent (lane 3), and their F_1 offspring (lanes 4 through 9).

TABLE 19.5
Allele Size of the Parents (Nax Lines and Selim) and Their F_1 Offspring

Sample	Variety/Cross	Allele Size (bp)
1	Nax1 line	199
2	Nax1 line	199
3	Selim	236
4	C31	199, 236
5	C31	199, 236
6	C32	199, 236
7	C32	199, 236
8	C34	199, 236
9	C34	199, 236

REFERENCES

Alam, M.Z., T. Stuchbury, R.E.L. Naylor, and M.A. Rashid. 2004. Effect of salinity on growth of some modern rice cultivars. *J Agron* 3:1–10.

Asadi, A.A. and B.N. Khiabani. 2007. Evaluation of salt tolerance based on morphological and yield traits in wheat cultivars and mutants. *Int J Agr Biol* 9:693–700.

Ashraf, M. and A. Bashir. 2003. Salt stress induced changes in some organic metabolites and ionic relations in nodules and other plant parts of two crop legumes differing in salt tolerance. *Flora* 198:486–498.

Ashraf, M. and M. Shahbaz. 2003. Assessment of genotypic variation in salt tolerance of early CIMMYT hexaploid wheat germplasm using photosynthetic capacity and water relations as selection criteria. *Photosynthetica* 41(2):273–280.

Ashraf, M.Y. and A.S. Bhatti. 2000. Effect of salinity on growth and chlorophyll content of rice. *Pakistan J Scientif Indust Res* 43:130–131.

Bchini, H., M. Ben Naceur, R. Sayar, H. Khemira, and L. Ben Kaab-Bettaeïb. 2010. Genotypic differences in root and shoot growth of barley (*Hordeum vulgare* L.) grown under different salinity levels. *Hereditas* 147:114–122.

Bouksila, F. 2011. Sustainability of irrigated agriculture under salinity pressure—A study in semiarid Tunisia. PhD Thesis. Department of Water Resources Engineering. Faculty of Engineering, Lund University, Sweden. Report N°1053, code: Lutvdg/Tvvr-1053.

Byrt, C., J.D. Platten, W. Spielmeyer et al. 2007. HKT1;5-like cation transporters linked to Na^+ exclusion loci in wheat, *Nax2* and *Knal*. *Plant Physiol* 143:1918–1928.

Chaabane, R., H. Bchini, H. Ouji et al. 2011. Behaviour of Tunisian durum wheat (*Triticum turgidum* L.) varieties under saline stress. *Pakistan J Nutrition* 10(6):539–542.

Chaabane, R., K. Sahari, K. Khamassi et al. 2012. Molecular and agro-physiological approach for parental selection before intercrossing in salt tolerance breeding programs of durum wheat. *Int J Pl Breed* 6(2):100–105.

Deinlein, U., A.B. Stephan, T. Horie et al. 2014. Plant salt-tolerance mechanisms. *Trends Plant Sci* 19:371–379.

Dubcovsky, J., G. Santa María, E. Epstein, M.C. Luo, and J. Dvorak. 1996. Mapping of the K^+/Na^+ discrimination locus *Knal* in wheat. *Theor Appl Genet* 92:448–454.

El-Hendawy, S.E., Y. Hu, J.I. Sakagami, and U. Schmidhalter. 2011. Screening Egyptian wheat genotypes for salt tolerance at early growth stages. *Int J Pl Prod* 5:283–298.

El-Hendawy, S.E., H. Yuncai, M. Gamal et al. 2005. Evaluating salt tolerance of wheat genotypes using multiple parameters. *Eur J Agron* 22:243–253.

Ghogdi Akbari, E., A. Izadi-Darbandi, and A. Borzouei. 2012. Effects of salinity on some physiological traits in wheat (*Triticum aestivum* L.) cultivars. *Indian J Sci Technol* 5:1901–1906.

Gorham, J., R.G. Wyn Jones, and A. Bristol. 1990. Partial characterization of the trait for enhanced K1-Na1 discrimination in the D genome of wheat. *Planta* 180:590–597.

Goudarzi, M. and H. Pakniyat. 2008. Evaluation of wheat cultivars under salinity stress based on some agronomic and physiological traits. *J Agr Forest Soc Sci* 4:35–38.

Gupta, B. and B. Huang. 2014. Mechanism of salinity tolerance in plants: Physiological, biochemical, and molecular characterization. *Int J Genomics* 2014: Article ID 701596.

Hasegawa, P.M., R.A. Bressan, J.K. Zhu, and H.J. Bohnert. 2000. Plant cellular and molecular responses to high salinity. *Annu Rev Plant Physiol Plant Mol Biol* 51:463–499.

Hochman, Z.V.I. 1982. Effect of water stress with phasic development on yield of wheat grown in a semi-arid environment. *Field Crop Res* 5:55–67.

Huang, S., W. Spielmeyer, E.S. Lagudah et al. 2006. A sodium transporter (HKT7) is a candidate for *Nax1*, a gene for salt tolerance in durum wheat. *Plant Physiol* 142:1718–1727.

Jaleel, C.A., A. Manivannan, and G.M.A. Lakshmanan. 2007. NaCl as a physiological modulator of proline metabolism and antioxidant potential in *Phyllanthus amarus*. *CR Biol* 330:806–813.

Jamal, Y., M. Shafi, and J. Bakht. 2011. Effect of seed priming on growth and biochemical traits of wheat under saline conditions. *African J Biotechnol* 10(75):17127–17133.

James, R.A., C. Blake, A.B. Zwart, R.A. Hare, A.J. Rathjen, and R. Munns. 2012. Impact of ancestral wheat sodium exclusion genes *Nax1* and *Nax2* on grain yield of durum wheat on saline soils. *Funct Pl Biol* 39:609–618.

James, R.A., R.J. Davenport, and R. Munns. 2006. Physiological characterisation of two genes for Na^+ exclusion in durum wheat: *Nax1* and *Nax2*. *Plant Physiol* 142:1537–1547.

Jones, H.G. and E.J.M. Kirby. 1977. Effects of manipulation of number of tillers and water supply on grain yield in barley. *J Agr Sci Cambridge* 88:391–397.

Khan, M.A., S. Yasmin, R. Ansari, M.U. Shirazi, and M.Y. Ashraf. 2007. Screening for salt tolerance in wheat genotypes at an early seedling stage. *Pakistan J Bot* 39:2501–2509.

Lindsay, M.P., E.S. Lagudah, R.A. Hare, and R. Munns. 2004. A locus for sodium exclusion (*Nax1*), a trait for salt tolerance, mapped in durum wheat. *Funct Pl Biol* 31:1105–1114.

Mougou, R., M. Mansour, A. Iglesias, R.Z. Chebbi, and A. Battaglini. 2011. Climate change and agricultural vulnerability. *Reg Environ Change* 11:137–142.

Munns, R. 2005. Genes and salt tolerance: Bringing them together. *New Phytol* 167:645–663.

Munns, R., R.A. Hare, R.A. James, and G.J. Rebetzke. 2000. Genetic variation for improving the salt tolerance of durum wheat. *Aust J Agric Res* 51:69–74.

Munns, R. and R.A. James. 2003. Screening methods for salinity tolerance: A case study with tetraploid wheat. *Plant Soil* 253:201–218.

Munns, R., R.A. James, B. Xu et al. 2012. Wheat grain yield on saline soils is improved by an ancestral Na^+ transporter gene. *Nat Biotechnol* 30:360–364.

Munns, R.M., G.J. Rebetzke, S. Husain, R.A. James, and R.A. Hare. 2003. Genetic control of sodium exclusion in durum wheat. *Aust J Agric Res* 54:627–635.

Nicolas, M.E., R. Munns, A.B. Samarakoon, and R.M. Gifford. 1994. Elevated CO_2 improves the growth of wheat under salinity. *Aust J Plant Physiol* 20:349–360.

Qureshi, R.H. and E.G. Barrett-Lennard. 1998. Three approaches for managing saline, sodic and waterlogged soils. In *Saline agriculture for irrigated land in Pakistan*, eds. N. Elbasam, M. Damborth, and B.C. Laugham, 8–19, Dordrecht, The Netherlands: Kluwer Academic Publishers.

Raza, S.H., H.R. Athar, and M. Ashraf. 2006. Influence of exogenously applied glycinebetaine on the photosynthetic capacity of two differently adapted wheat cultivars under salt stress. *Pakistan J Bot* 38(2):341–351.

Rezgui, S., M.M. Fakhfakh, S. Boukef et al. 2008. Effect of common cultural practices on septoria leaf blotch disease and grain yield of irrigated durum wheat. *Tunisian J Plant Protect* 3(2):59–68.

Shahzad, A., M. Ahmad, M. Iqbal, I. Ahmed, and G.M. Ali. 2012. Evaluation of wheat landrace genotypes for salinity tolerance at vegetative stage by using morphological and molecular markers. *Genet Mol Res* 11(1):679–692.

Singh, M.P., D.K. Singh, and M. Rai. 2007. Assessment of growth, physiological and biochemical parameters and activities of antioxidative enzymes in salinity tolerant and sensitive basmati rice varieties. *J Agron Crop Sci* 193:398–412.

The, T.T. 1973. Transference of resistance to stem rust from *Triticum monococcum* L. to hexaploid wheat. PhD thesis, University of Sydney, Sydney, Australia.

Turki, N., T. Shehzad, M. Harrab, M. Tarchi, and K. Okuno. 2014. Variation in response to salt stress at seedling and maturity stages among durum wheat varieties. *J Arid Land Studies* 24(1):261–264.

20 Gene Flow as a Source of Adaptation of Durum Wheat to Changing Climate Conditions

Double Gradient Selection Technique

M. Nachit, J. Motawaj, Z. Kehel, D.Z. Habash, I. Elouafi, M. Pagnotta, E. Porceddu, A. Bari, A. Amri, O.F. Mamluk, and M. El-Bouhssini

CONTENTS

INTRODUCTION

Fluctuating environments are recurring scenarios in dryland agriculture with detrimental effects to crop production such as durum wheat crop that could lead to losses in yield. The dryland Mediterranean type of environment in particular vis-à-vis climate is characterized by low and highly erratic annual rainfall varying from 200 to 800 mm, with usually a poor rainfall distribution coupled with intermittent periods of drought and temperature extremes (cold and heat) that can occur at any stage of durum wheat plant development (Nachit et al. 1992a, 1992b).

These environmental fluctuations and stresses are expected to amplify as climate models show relatively consistent predictions that the Mediterranean basin will become hotter and drier over the next century with annual precipitation likely to decrease by 4%–27% and 3–5°C increase in temperature (IPCC 2007, Habash et al. 2009). The assessment of the fifth report remains in agreement with the later projections including the probable increase in extreme weather events.

Because of limitation of irrigation on a large scale for durum wheat commercial production, the only applicable alternative is the improvement of drought tolerance and yield stability through genetics and plant breeding supported by an understanding of the environment fluctuations, stress physiology and the use of molecular markers.

This chapter gives a broad review of the durum wheat research program, with emphasis on the challenges involved, particularly drought research to cope with water limitations and environmental fluctuations. Overall, the chapter focuses on strategies to breed for fluctuating environment in the context climate changing conditions.

STRATEGY OF ADAPTATION

The ICARDA dryland durum wheat-breeding program was initiated in 1977 in northern Syria. The ICARDA main research station (Tel Hadya) and its related research sites (Latakia, Terbol, Kfardan, and Breda) are located in dryland area where agriculture originated and evolved across different environmental climatic gradients over millennia. This region, known as the Fertile Crescent harbors, as a result, has a wealth of wheat landraces along with their wild relatives spanning these different agro-ecological zones, from lowland plains to highland plateaus, and from favorable to stressed environmental conditions (Nachit 1992, 1998a). There are thousands of landraces that were grown over long periods and are still found along with their related wheat *Triticum* parental species (Nachit and Elouafi 2004). These genetic resources grow under drought, cold, and heat as well as under multiple biotic stresses with the highest virulence of disease races and insect biotypes along the gradient environmental conditions.

The combinations of these abiotic and biotic stresses made the breeding work in dryland, such as West Asia and North Africa (WANA), both complex and challenging. Thus these wheat landraces and their wild relatives have been strategically incorporated in germplasm renewal continuously to buffer the environmental change and fluctuating conditions (Nachit and Ouassou 1988, Nachit 1992). This sustained gene flow from landraces and wild relatives to the improved varieties has been adopted strategically to enable durum wheat crop to withstand environmental fluctuations, abiotic stresses, and recurrent biotic virulence.

Modeling of gene flow has shown that not only does it enhance genetic variation, and thus the capacity to adapt, but it could also increase directly the population's adaptation as it introduces alleles that are newly adaptive in the recipient population (Shaw and Etterson 2012). Gene flow may involve a large number of networked genes that have been found to be involved with traits, such as those associated with water-use efficiency (WUE). The expression of some genes may increase or decrease the expression of other genes, thus forming a complex network of interactions (Liebovitch et al. 2006). There are a number of such complex networks of genes that involve genotype–environment interactions and also epistatic interactions between genes regulating variation for traits (Cooper et al. 2000, Dennis 2002). The interaction and functioning among these networks of genes could be predicted if these networks were understood and appropriately quantified (van Oosterom et al. 2004). The models of gene networks are based on the expression of gene i as a level of mRNA, with value X, among N genes, represented by a connection matrix M. The relationship of a new value X' at a subsequent time with the original X is given as

$$X' = MX$$

The crosses that have been made with the sustained gene flow have led to the development of germplasm with essentially durum genotypes performing well under stressed environments with abiotic constraints such as drought and temperature extremes.

The assessment of the extent of genetic variation created and acquired was based on the use of trait measurements such as carbon isotopic composition ($\delta^{13}C$), which is a useful surrogate for integrated plant WUE. Negative $\delta^{13}C$ is indicative of lower WUE. The definition of $\delta^{13}C$, expressed in per thousand units (‰), is given as follows:

$$\delta^{13}C = [(R_{sample}/R_{standard}) - 1] \times 1000 \tag{20.1}$$

where R is the ratio between carbon 13 (^{13}C) and carbon 12 (^{12}C) isotopes.

The $^{13}C/^{12}C$ ratio of plant biomass is used as a natural indicator of photosynthetic WUE to measure the variation of $^{13}C/^{12}C$ in plant tissue with respect to the atmospheric source of carbon during growth (Farquhar et al. 1989, Nachit 1998b, Xu et al. 2004).

In C3 plant species, there is discrimination against ^{13}C, in comparison to the more abundant ^{12}C in the atmosphere during photosynthetic carbon capture. Genetic variation in transpiration efficiency has been found to be negatively correlated in C3 species (Xu et al. 2004).

Targeted Environment

Before developing a selection procedure for tolerance to abiotic stresses, it is crucial to determine the frequency of occurrence of a particular stress and its timing in relation to crop development in each agroecological zone. Results of our earlier selection work under contrasting environments show that genotypes selected under favorable conditions do not necessarily do so under less favorable conditions and vice versa. It may be difficult to select for genotypes with a high-yield potential in dryland environments, but it is far more difficult to select for moisture-stress tolerance under high-input environments (Nachit 1992). However, it appears that selection only under extreme environmental conditions (too favorable or too dry) is not an efficient way to identify cultivars for Mediterranean drylands, which are characterized by a high year-to-year variability and by an unpredictable alternation between favorable and less favorable seasons. Breeding cultivars that combine yielding ability with stress tolerance and yield stability are therefore a prerequisite to the adaptation to the Mediterranean dryland conditions (Nachit 1989).

In our selection strategy, all early segregating populations are subjected to the stresses encountered in the Mediterranean dryland, for example, drought, cold, heat, rusts, *Septoria tritici*, root rot, Hessian fly, and stem sawfly, with the aim of identifying the populations that do particularly well in certain environments and are not sensitive to the stresses of other environments (Nachit 1996). The pedigree method of selection is used to select individual plants from the populations that were selected across sites/environments. As for the bulk method, it is more extensively used to select among populations across sites during the winter and summer testing. This method presents the double advantage of testing more crosses at several sites/environments. The promising bulked segregating populations are tested in several agroecological zones of the Mediterranean region in collaboration with national programs in the region. With this selection procedure, it is possible to identify at early stages the populations that combine productivity and stability with tolerance to biotic and abiotic stresses. Disease resistance requires constant surveillance, as it can be broken by new races, as in the case of rusts.

Double Gradient Selection Technique

Dryland productivity can fluctuate from null to 6 tons per hectare. This fluctuation may also be accompanied by variations in attacks by biotic stresses. In relatively favorable rainfall seasons, attacks by rusts and *Septorai tritici* are a threat to the crop, whereas, in the dry seasons, Hessian fly and wheat stem sawfly are the major biotic constraints with devastating effects when the environment is conducive for these diseases to thrive. Thus, the basis of our breeding work is to select durum wheat populations and advanced lines with resistance to abiotic and biotic stresses, and to test for yield stability and productivity under Mediterranean dryland conditions. The cornerstone of this strategy is the introgression of resistance genes from landraces and wild relatives to durum advanced genotypes and the utilization of contrasting and representing environments in the Mediterranean region. A double gradient selection technique (DGST) was developed in the 1980s. The DGST sites are representatives of abiotic and biotic stresses in WANA for temperature extremes, varying from cold to hot, and for water regimes, varying from severe drought to irrigated conditions. The DGST reflects the environmental fluctuations encountered in the Mediterranean drylands.

The DGST employs five environments that are extensively used during the various phases of selection in segregating populations and testing of advanced lines. The DGST approach has its origin from practice, and later in the program its conceptual framework has been established. Under the DGST, the environments/sites used were as follows:

1. Tel Hadya, the main research station at ICARDA-HQ, located in Syria at 36°01′ N latitude, 36°56′ E longitude, and 284 masl. It has a Mediterranean continental climate with an average annual precipitation of 335 mm.
2. Breda station (Br), located also in Syria at 35°56′ N latitude, 37°10′ E longitude, and 300 masl, with clay soil, and an average annual precipitation of 260 mm.

The Breda station is characterized by drought and harsh continental climatic conditions, and it provides a dry site with cold winters and high natural infestation of wheat stem sawfly.

3. Latakia station, located at 35°32′ N latitude, 35°46′ E longitude, and 7 masl, with an annual average rainfall of 784 mm, mild winters, and severe disease pressure. Latakia is a high rainfall site used to test for resistance to diseases under natural and artificial infestation, particularly to *S. tritici* and BYDV.
4. Terbol station, located in the Bekaa valley in Lebanon at 33°33′ N latitude, 35°59′ E longitude, and 890 masl. It is characterized by cold winters but favorable growing conditions, and it has a fine clay soil and an average annual precipitation of 524 mm. It is used during the winter season to test for yield potential and resistance to yellow rust and cold, and during the summer to screen for resistance to heat and stem and leaf rusts.
5. Kfardan (Kf) station, located also in Lebanon in the Bekaa valley at 34°01′ N latitude, 36°06′ E longitude, and 1080 masl and characterized by extreme temperature fluctuations. It has a fine clay soil and an average annual precipitation of 402 mm.

Further, with the DGST at the Tel Hadya main station, six effectively different environments were created (Nachit et al. 1995). Realizing that an important part of our work, particularly in the early stage, had to be done on a research station, we developed a stress-screening technique using simulated environments (Nachit 1983, Nachit and Ketata 1986). The same germplasm is thus subjected to different stresses (cold, drought, terminal stress, and heat) according to date of sowing. Thus, at the Tel Hadya station, durum wheat germplasm is sown at these times and conditions:

- Mid-October (early sowing) with supplemental irrigation (including rainfall of 450 mm) to simulate a crop cycle with long duration and favorable growing conditions. The early sowing conditions subject the plants to cold damage (winterkill) during the tillering stage, to frost during the anthesis stage, and to attacks of yellow rust and *S. tritici*.
- Normal date of sowing, which here presents the plants with a typical Mediterranean continental dryland condition.
- Early April (late sowing) to simulate a short growing season, subjecting the plants to terminal heat and drought stresses, particularly during the grain-filling stage. Late sowing conditions increase the attacks by aphids, barley yellow dwarf virus (BYDV), and Hessian fly.
- Early July to mid-October (summer sowing) to test for high temperature conditions during all crop-growth stages. The summer sowing conditions subject the durum wheat plants to extreme high temperatures and sirocco winds.
- Under irrigated conditions for high-yield potential selection.
- After rotation with vetch to induce dryland-favorable conditions and slow release of nitrogen, mobilized from the vetch root decomposition.

Overall, the testing sites/environments of the DSGT provide two interacting selection gradients for rainfall and temperature regimes, and the environmental conditions of the DGST encompass the main abiotic and biotic stresses that prevail in the Mediterranean dryland.

Use of Landraces to Broaden the Genetic Base for Drought Resistance in Durum

In terms of genetic variation, the Mediterranean durum landraces were found to possess also desirable traits lacking in other materials, such as resistance to drought and cold, early growth vigor, long peduncle, and high fertile tillering. Our results on the use of durum wheat landraces in the hybridization program show that substantial progress can be achieved in developing improved cultivars for dry areas. Several genotypes were developed for rainfed conditions with reduced height (80–100 cm). It was clearly demonstrated that grain yield in dryland does not correlate with the height of plant.

Furthermore, the knowledge of physiological mechanisms involved in drought tolerance is a prerequisite to increasing durum dryland productivity. High relative water content and high capacity of osmotic adjustment were identified as important traits for drought tolerance (Habash et al. 2014). Selection for several morpho-physiological traits related to drought tolerance has been performed in populations issued from crosses between durum wheat and its wild relatives (Nachit 1996). Crosses initiated in the mid-1980s at our program (Nachit 1996) are now generating several advanced lines with better performance under environments with abiotic constraints such as drought and temperature extremes (heat and drought).

GENOMIC TECHNOLOGIES DISSECT SYSTEM RESPONSES TO WATER STRESS

Genomic tools and "omic" technologies are increasingly employed in screening for genes and metabolic pathways responsive to water stress to help identify key molecular signatures for stress resistance in adapted material. It is hoped that such tools could complement other genetic and physiological approaches to gain a better insight into mechanisms defining resistance to water stress and to identify genes that could be used as markers in molecular genetic efforts. A recent study applied transcriptome technologies to RILs from the Lahn × Cham1 mapping population known to segregate for yield stability and drought resistance established in extensive field trials. The study designed a unique water stress time transient that enabled a suite of physiological, biochemical, and molecular measurements to be applied (Habash et al. 2014). This enabled the identification of one RIL (#2219) as having constitutively higher stomatal conductance, photosynthesis, transpiration, abscisic acid content, and enhanced osmotic adjustment at equivalent leaf water compared to parents, thus defining a possible physiological strategy for high-yield stability under water stress. Various statistical models were applied to the extensive flag leaf transcriptome datasets and uncovered:

1. Global and similar trends of early changes in regulatory pathways, reconfiguration of primary and secondary metabolism, and lowered expression of transcripts in photosynthesis in all lines subjected to water stress.
2. Statistically significant differences in a large number of genes among the genotypes, in terms of gene expression magnitude and profile under stress, with a high number belonging to regulatory pathways.
3. Constitutive differences in a large number of genes between the genotypes, demonstrating the uniqueness of each line transcriptome.
4. A high level of structure in the transcriptome response to water stress in each wheat line, suggesting genome-wide co-ordination of transcription.

Applying a systems-based analytical framework, in terms of biological robustness theory, the findings suggest that each durum line transcriptome responded to water stress in a genome-specific manner, which may contribute to an overall different strategy of resistance to water stress integrated over levels of function and time. Such studies produce large dataset that can be used in meta-analysis to identify fundamental features of plant responses to water stress and most immediately can be exploited in the search for new molecular markers and breeding strategies for drought resistance in durum wheat.

Implications of the DGST Approach vis-à-vis Durum Wheat Improvement

During the past 10 years, the contribution made by the stress-tolerant and productive durum wheat genotypes developed by ICARDA is reflected in the significant production increase, from less than 1–3.4 million tons in some areas, where the new varieties were adopted, without any significant increase in the cropped area (1.2 mha). However, with unrelenting population growth in most countries such as WANA region where nearly all of countries are experiencing food deficits, the major challenge for researchers is to increase food production, mainly wheat.

Over the past three decades, ICARDA has developed durum wheat cultivars with high tolerance to drought and WUE, such as Cham1, Cham3, Cham5, and Cham7. These cultivars were released in some areas where they have contributed significantly to increase durum wheat production. For example, in Syria, these cultivars also have a high standard of grain quality. Through the enhanced production of durum grain and good quality, Syria not only achieved self-sufficiency in durum production, but it rose to become the third largest exporter of durum wheat in the world. This achievement (Figure 20.1) was mainly because of variety improvement, which has also prompted the farmers to use improved agricultural practices.

FUTURE WORK

The increase in environmental fluctuation as a result of climate change will negatively affect yield and increase yield variation from year to year. To develop a germplasm with

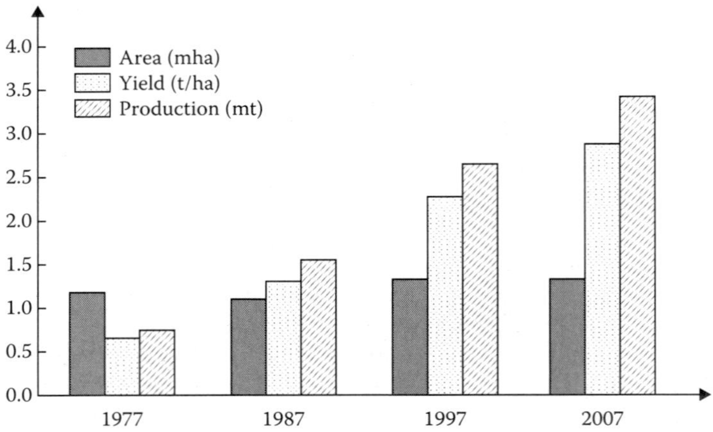

FIGURE 20.1 Area (*left bars*), grain yield (*middle bars*), and production (*right bars*) of the major varieties of durum wheat in Syria over the past 30 years (Haurani/Gezira in 1977, Haurani/Cham1 in 1987, Cham3/Cham1 in 1997, and Cham3/Cham5 in 2007).

buffering ability in durum wheat against these environmental and climatic changes constraints, more gene introgression will be carried out to improve wheat cultivars through the use of landraces and *Triticum* wild relatives. Gene flow from landraces and wild relatives to the improved varieties has been adopted to enable the durum crop to withstand environmental fluctuations, abiotic stresses, and recurrent biotic virulence.

The use of modeling of gene flow is of paramount importance to broaden the genetic base of wheat and to capture complex traits based on quantitative models for predicting functional gene networks.

Conceptual mathematical frameworks will also continue to be used to seek ways to speed up the process of the DGST approach as part of the strategy. The process will involve modeling dynamics that capture complexity such as the predator–prey–predator model to study the infection of a plant when a virus (the initial predator) attacks the plant cells (the initial prey). In such models, the goal is the recovery of solutions of the various inverse problems, each of which highlights a specific feature of the underlying predator–prey–predator dynamics.

The future work will continue to involve capturing complex traits along with modeling *in silico* of gene flow and gene introgression of useful genes in anticipation to buffer the environmental fluctuations and tolerance to stresses.

REFERENCES

Cooper, M., D.W. Podlich, and S.C. Chapman. 2000. Computer simulation linked to gene information databases as a strategic research tool to evaluate molecular approaches for genetic improvement of crops. In *Molecular approaches for the genetic improvement of cereals for stable production in water-limited environments*, A strategic planning workshop held at CIMMYT, El Batan, Mexico, June 21–25, 1999, eds. J.-M. Ribaut and D. Poland, 192–166. Mexico, DF, Mexico: CIMMYT.

Dennis, C. 2002. Gene regulation: The brave new world of RNA. *Nature* 418:122–124.

Farquhar, G.D., J.R. Ehleringer, and K.T. Hubick. 1989. Carbon isotope discrimination and photosynthesis. *Annu Rev Plant Physiol Plant Mol Biol* 40:503–537.

Habash, D.Z., M. Baudo, M. Hindle, S.J. Powers, M. Defoin-Platel, R. Mitchell, M. Saqi, C. Rawlings, K. Latiri, J.L. Araus, A. Abdulkader, R. Tuberosa, D.W. Lawlor, and M. Nachit. 2014. Systems responses to progressive water stress in durum wheat. *PLoS One* 29:9(9).

Habash D.Z., Z. Kehel, and M. Nachit. 2009. Genomic approaches for designing durum wheat ready for climate change with a focus on drought. *J Exp Bot* 60(10):2805–15.

IPCC. 2007. Climate change 2007: The physical science basis. In *Contribution of Working Group I to the Fourth Assessment Report of the Intergovernmental Panel on Climate Change*, eds. S. Solomon, D. Qin, M. Manning, Z. Chen, M. Marquis, K.B. Averyt, M. Tignor, and H.L. Miller. Cambridge, UK, New York: Cambridge University Press.

Kehel, Z., A. Garcia-Ferrer, and M. Nachit. 2013. Using Bayesian and Eigen approaches to study spatial genetic structure of Moroccan and Syrian durum wheat landraces. *Am J Mol Biol* 3:17–31. doi:10.4236/ajmb.2013.31003.

Liebovitch, L.S., V.K. Jirsa, and L.A. Shehadeh. 2006. Structure of genetic regulatory networks: Evidence for scale free networks. In *Complexus mundi: Emergent patterns in nature*, ed. M. Novak, 1–8. Singapore: World Scientific.

Nachit, M.M. 1983. Use of planting dates to select stress tolerant and yield stable genotypes for the rainfed Mediterranean environment. *Rachis* 3:15–17.

Nachit, M.M. 1989. Moisture stress tolerance in durum wheat (*T. turgidum* var. L. *durum*) under Mediterranean dryland conditions. In *Proceedings of the 12th EUCARPIA congress: Science for plant breeding*, February 27–March 4, 1989, Göttingen, Germany, 2.

Nachit, M.M. 1992. Durum breeding for Mediterranean drylands of North Africa and West Asia. In *Durum wheats: Challenges and opportunities*. Wheat Special Report, No. 9, 14–27. Mexico, DF, Mexico: CIMMYT.

Nachit, M.M. 1996. Durum wheat breeding. In *Cereal Improvement Program Annual report*, 84–123. Aleppo, Syria: ICARDA 010/1500.

Nachit, M.M. 1998a. Durum breeding research to improve dry-land productivity in the Mediterranean region. In *SEWANA Durum Research Network*, eds. M.M. Nachit, M. Baum, E. Porceddu, P. Monneveux, and E. Picard, 1–15. Aleppo, Syria: ICARDA.

Nachit, M.M. 1998b. Association of grain yield in dryland and carbon isotope discrimination with molecular markers in durum (*Triticum turgidum* L. var. *durum*). In *Proceedings of 9th International Wheat Genetics Symposium*, Saskatoon, ed. A.E. Slinkard, 218–223. Saskatchewan, Canada.

Nachit, M.M., M. Baum, A. Impiglia, and H. Ketata. 1995. Studies on some grain quality traits in durum wheat grown in Mediterranean environments. In *Durum wheat quality in the Mediterranean region, Proceedings of the seminar held in Zaragoza, Spain*, November 17–19, 1993, *Options Méditerranéenes*, eds. N.D. Fonzo, F. Kaan, and M. Nachit, Series A, 22:181–187. Zaragoza, Spain: CIHEAM. om.ciheam.org/article.php?IDPDF=95605369.

Nachit, M.M. and I. Elouafi. 2004. Durum wheat adaptation in the Mediterranean dryland: Breeding, stress physiology, and molecular markers. In *Challenges and strategies for dryland agriculture*, eds. S.C. Rao and J. Ryan, CSSA Special Publication no. 32, 203–218. Madison, WI: Crop Science of America and American Society of Agronomy.

Nachit, M.M. and H. Ketata. 1986. Breeding strategy for improving in Mediterranean rainfed conditions. In *Proceedings of the International Wheat Conference*, Rabat, Morocco, May 2–9, 1983.

Nachit, M.M. and A. Ouassou. 1988. Association of yield potential, drought tolerance and stability of yield in *Triticum turgidum* var. *durum*. In *Proceedings of the 7th International Wheat Genetics Symposium*, eds. T.E. Miller and R.M.D. Koebner, 867–870, Cambridge: Institute of Plant Science Research.

Nachit, M.M., M.E. Sorrells, R.W. Zobel, H.G. Gauch, W.R. Coffman, and R.A. Fischer. 1992a. Association of morpho-physiological traits with grain yield and components of genotype-environment interaction in durum wheat. Pt. 1. *J Genet Breed* 46:363–368.

Nachit, M.M., M.E. Sorrells, R.W. Zobel, H.G. Gauch, W.R. Coffman, and R.A. Fischer. 1992b. Association of morpho-physiological traits with grain yield and components of genotype-environment interaction in durum wheat. Pt. 2. *J Genet Breed* 46:369–372.

Shaw, R.G. and J.R. Etterson. 2012. Rapid climate change and the rate of adaptation: Insight from experimental quantitative genetics. *New Phytol* 195(4):752–765.

van Oosterom, E., G. Hammer, and S. Chapman. 2004. Can transition to flowering be modelled dynamically from the gene level? In *New directions for a diverse planet: Proceedings of the 4th International Crop Science Congress*, eds. T. Fischer, N. Turner, J. Angus, L. McIntyre, M. Robertson, A. Borrell, and D. Lloyd, 647. Gosford, Australia: The Regional Institute. http://www.cropscience.org.au/icsc2004/poster/3/2/1/647_vanoosteromej.htm.

Welch, S.M., Z. Dong, and J.L. Roe. 2004. Modelling gene networks controlling transition to flowering in *Arabidopsis*. In *New directions for a diverse planet: Proceedings for the 4th International Crop Science Congress*, eds. T. Fischer, N. Turner, J. Angus, L. McIntyre, M. Robertson, A. Borrell, and D. Lloyd, 158. Gosford, Australia: The Regional Institute. http://www.cropscience.org.au/icsc2004/symposia/3/2/158_welchsm.htm.

Welch, S.M., J.L. Roe, and Z. Dong. 2003. A genetic neural network model of flowering time control in *Arabidopsis thaliana*. *Agron J* 95:71–81.

Xu, Y., J.R. Coburn, B.E. Gollands et al. 2004. *Mapping quantitative trait loci for water use efficiency in rice*. Paper presented at the International Plant and Animal Genome XII Conference, January 10–14, 2004, San Diego, CA.

21 Addressing Diversity of Ethiopian Durum Wheat Landrace Populations Using Microsatellite Markers

A.G. Tessema, G.G. Venderamin, and E. Porceddu

CONTENTS

Ethiopia possesses the second largest wheat-growing area (approximately 877,000 ha) of the sub-Saharan African countries. Wheat provides energy, protein, calcium, vitamins, and iron to the diet of the local people. An understanding of the distribution of genetic variation within landrace populations is crucial for efficient germplasm preservation and subsequent distribution of material to wheat researchers.

Durum wheat (*Triticum turgidum* L. ssp. *durum* (Desf.) Husn.) is the only tetraploid ($2n = 4x = 28$) wheat of commercial importance. It is a staple food for people, providing protein and energy in different forms. It was probably introduced into

Ethiopia by Hammite immigrants some 5000 years ago. Durum wheat is found growing under highly diverse agroecological conditions, which results in further genetic diversification, making the country one of the most important centers of diversity of this crop (Vavilov 1926). Wheat is traditionally grown in Ethiopia on the heavy black soils (vertisols) of the highlands and at altitudes ranging between 1500 and 3000 masl under rainfed conditions (Tesemma and Belay 1991). Landraces still form the backbone of cultivated material.

Diversity studies on Ethiopian wheat germplasm have been performed at morphological and agronomical levels (Vavilov 1926, 1951, 1964, Bekele 1984, Belay et al. 1993, Bechere and Tesemma 1997), providing insights into population structure and breeding systems. However, the morphological and other field measurements present several inconveniences; not only are morphological characters an indirect measure of the genetic variation, but they also are prone to environmental conditions and are controlled by multiple genes (Newbury and Ford-Lloyd 1997), making assessment difficult. Analysis of variance (ANOVA) has been used to assess the potential of biochemical markers such as isozymes and protein subunits, but, again, they are phenotypes, with the aforementioned difficulties (Hamrick and Godt 1990, Eagles et al. 2001). Molecular markers are the last comers; they offer the potential to assess diversity directly at the DNA level, completely independent of environmental effects, although such markers may represent variation at noncoding regions, thus not affected by selection. Molecular markers such as microsatellites have a high potential use for genetic analysis because of their high degree of polymorphism and the fact that they are co-dominantly inherited (Akkaya et al. 1992, Plaschke et al. 1995, Röder et al. 2004, Al Khanjari et al. 2007). Their polymorphism is due to the variation in numbers of repeats, good distribution, and hypervariability in the whole genome (Morgante and Olivieri 1993). Consequently, microsatellite markers are able to distinguish between closely related genotypes. The present study was aimed at evaluating genetic diversity of durum wheat landraces from Ethiopia.

MATERIAL AND METHODS

Genetic Material

A total of 790 accessions of durum wheat were used, belonging to 30 landrace populations from 10 regions of Ethiopia, representing three agroecological zones, with elevations ranging from 1500 to 3020 masl. All seeds were supplied by the Genebank of the Ethiopian Institute of Biodiversity Conservation (IBC).

DNA Isolation

Leaf samples were collected from single seedlings 7–15 days old, frozen in liquid nitrogen, and stored at −80°C until used. DNA was extracted manually by DNeasy plant mini kits (Qiagen). DNA quality and quantity was determined on agarose gels (1% in TBE buffer) by comparing bands to known concentrations of stock lambda DNA.

Microsatellite Markers and Polymerase Chain Reaction Amplification

Fourteen primer pairs, one per chromosome, were used in the analysis. For microsatellite analysis, DNA concentration was adjusted to 50 ng/μl. Polymerase chain reaction (PCR) amplifications were performed in a 10-μl reaction volume. Each reaction contained 0.15 μl Taq polymerase, 2 μl of 5× buffer, 2 μl of 10 mM of each of the four nucleotides (dNTPs), 0.03 μl of 10 mM forward primers, 0.2 μl of 10 mM reverse primers, and 0.2 μl of 10mM (M13) fluorescent dyes. Fluorescently labeled forward and unlabeled reverse primers were used. The amplification protocol of 2 min at 94°C with 32 cycles was followed with 50 s at 94°C, 1 min at either 50°C, 51°C, 55°C, 59°C, 60°C, or 61°C, 2 min at 72°C, and a final extension step of 10 min at 72°C (Plaschke et al. 1995, Röder et al. 1998).

Genotyping

The mix was prepared from 1 μl of PCR product, 0.5 μl of sizer, and 18 μl of formamide (DNA analyzer) per well, in 96- as well as in 384-well plates. The mixed product was centrifuged for 20 s at 20,000× g (14,000 rpm) and denatured for 3–5 min. Microsatellite fragments were detected on an automated laser fluorescence sequencer. All genotyping samples up to 340 bp were labeled with fluorescent dyes (6-FAM, VIC, NED, and PET), then the labeled primers and the 500 LIZ-size standard were run on an Applied Biosystems Prism® 3130xl Genetic Analyzer using GeneScan. After the fragment analysis was performed, confirmed alleles were processed, analyzed, and reviewed by use of Genemapper software v. 3.7. Alleles were coded in base-pair sizes.

Statistical Analysis

The absence of specific microsatellite alleles were coded as (0) The allelic richness was computed, based on available alleles per each locus detected within their base-pair ranges. Analysis of molecular variance (AMOVA) was used to investigate the hierarchical partitioning of genetic variation among the populations and regions. Within AMOVA, principal coordinate analysis (PCoA) was used to find and plot the major patterns within a multivariate dataset (multiple loci and multiple samples). This analysis is based on an algorithm published by Orlóci (1978). The analysis for effective alleles yielded values for observed heterozygosity, expected heterozygosity, unbiased heterozygosity, and F-fixation value (Peakall and Smouse 2006).

RESULTS AND DISCUSSION

Allelic Variation of Microsatellite Markers

Microsatellite markers have been shown to be more variable than most of the other molecular markers (Röder et al. 1995, Huang et al. 2002), and it has been found that microsatellites are preferentially associated with low-copy regions of plant genomes (Morgante et al. 2002). Microsatellites are highly mutable loci, which may be present at many sites in a genome (Morgante et al. 1998), and they provide highly informative markers because they are co-dominant and generally have high polymorphic information content (Gupta et al. 1996). In this study, microsatellite markers have been used based on PCR systems and compared for studying the genetic diversity between and within 30 populations of 790 individuals of tetraploid wheat. Compared to the source of simple sequence repeat (SSR), wheat microsatellites produced the highest number of alleles per locus, 34. The maximum number of alleles was detected for marker *wms170* which maps to a single locus on the long arm of chromosome 2A (Röder et al. 1998). Marker *wms375*, located on the long arm of chromosome 4B, showed lower polymorphism for all durum wheat genotypes. The total number of alleles found in the A genome was 190, whereas in the B genome, it was 157. At the genome level, a larger number of alleles per locus occurred in the A genome (27 per locus) than in the B genome (22.4 per locus), in agreement with results of Teklu et al. (2005). In contrast, other researchers detected the highest number of alleles per locus in the B genome (Figliuolo and Perrino 2004, Alamerew et al. 2003), and Al Khanjari et al. (2007) reported an average of 7.9 alleles per locus for the B genome compared to 6.5 alleles per locus for the A genome.

A total of 12 and 65 durum wheat accessions originating from Ethiopia were investigated by Sentayehu et al. (2003) and Teklu et al. (2005), respectively; they discovered an average number of alleles per locus of 7.9 and 11.03, respectively. This is comparable to the 7.9 alleles per locus (B genome) and 6.54 alleles per locus (A genome) detected on 38 Omani tetraploid wheat landraces (Al Khanjari et al. 2007).

Comparisons of diversity were also made among regions and the averages of allele numbers were different for each region. Allele number was highest in the Arssi region with 184, followed by Wello (156), Bale (136), Wellega (127), Hararghe (126), Tigray (125), Shewa (125), Gonder (120), Gojam (119), and Jimma (87). The highest number of alleles per population was found in the Jimma region. However, all regions exhibited their own uniqueness and specific alleles. Rare alleles were detected in most of the microsatellite loci.

Patterns of Allelic Variation among and within Populations and among Regions

Analysis of microsatellite allelic diversity within the 30 durum wheat populations revealed that 72% of the genetic variation resided within populations, while 12% was among populations. These populations belong to 10 different regions, and the variation among regions was 16%. Diversity was also evaluated between the genomes, with the A genome showing 75% within and 10% among population diversity versus 74% and 10%, respectively, for the B genome. Among regions, higher variability was detected for the B genome (16%) than for the A genome (14%). The populations with the highest among-population diversity were those from Shewa (21%), Bale (15%), and Wello (14%) regions. The regions that had the highest within-population diversity were Tigray, Jimma, Hararghe, and Gojam, all at 91%. The large within-population allelic diversity found in this study was unexpected for a self-pollinating species. Self-pollination is known to decrease the effectiveness of pollen flow, and is therefore expected to reduce genetic diversity within populations (Jarne 1995, Hamrick and Godt 1996, Ingvarsson 2002, Nybom 2004). The present results were consistent with earlier observations, which showed high genetic diversity within populations (Fahima et al. 1999, 2002). Environmental gradients are shaping biological diversity at all levels of organization (e.g., Darwin 1859, Connell 1978) and we suggest that the farmer's ingenuity coupled with diverse environments and different selection pressures are acting on the various populations to increase within-population genetic diversity as shown in the current study.

Principal Coordinate Analysis

In order to illustrate the relatedness between the 30 populations of 10 different regions, PCoA was exploited. The PCoA results for the individual populations showed the distribution of the landrace populations spreading in two quadrants only. Since the analysis was a multivariate technique, it plotted the major patterns within a multivariate data set of every population with respect to their regions. Genetic differentiation between populations in the Bale region was higher in the PCA1 quadrant (35.78%) than in the PCA2 quadrant (16.72%). The Arssi region comprised eight populations and the PCoA was organized clearly in two groups. The first (PCA1) and second (PCA2) principal coordinates accounted for 33.7% and 16.97% of the total variation, respectively. The analysis differentiated the highest variation in the PCA1 quadrant from Bale, Arssi, Shewa, and Tigray regions (35.78%, 33.7%, 29.67%, and 24.8%, respectively). This was demonstrated by the scattered presence for the four selected regions discriminated using PCoA analysis, and the clear separation of all regions indicated that the genetic differentiation was significant.

Genetic Structure Analysis

The data were analyzed to resolve the genetic structure and to interpret the genetic relationships between groups. The genetic structuring, showing the genetic relationships among the populations, was constructed based on the allelic frequency. Three main groups (I, II, and III) were detected. Group I comprised 13 populations of Arssi, Bale, Hararghe, and Wellega regions; group II comprised nine populations of Wello, Gojam, Gonder, and Jimma regions; and group III comprised eight populations of Tigray and Shewa regions. Individuals from all the different regions of origin were included in the above-described groups. The genetic relationships among the durum wheat landrace populations from Arssi and Bale were more closely related with each other, and both were from similar geographical origins. But, interestingly, populations from Gonder formed close relationships with those from Arssi and Bale. This observed relationship may reflect perhaps more on the genomic locations of the markers, indicating that landraces with similar traits occur in several geographic locations. All western geographical origins were grouped together in group II. But, the analysis also related these with two regions of northern geographical origins. However, this effect might be the result of the displacement of farmers, moving from their place of origin to another region. Finally, Tigray and Shewa regions were grouped together, in spite of the fact that within these two regions more variation was exhibited. However, when the analysis was compared with all regions, these two regions were more closely related to each other than to any of the others.

Genetic Diversity

Durum wheat is predominantly self-pollinating (Golenberg 1998, Felsenburg et al. 1991) with an outcrossing rate estimated at 4%–5%. Consequently, a very low proportion of heterozygosity is expected. The present study was initiated to quantify the level of outcrossing within several dynamically managed populations. Based on the analysis of heterozygosity, *F*-statistics, and polymorphism, the mean observed heterozygosity (measured with microsatellites) was high for all microsatellite loci except one locus, for which heterozygosity was much higher than expected under purely self-pollination. This is an indicator that outcrossing does play a role in natural wheat populations and may influence fitness. The levels of heterozygosity varied among microsatellite markers. The mean observed heterozygosity per population (HO) ranged from 1.9% to 9%. The mean observed heterozygosity per region ranged from 2% to 9%. The mean observed heterozygosity for the A genome ranged from 2% to 18%, whereas for the B genome, it ranged from 1% to 10%. Comparison was also made among four chromosome arms; the highest mean of observed heterozygosity was found for short arms of the A genome (15%). The highest mean of observed heterozygosity for regions was for the Tigray region (9%). Mean observed heterozygosity was highest at locus *wms124* in the long arm of chromosome 1B (8%) and the lowest was at locus *wms375* in the long arm of chromosome 4B (1%). The highest grand mean of observed heterozygosity was detected in Tigray, Hararghe, Shewa, and Wellega regions (9%, 7%, 7%, and 6%, respectively). Similarly, in previous studies, the highest observed heterozygosity observed from Ethiopian-origin germplasm was 10%–15% (Sentayehu et al. 2003). The highest grand mean diversity (effective alleles) was found in Gojam, Wello, Hararghe, Arssi, and Wellega regions (4.1%, 4%, 3.9%, 3.8%, and 3.7%, respectively), while the lowest was in the Shewa region (2.8%). In general, for all parameters, every region exhibited significantly higher polymorphism. According to Hamrick et al. (1991), selfing species can have fivefold more genetic diversity among populations and less than half the genetic diversity within populations than have wind-pollinated, outcrossing species. In agreement with a previous study, the Shewa region exhibited higher variation among populations (21%) than did other regions. In contrast, in the current study of nine regions, high variation was found within populations (85%–91%).

CONCLUSION AND RECOMMENDATIONS

- An understanding of the extent and distribution of genetic variation within landrace populations is a key to efficient germplasm preservation and subsequent distribution of material to researchers.
- High priority should be given to in-situ and ex-situ conservation of genetic diversity. Moreover, the breeding and conservation programs must be interlocked with the understanding of genetic variability present in genepools for the success in the present and in the future.
- This current molecular study confirmed earlier work showing a high variation in farmers varieties conserved in the IBC ex-situ genebank. This diversity reflects the effects of the well-developed and cultured traditional way of farmers' seed-selection criteria.

- Molecular evaluation of durum wheat landraces should be followed by an agronomic and morphologic evaluation in order to develop DNA markers linked to traits adaptive to biotic and abiotic stress conditions.

REFERENCES

Akkaya, S.M., A.A. Bhagwat, and P.B. Cregan. 1992. Length polymorphism of simple-sequence repeat DNA in soybean. *Genetics* 132:1131–1139.

Al Khanjari, S., K. Hammer, A. Buerkert, and M.S. Röder. 2007. Molecular diversity of Omani wheat revealed by microsatellites: I. Tetraploid landraces. *Genet Resour Crop Ev* 54(6):1291–1300.

Bechere, E. and T. Tesemma. 1997. Enhancement of durum wheat landraces in Ethiopia. In *Proceedings of the 1st Joint Meeting of the National Project Advisory and Overseeing Committee and Partner Institutions*, February 24, 1997, eds. F. Kebebew, A. Demissie, and S. Ketema, 35–48. Addis Ababa, Ethiopia: Community Biodiversity Development and Conservation (CBDC), Ethiopian Biodiversity Institute.

Bekele, E. 1984. Analysis of regional patterns of phenotypic diversity in the Ethiopian tetraploid and hexaploid wheats. *Hereditas* 100(1):131–154.

Belay, G., T. Tesemma, H.C. Becker, and A. Merker. 1993. Variation and interrelationships of agronomic traits in Ethiopian tetraploid wheat landraces. *Euphytica* 71(3):181–188.

Connell, J.H. 1978. Diversity in tropical rain forests and coral reefs. *Science* 199:1302–1310.

Darwin, C. 1859. *On the origin of species by means of natural selection*. London: John Murray.

Eagles, H., H. Bariana, F. Ogbonnaya et al. 2001. Implementation of markers in Australian wheat breeding. *Aust J Agric Res* 52:1349–1356.

Fahima, T., M.S. Röder, K. Wendehake, V.M. Kirzhner, and E. Nevo. 2002. Microsatellite polymorphism in natural populations of wild emmer wheat, *Triticum dicoccoides*, in Israel. *Theor Appl Genet* 104:17–29.

Fahima, T., G.L. Sun, A. Beharav, T. Krugman, A. Beiles, and E. Nevo. 1999. RAPD polymorphism of wild emmer wheat populations, *Triticum dicoccoides*, in Israel. *Theor Appl Genet* 98:434–447.

Felsenburg, T., A.A. Levy, G. Galili, and M. Feldman. 1991. Polymorphism of high- molecular weight glutenins in wild tetraploid wheat: Spatial and temporal variation in a native site. *Israel J Bot* 40:451–479.

Figliuolo, G. and P. Perrino. 2004. Genetic diversity and intra-specific phylogeny of *Triticum turgidum* L. subsp. *dicoccon* (Schrank) Thell. Revealed by RFLPs and SSRs. *Genet Resour Crop Ev* 51:519–527.

Golenberg, E.M. 1998. Outcrossing rates and their relationship to phenology in *Triticum dicoccoides*. *Theor Appl Genet* 75:937–944.

Gupta, P.K., H.S. Balyan, P.C. Sharma, and B Ramesh. 1996. Microsatellites in plants: A new class of molecular markers. *Curr Sci India* 70:45–54.

Hamrick, J.L. and M.J.W. Godt. 1990. Alloenzyme diversity in plant species. In *Plant population genetics, breeding, and genetic resources*, eds. A.H.D. Brown, M.T. Clegg, A.L. Kahler, and B.S. Weir, 43–63. Sunderland, MA: Sinauer Associates.

Hamrick, J.L. and M.J.W. Godt. 1996. Effect of life history traits on genetic diversity in plant species. *Philos T Roy Soc B* 351:1291–1298.

Hamrick, J.L., M.J.W. Godt, D.A. Murawski, and M.D. Loveless. 1991. Correlations between species traits and allozyme diversity: Implications for conservation biology. In *Genetics and conservation of rare plants*, eds. D.A. Falk and K.E. Holsinger, 75–86. Oxford: Oxford University Press.

Huang, X.Q., A. Borner, M.S. Röder, and M.W. Ganal. 2002. Assessing genetic diversity of wheat (*Triticum aestivum* L.) germplasm using microsatellite markers. *Theor Appl Genet* 105:699–707.

Ingvarsson, P.K. 2002. A metapopulation perspective on genetic diversity and differentiation in partially self-fertilizing plants. *Evolution* 56:2368–2373.

Jarne, P. 1995. Mating system, bottleneck and genetic polymorphism in hermaphroditic animals. *Genet Res* 65:193–207.

Morgante, M., M. Hanafey, and W. Powell. 2002. Microsatellites are preferentially associated with nonrepetitive DNA in plant genomes. *Nat Genet* 30:194–200.

Morgante, M. and A.M. Olivieri. 1993. PCR-amplified microsatellites as markers in plant genetics. *Plant J* 3:175–182.

Morgante, M., A. Pfeiffer, I. Jurman, G. Paglia, and A.M. Olivieri. 1998. Isolation of microsatellite markers in plants. In *Molecular tools for screening biodiversity, plants and animals*, eds. A. Karp, P.G. Isaac, and D.S. Ingram, 75–134. London: Chapman & Hall.

Newbury, H.J. and B.V. Ford-Lloyd. 1997. Estimation of genetic diversity. In *Plant genetic conservation*, eds. N. Maxted, B.V. Ford-Lloyd, and J.G. Hawkes, 192–206. London: Chapman & Hall.

Nybom, H. 2004. Comparison of different nuclear DNA markers for estimating intraspecific genetic diversity in plants. *Mol Ecol* 13:1143–1155.

Orlóci, L. 1978. *Multivariate analysis in vegetation research*, 2nd ed. The Hague, The Netherlands: Dr. W. Junk Publishers.

Peakall, R. and P.E. Smouse. 2006. GenAlEx 6: Genetic analysis in excel. Population genetic software for teaching and research. *Mol Ecol Notes* 6:288–295.

Plaschke, J., M.W. Ganal, and M.S. Röder. 1995. Detection of genetic diversity in closely related bread wheat using microsatellite markers. *Theor Appl Genet* 91:1001–1007.

Röder, M.S., X.Q. Huang, and M.W. Ganal. 2004. Wheat microsatellite in plant breeding-potential and implications. In *Molecular markers in plant breeding*, eds. H. Loerz and G. Wenzel, 255–266. Berlin, Germany: Springer.

Röder, M.S., V. Korzun, K. Wendehake et al. 1998. A microsatellite map of wheat. *Genetics* 149:2007–2023.

Röder, M.S., J. Plaschke, S.U. König et al. 1995. Abundance, variability and chromosomal location of microsatellites in wheat. *Mol Gen Genet* 246:327–333.

Sentayehu, A., S. Chebotar, X. Huang et al. 2003. Genetic diversity in Ethiopia hexaploid and tetraploid wheat germplasm assessed by microsatellite markers. *Genet Resour Crop Ev* 51:559–567.

Teklu, Y., K. Hammer, and M.S. Röder. 2005. Comparative analysis of diversity indices based on morphological and microsatellite data in tetraploid wheats. *J Genet Breed* 59:121–130.

Tesemma, T. and G. Belay. 1991. Aspects of Ethiopian tetraploid wheats with emphasis on durum wheat genetics and breeding research. In *Wheat research in Ethiopia: A historical perspective*, eds. H. Gebre-Mariam, D.G. Tanner, and M. Hulluka, 47–71. Addis Ababa, Ethiopia: IAR/CIMMYT.

Vavilov, N.I. 1926. Studies on the origin of cultivated plants (Russian with English summary). *Bull Appl Bot Plant Breed* (Leningrad) 16(2):1–128.

Vavilov, N.I. 1951. The origin, variation, immunity and breeding of cultivated plants. (Transl. from Russian by K.S. Chester). *Chronica Botanica* 13:1–366.

Vavilov, N.I. 1964. *World resources of cereals, legumes, flax cultivars and their utilization in breeding. Wheat.* Moscow, Russia: Nauka (In Russian).

Index

Note: Locator followed by '*f*' and '*t*' denotes figure and table in the text

E

F